WIRELESS AND GUIDED WAVE ELECTROMAGNETICS

Fundamentals and Applications

Optics and Photonics

Series Editor

Le Nguyen Binh

Huawei Technologies, European Research Center, Munich, Germany

Wireless and Guided Wave Electromagnetics

Fundamentals and Applications

Le Nguyen Binh

CRC Press
Taylor & Francis Group
Boca Raton London New York

CRC Press is an imprint of the
Taylor & Francis Group, an **informa** business

CRC Press
Taylor & Francis Group
6000 Broken Sound Parkway NW, Suite 300
Boca Raton, FL 33487-2742

First issued in paperback 2017

© 2013 by Taylor & Francis Group, LLC
CRC Press is an imprint of Taylor & Francis Group, an Informa business

No claim to original U.S. Government works

Version Date: 20130206

ISBN 13: 978-1-138-07787-4 (pbk)
ISBN 13: 978-1-4398-4753-4 (hbk)

To the memory of my father, and to my mother

To Phuong and Lam

Contents

Preface

In the 19th century, Maxwell combined the laws of electricity and magnetism to speculate on the evolution of and intertwining between the time-variant electric and magnetic fields to form the foundation of electromagnetism (EM). Hertz, in 1885, conducted the first experiment on the radiation of such electromagnetic waves to test Maxwell's equations and concept of time-dependent excitation of electromagnetic waves, the Hertz dipole antenna. Thus, he conclusively proved the existence of electromagnetic waves by engineering instruments to transmit and receive radio pulses using experimental procedures that ruled out all other known wireless phenomena. Thus began the science of generation and radiation of electromagnetic waves for communication, especially during the Second World War, when extensive use of EM waves was employed by both sides to track, code, and decode information.

Radiation of EM waves was also detected from the creation of stars and galaxies in the universe. Extensive research and development work were pursued to achieve such generation, transmission, and detection of EM waves, with frequencies increasing with time, even to the terahertz region currently, for example, tera-waves.

EM waves are considered to extend from less than 1 Hz to several hundred terahertz, the visible, near- and central-infrared, and infrared spectral regions. Not until 1960, when the laser was invented, could optical waves be considered the carrier for communications. Charles Kao and George Hockham* first proposed a structure of dielectric waveguides for guiding lightwaves, using the evolutionary concept of metallic waveguides. Guiding EM waves at ultra-high and microwave frequency was also extensively investigated at the same time. The invention of semiconductor junctions, and then devices, allowed tremendous development of microwave and millimeter-wave technologies, as well as lightwave sources in the visible and near-infrared regions.

Since the 1970s, the tremendous investments of several traditional glass corporations, such as Corning, Schott Glass, and Sumitomo Cement Company, developed techniques for a fabrication optical guiding medium, the circular optical waveguides, and optical fibers. From a very high propagation loss of 100 dB/m, the attenuation coefficient was reduced to 0.2 dB/km, and today's fiber attenuation coefficient in the 1550 nm wavelength window was achieved in 1970. Initially the circular waveguides supported several hundred guide modes, until the diameter of the fibers was reduced, along with the refractive index difference between the core and cladding regions. Single-mode optical fibers can be produced and proven to the best-guided media for ultra–high-speed, ultra–long-reach transmission. Only one guided mode of lightwaves is propagated in such fibers; however, it is composed of two polarized modes—hence, they are not a monomode type.

---

* K.C. Kao and G.A. Hockham, "Dielectric-Fibre Surface Waveguides for Optical Frequencies," *Proc. IEEE*, 113(7), 1151–1156, 1966.

In parallel with the development of optical communications, wireless technologies have been progressing and penetrating into almost everyone's lives. Information capacity has reached its maximum limit in the frequency bands for mobile networks. Smart antennae have been employed with multiple sources and multiple detection devices; hence, multiple-input multiple-output (MIMO) techniques further increase transmission capacity.

Thus in modern communication technologies, both wireless and guided wave technologies dominate global societies with the employment of EM waves in ultra-high frequency to microwave to millimeter wave to lightwave. At the same time, tremendous progress has been made in digital signal processing with ultra-dense and ultra–high-speed digital electronic processors. This digital technology has pushed processing techniques to all-time complex algorithms for wireless communication systems and increased the channel capacity with spectral efficiency of several tens of bits per Hertz. These algorithms have been extensively applied to ultra–high-speed optical communications employing coherent detection techniques in which the polarization and phase of the optical carriers can be recovered. Thus, the roles of the optical fields in terms of electric and magnetic components of the lightwaves are very important.

This book proposes to introduce the concept of EM waves with applications oriented to wireless and guided waves in both electrical and optical domains. The contents are appropriate for students at the senior level of electrical and electronic engineering after some introduction to electromagnetism, but before introduction to communication techniques and systems, especially the wireless and optical transmission technologies.

Thus, we give a brief introduction to the mathematical relationships between the electric and magnetic fields, which are time varying, and the intertwining between them to radiate EM waves in free space or bounded under certain boundary conditions. In Chapter 2 we introduce transmission lines in the electrical domain; in modern days this topic is commonly faced by engineers in transmission systems operating from the ultra–high-frequency to microwave and millimeter-wave ranges and optical frequency with line rates reaching several Gbits/s. The excitation of time-varying currents into conducting elements to produce radiation of EM waves into free space—the antennae—is introduced in Chapter 3. Only fundamental issues of some basic antennae are given, with a brief introduction of electrical waveguides, in which the high-frequency waves can be confined and vanished at the electric conducting walls.

In the 1960s, when lasers in the visible range, specifically the HeNe gas laser, were invented, the thought of using such waves for ultra-wideband communications was proposed. The concept of guiding such EM waves was first developed and proposed for guiding optical waves propagated by the zigzag ray model. This was later proven to be impossible for propagation over very long distances. Thence, all dielectric waveguides were introduced, leading to present-day optical communication technologies, even though the fibers were to be theoretically analyzed and numerically confirmed with manufactured fibers. The applications of Maxwell's equations to arrive at the wave equations, and then the eigenvalue equation due to boundary conditions, offer the conditions for the number of guided modes. However, if only one mode can be guided, the interference between the guided modes would be eliminated, thus the wideband response of the transmission medium. When single-mode optical fibers were proven to be practical, several measurements of the distribution profile of the field intensity were made, and conclusions were obtained that the field profile of the single mode follows that of a Gaussian shape when the refractive index difference between the core guiding region and the cladding region is very small. This known empirical function leads to the simplification of the application of electromagnetism in the understanding of lightwave confinement and guiding in optical fibers, the *gentle* or *weakly* guiding phenomena for extremely long-reach transmission and detection of modulated optical signals. These conceptual developments are still holding and playing critical roles in modern optical fiber communication systems and networks. We thus introduce the phenomena of guided wave optical transmission lines in Chapter 7, followed by Chapter 8, which introduces the eventual applications of such guided waves in the transmission or propagation of modulated lightwave signals over long-reach optical transmission lines, especially the attenuation, dispersion, and broadening of modulated signals due to the interference of lightwaves of different spectral components of the sidebands of the passband signals.

However, the basic concept of lightwave guiding in optical waveguides is introduced in Chapter 4, and planar and three-dimensional structures in Chapter 5, which are commonly employed in the design and implementation of guided media for lasers and optical modulators, hence the term *lightwave circuit technology*.

Chapter 9 introduces somewhat more advanced applications of guided waves in the optical domain, for example, the Fourier guided wave optics in which the uses of Fourier transformation can be implanted by using guided wave components. This chapter concludes this book on the introduction of guided waves using electromagnetics.

Lecturers can elect to introduce these chapters in different sequences after introducing the fundamental parts of electromagnetism with time-varying fields. Guided waves in optical fibers can be introduced, bypassing the EM mathematical concept, but with the weakly guiding phenomena (Chapters 7 and 8), interest can be created for students to appreciate the importance of guiding lightwaves. Formal mathematical analyses (Chapter 1) can be introduced regarding radiation of EM waves and the confinement of lightwaves in optical devices (Chapters 4 and 5). After the introduction in Chapter 1, the introduction of electrical and guided wave optical sections (Chapters 2 and 4) can be taught in parallel, if preferred.

There are several textbooks that introduce courses in time-varying EM waves, but not one, to my best knowledge, gives specific uses of EM waves focusing on wireless and optical guided waves, which I think are very important for electrical engineers to appreciate for practical purposes, thus reducing the possible boredom that modern engineers may face in practice. Hence, the title and contents of this book are proposed.

The traditional teaching of electromagnetics must evolve toward technological applications, especially under the transmission of modulated optical signals over very long distances, the weakly guiding and confinement of the EM phenomena.

Le Nguyen Binh
Munich, Germany

Acknowledgments

I thank my former undergraduate students at Monash University, especially those who have passed through the course in Level 3 over a number of years and may have had some difficult periods trying to grasp the EM concepts in laboratory practices and theoretical lectures, while still maintaining much interest in and challenging the contents of the lectures. I also thank my colleagues in the Optical Technology Department of the Fixed Network Division of the European Research Center of Huawei Technologies for fruitful discussions, especially on several practical issues related to the conceptual understandings of guided waves, particularly when pushing them into the nonlinear region for ultra–long-reach, non–CD-compensating transmission.

Throughout my teaching years I have worked and taught electromagnetism together with Dr. John Bennett and Dr. G.K. Cambrell of Monash University, and I appreciated fruitful discussions with them during several teaching sessions.

I also thank CRC Press acquisitions editor Ashley Gasque for her encouragement throughout the development of the book's chapters.

I dedicate this book to my parents, my late father and my mother, Nguyen Thi Huong, who will not understand a word in it, but her teaching endeavors and learning in life have always been alive in my heart. My wife, Phuong, and son, Lam, have strongly supported my dedication of time to writing the chapters at home during late nights and weekends.

About the Author

Le Nguyen Binh received a BE (Hons) and PhD in electronic engineering and integrated photonics in 1975 and 1980, respectively, both from the University of Western Australia in Nedlands, Western Australia. In 1980 he joined the Department of Electrical Engineering at Monash University after a three-year period as a research scientist with CSIRO Australia. He was appointed as a reader at Monash University in 1995.

Dr. Binh has worked on several major advanced world-class projects, especially including the DWDM optical transmission systems and networks for the Department of Optical Communications of Siemens AG Central Research Laboratories in Munich and the Advanced Technology Centre of Nortel Networks in Harlow, UK. He was the Tan-Chin Tuan Professorial Fellow of Nanyang Technological University of Singapore and a visiting professor of the Faculty of Engineering at Christian Albrechts University of Kiel, Germany. He was, for a short period of time, the director of the Center for Telecommunications and Information Engineering of Monash University.

Dr. Binh has published more than 250 papers in leading journals and refereed conferences as well as four books in the fields of photonic signal processing and digital optical communications: *Photonic Signal Processing* (2008), *Digital Optical Communications* (2009), *Optical Fiber Communications Systems* (2010), and *Ultra-Fast Fiber Lasers* (2010), all published by CRC Press/Taylor & Francis Group of Boca Raton, Florida (USA). His current research interests include advanced modulation formats for long haul optical transmission, electronic equalization techniques for optical transmission systems, ultra-short pulse lasers and photonic signal processing, and optical transmission systems and network engineering. Dr. Binh has developed and taught several courses in engineering, including Fundamental of Electrical Engineering, Physical Electronics, Small Signal Electronics, Large Signal Electronics, Signals and Systems, Signal Processing, Digital Systems, Micro-Computer Systems, Electromagnetism, Electromagnetic Wave Propagation, Wireless and Guided Wave Electromagnetics, Communications Theory, Coding and Communications, Optical Communications, Advanced Optical Fiber Communications, Advanced Photonics, Integrated Photonics, and Fiber Optics. He has also led several course and curriculum developments in electrical and computer systems engineering (E&CSE) and joint courses in physics and E&CSE.

From 2001–2009, Dr. Binh chaired Commission D: Photonics and Electronics for the National Committee for Radio Science of the Australian Academy of Science. In January 2011 he was appointed technical director of the advanced optical transmission laboratories for the European research center of Huawei Technologies in Munich, Germany. His team has recently demonstrated the world record in transmission of Tb/s channels over more than 3500 kilometers of installed fiber in European Networks.

1 Electric and Magnetic Fields and Waves

1.1 BRIEF OVERVIEW

Electromagnetic waves always consist of two orthogonal vectorial components, the electric and magnetic fields, whether they are in wireless or guided media. They are normally induced by the excitation from a time-varying source. Thus, in this chapter the time-varying electric and magnetic (EM) fields are described. That means the interdependence and dynamics of the electromagnetic fields and the formation and propagation of plane waves are assumed, which is important for the radiation, guiding, and propagation through either wireless or guided media. Essential fundamental understandings of the vector analyses are given.

The word *field* has been commonly used in several textbooks and published works since the generalization works by Maxwell. It indicates the strength of the electric or magnetic force and the spatial region under its influence.

1.2 WAVE REPRESENTATION

1.2.1 OVERVIEW

EM radiation is the electrodynamics of the electric and magnetic fields. Both components obey the principle of superposition when the medium in which they behave is linear. Thus, a field due to the time-varying electric or magnetic field will influence the property of the EM field strength of the medium. So what is the relationship between the variations of these two fields under the time-varying condition?

Maxwell derived a waveform of the electric and magnetic equations that constitute the wave-like nature of the electric and magnetic fields and their symmetric property. The speed of EM waves concurs with that of the measured speed of light; thus Maxwell concluded that the light itself is an EM wave. This conclusion has been extensively employed over the last century and currently is used in the guiding of lightwaves in optical fibers for ultra-high-speed optical communications. Later sections of this chapter will give insight into the derivation of the wave propagation equations and Maxwell's equations.

Accordingly, based on Maxwell's equations a spatially varying electric field will enforce the variation of the magnetic field over time. Likewise for the change relationship, when there exists a spatial change of the magnetic field leading to time variation in the electric field. In an EM wave, the electric field component under an influential change would shift the magnetic field in one direction, which in turns shifts the electric field in the same direction. Together these fields form the propagation mechanism of an EM wave that moves into space and never again affects the sources. However, this would be true in a vacuum or free space. If the EM waves are propagating in a charged medium, then depending on the impedance of the line, the waves can be reflected or absorbed or confined in a waveguide traveling at a reduced speed. The issues of confinement of EM waves in an optical or electrical waveguide will be dealt with in later chapters.

1.2.2 GENERAL PROPERTY

We consider in this book the continuous and time-harmonic waves represented by sine waves under steady-state conditions. An electric field propagating in the z-direction can be represented by

$$\underline{E}(z,t) = E_0 e^{-\alpha z} \cos(\omega t - \beta z + \phi)\underline{a}_z \tag{1.1}$$

where ω is the angular frequency of the wave ($\omega = 2\pi f$), where f is the frequency of the wave, β is the propagation of the waves along the z-direction, ϕ is the initial phase, and α is the attenuation constant of the propagation medium. It is noted here that the propagation/phase constant β can be estimated very straightforwardly if the waves propagating in a wireless medium are assumed to be air or in a vacuum. However, this is more complex if the waves propagate in a guided medium in which the propagation/phase constant depends on the modes of the waves confined in the transverse directions, and how tight the waves are confined in this plane. We will deal with these issues in the chapters following. Thus, this constant indicates how fast or how slow the waves are propagating through the medium. Sometimes this is called the wavenumber.

Assuming that the initial phase can be adjusted to zero and at the initial instant, i.e., at $t = 0$, in a lossless ($\alpha = 0$) medium the waves can be rewritten as

$$\underline{E}(z,0) = E_0 \cos(-\beta z)\underline{a}_z \tag{1.2}$$

This indicates that the phase/propagation constant β is related to the EM wavelength λ of such a wave in the propagation media by

$$\beta = \frac{2\pi}{\lambda} \tag{1.3}$$

with the wavelength taking the value in vacuum if the EM waves propagate in air or vacuum, and a lower value when under a guiding condition by an effective permittivity or refractive index.

Note that when the EM waves propagate in a medium, the frequency of the waves remains unchanged, and only the wavelength changes as the propagating velocity is slowed down by the permittivity constant of the medium. Thus, the wavelength varies accordingly. If in a guided medium, the wave velocity is slowed down by the effective permittivity of the medium, as seen by the waves under a guiding condition.

Under an attenuated medium with the coefficient along the z-direction, the electric field vector at the initial instant $t = 0$ can be represented as

$$\underset{\sim}{E}(z,0) = E_0 e^{-\alpha z} \cos(-\beta z)\underset{\sim}{a}_z \tag{1.4}$$

Thus, we could see that the amplitude of the wave is exponentially decreased with respect to the propagation distance z, indicating an oscillatory wave as a function of time, with its amplitude multiplied by the exponentially decreased coefficient $e^{-\alpha z}$.

Exercise

Plot the amplitude of an EM wave as a function of time represented by

$$\underset{\sim}{E}(z,t) = E_0 e^{-\alpha z} \cos(\omega t - \beta z + \phi)\underset{\sim}{a}_z$$

under the following conditions:

    a. $E_0 = 1.0$; $\alpha = 0$ dB/km; $\omega = 2\pi \cdot 10^9$ Hz, $\lambda = 1000$ nm; initialphase $= 0$
    b. $E_0 = 10$; $\alpha = 0$ dB/km; $\omega = 2\pi \cdot 10^9$ Hz, $\lambda = 1000$ nm; initialphase $= \pi/2$
    c. $E_0 = 10$; $\alpha = 1$ dB/km; $\omega = 2\pi \cdot 10^9$ Hz, $\lambda = 1000$ nm; initialphase $= 0$

with the total traveled distance of 10 km.

1.2.3 WAVES BY PHASOR REPRESENTATION

The wave represented in (1.1) can be rewritten, in the case where the attenuation constant is negligible, as

$$\underset{\sim}{E}(z,t) = \mathrm{Re}(E_0 e^{-\alpha z} e^{j\phi} e^{j(\omega t - \beta z)})\underset{\sim}{a}_z$$

$$\underset{\sim}{E}(z,t) = \mathrm{Re}(\vec{E}_s e^{j\omega t})\underset{\sim}{a}_z \tag{1.5}$$

$$\underset{\sim}{E}_s = \mathrm{Re}(E_0 e^{j\phi})\underset{\sim}{a}_z$$

This uses the property of the polar form representation of the sinusoidal wave as follows:

$$re^{j\phi} = r\cos\phi + jr\sin\phi = r \prec \phi \tag{1.6}$$

In general form we have

$$\underset{\sim}{E}_s = E_0(x,y,z)e^{j\phi}\underset{\sim}{a}_r \tag{1.7}$$

where $\underset{\sim}{a}_r$ is the unit vector along the radial direction of the phasor. That means that the sinusoidal representation is now transformed to complex form and taking the real

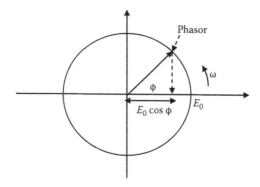

FIGURE 1.1   Phasor representation of the wave of amplitude E_0 and frequency ω with amplitude E_0. The vector rotates around the circle with an angular frequency ω.

part. This complex representation assists the manipulation of the waves using vectorial operation and, hence, simplifies the algebraic steps. The waves expressed in (1.5) can be represented in vector form as shown in Figure 1.1, in which the initial phase and angular frequency are represented by a vector of an initial angle ϕ. This vector is rotating around the circle of amplitude E_0 with an angular frequency of ω. The projection of this vector to the real horizontal axis of the complex plane is the sinusoidal wave. The attenuation factor can be incorporated according to the length of the propagation along the z-direction. This vector representation allows us to add or subtract several other wave vectors provided that they are oscillating at the same frequency.

1.2.4   Phase Velocity

Now consider a traveling wave in a lossless medium expressed by

$$\underset{\sim}{E}(z,t) = E_0 \cos(\omega t - \beta z + \phi)\underset{\sim}{a}_z \tag{1.8}$$

This wave is propagating in the $+z$-direction with a constant phase of

$$\omega t - \beta z = C \tag{1.9}$$

in which we assume a zero *initial* phase.

Then the phase velocity or the speed of the wave propagating along the z-direction is given by differentiating (1.9):

$$v_p = \frac{dz}{dt} = \frac{\omega}{\beta} = \lambda f \tag{1.10}$$

Now if the wave angular frequency is not a single frequency but composed of several other spectral components, such as in the case that the wave is modulated by a pulse envelope, then these spectral components are propagated along the z-direction with a group velocity of

$$v_g = \frac{dv_p}{d\omega} \tag{1.11}$$

Indeed, this group velocity allows us to evaluate the dispersion of carrier-modulated signals after propagating through a distance L in a wireless or guided medium, which will be described in later chapters of the book.

1.3 MAXWELL'S EQUATIONS

Maxwell's equations are formed with the unity of four equations, Ampere's circuital law, Faraday's law, and Gauss's laws for electric and magnetic fields, as listed in Table 1.1.

1.3.1 FARADAY'S LAW

Faraday's law is a fundamental relationship between a generated voltage and the electric field in a changing magnetic field. The induced electromotive voltage V_{emf} is related to the electric field $\underset{\sim}{E}$ and the magnetic field strength $\underset{\sim}{B}$ by

$$\oint \underset{\sim}{E} \cdot d\underset{\sim}{L} = V_{emf} = -\frac{\partial}{\partial t} \oiint_S \underset{\sim}{B} \cdot d\underset{\sim}{S} \tag{1.12}$$

where $d\underset{\sim}{L}$ is the vectorial differential length within a close loop and $d\underset{\sim}{S}$ is the differential area perpendicular to the surface. This equation involves the interaction of charges and magnetic field. Note that the vectorial directions of the fields and the surface and differential lengths must follow the right-hand rule.

1.3.2 AMPERE'S LAW

The continuity of the current flowing is the original form of Ampere's circuital law. This represents the relationship between the electric current source and the magnetic field. The law can be written in integral form, and its equivalent differential form via the Stokes' theorem is as follows:

$$\oint_C \underset{\sim}{H} \cdot d\underset{\sim}{L} = \oiint_S \underset{\sim}{J}_c \cdot d\underset{\sim}{S} + \frac{\partial}{\partial t} \oiint_S \underset{\sim}{D} \cdot d\underset{\sim}{S} \Leftrightarrow \nabla \times \underset{\sim}{H} = \underset{\sim}{J}_c + \frac{\partial \underset{\sim}{D}}{\partial t} \tag{1.13}$$

where

- $\oint_C$ is the closed line integral around the closed curve C.

TABLE 1.1
Fundamental Electromagnetic Equations

Equations	Differential Form	Integral Form
Gauss's law: Divergence of the displacement vector $\underset{\sim}{D}$ equals the charge density per unit volume ρ, or the total charge enclosing the product of the vector and the differential surface vector	$\nabla \cdot \underset{\sim}{D} = \rho$	$\oint\!\!\!\oint_S \underset{\sim}{D} \cdot d\underset{\sim}{S} = Q_{enc}$
Gauss's magnetic law: Likewise for the magnetic field density	$\nabla \cdot \underset{\sim}{B} = 0$	$\oint\!\!\!\oint_S \underset{\sim}{B} \cdot d\underset{\sim}{S} = 0$
Faraday's law: Differential variation of the electric field and temporal evolution of the magnetic field	$\nabla \times \underset{\sim}{E} = -\dfrac{\partial \underset{\sim}{B}}{\partial t}$	$\oint \underset{\sim}{E} \cdot d\underset{\sim}{L} = -\dfrac{\partial}{\partial t}\oint\!\!\!\oint_S \underset{\sim}{B} \cdot d\underset{\sim}{S}$
Ampere's circuital law: Relationship between magnetic field density and the charge flow rate and temporal variation of the electric field displacement	$\nabla \times \underset{\sim}{H} = \underset{\sim}{J}_c + \dfrac{\partial \underset{\sim}{D}}{\partial t}$	$\oint \underset{\sim}{H} \cdot d\underset{\sim}{L} = \oint\!\!\!\oint_S \underset{\sim}{J}_c \cdot d\underset{\sim}{S} + \dfrac{\partial}{\partial t}\oint\!\!\!\oint_S \underset{\sim}{D} \cdot d\underset{\sim}{S}$

Other EM Equations

Lorentz force equation

$$\underset{\sim}{F} = q(\underset{\sim}{E} + \underset{\sim}{v} \times \underset{\sim}{B})$$

E is the electric field, B is the magnetic field intensity, and q is the electronic charge moving with a velocity v through a magnetic field.

Constitutive equations: Electric/magnetic field density and the field with permittivity/permeability

$$\begin{cases} \underset{\sim}{D} = \varepsilon \underset{\sim}{E} \\ \underset{\sim}{B} = \mu \underset{\sim}{H} \\ \underset{\sim}{J} = \sigma \underset{\sim}{E} \end{cases}$$

ε, μ, σ = permittivity, permeability, and conductivity of the medium

Current continuity equation: 3D current density and rate of change of volume of charges

$$\nabla \cdot \underset{\sim}{J} = -\dfrac{\partial \rho_v}{\partial t}$$

- $\underset{\sim}{B}$ is the magnetic field in Weber.*
- The dot • is the vector dot product.
- $d\underset{\sim}{L}$ is an infinitesimal element (a differential) of the curve C (i.e., a vector with magnitude equal to the length of the infinitesimal line element, and direction given by the tangent to the curve C).

- $\displaystyle\oiint_S$ denotes an integral over the surface S enclosed by the curve C (the double integral sign is meant simply to denote that the integral is two-dimensional in nature.

- J_C is the free current density through the surface S enclosed by the curve C.
- $d\underset{\sim}{S}$ is the vectorial differential area of an infinitesimal element of the surface S (that is, a vector with magnitude equal to the area of the infinitesimal surface element, and direction normal to surface S). The direction of the normal must correspond with the orientation of C by the right-hand rule. I_{enc} is the net free current that penetrates through the surface S.

- $\displaystyle\oiint_S \underset{\sim}{J_c} \bullet d\underset{\sim}{S}$ represents the conduction current and $\dfrac{\partial}{\partial t}\displaystyle\oiint_S \underset{\sim}{D} \cdot d\underset{\sim}{S}$ the displacement current.

1.3.3 GAUSS'S LAW FOR ELECTRIC FIELD AND CHARGES

The electric flux through any closed surface is proportional to the enclosed electric charge Q_{enc}.

$$\oiint_S \underset{\sim}{D} \cdot d\underset{\sim}{S} = Q_{enc} \tag{1.14}$$

1.3.4 GAUSS'S LAW FOR MAGNETIC FIELD

The divergence of the magnetic field $\underset{\sim}{B}$ equates to zero; in other words, it is a solenoidal vector field. It is equivalent to the statement that magnetic monopoles do not exist.

$$\nabla \cdot \underset{\sim}{B} = 0 \tag{1.15}$$

1.4 MAXWELL EQUATIONS IN DIELECTRIC MEDIA

1.4.1 MAXWELL EQUATIONS

The general Maxwell equations can be written as

---

* The unit of the magnetic field intensity $\underset{\sim}{B}$ is 1 Weber $= \dfrac{Kg \cdot m^2}{C \cdot s}$.

$$\nabla \times E = -j\omega\mu H \tag{1.16}$$

$$\nabla \times H = J + j\omega\varepsilon E \tag{1.17}$$

$$\nabla \cdot D = \frac{\sigma}{\varepsilon_0} \tag{1.18}$$

$$\nabla \cdot B = 0 \tag{1.19}$$

Note that in the above equations $j = \sqrt{-1}$; $\mu \cong \mu_0$ is the magnetic permeability for nonmagnetic materials, which normally constitute an optical waveguide; $\varepsilon = \varepsilon_0 n^2$ is the dielectric constant of the material, where $e_{o\neg}$ is the dielectric constant of free space and n is the refractive index of the materials; J is the current density; and σ is the surface charge density, which is a possible source. The displacement vector D is related to the electric field via $D = \varepsilon E$, and the magnetic field induction B is related to the magnetic field via $B = \mu H$.

The first is Faraday's law of induction, the second is Ampere's law as amended by Maxwell to include the displacement current $\partial D/\partial t$, and the third and fourth are Gauss's laws for the electric and magnetic fields. The displacement current term $\partial D/\partial t$ in Ampere's law is essential in predicting the existence of propagating electromagnetic waves. Its role in establishing charge conservation is discussed in the previous section. The quantities E and H are the electric and magnetic field intensities and are measured in units of V/m and A/m, respectively. The quantities D and B are the electric and magnetic flux densities and are in units of Coulomb/m² and Weber/m², or Tesla. D is also called the electric displacement, and B the magnetic induction. The quantities ρ and J are the volume charge density and electric current density (charge flux) of any external charges (that is, not including any induced polarization charges and currents). They are measured in units of Coulomb/m³ and A/m². The right-hand side of (1.19) is zero because there are no magnetic monopole charges. The charge and current densities ρ, J may be thought of as the sources of the electromagnetic fields. For wave propagation problems, these densities are localized in space; for example, they are restricted to flow on an antenna. The generated electric and magnetic fields are radiated away from these sources and can propagate large distances to the receiving antennas—away from the sources, that is, in source-free regions of space.

In practice, problems of optical waveguide and couplers are often analyzed in the regions that are free of the above sources, i.e., $J = 0$ and $\sigma = 0$. In these cases, we have

$$\nabla \times E = -j\omega\mu H \tag{1.20}$$

$$\nabla \times H = j\omega\varepsilon E \tag{1.21}$$

$$\nabla \cdot (\varepsilon E) = 0 \tag{1.22}$$

$$\nabla \cdot H = 0 \tag{1.23}$$

1.4.2 WAVE EQUATION

With the usual expression of the time-dependent lightwave carrier-modulated signals $e^{j\omega t}$, the wave equation can thus be obtained as

$$\nabla x \nabla x \vec{E} - \varepsilon(\omega)\frac{\omega^2}{c^2}\vec{E} = 0 \qquad (1.24)$$

The refractive index of the medium can be related to the permittivity, including the nonlinear third-order effects. Using the relation:

$$\nabla x \nabla x \vec{E} = \nabla(\nabla.\vec{E}) - \nabla^2\vec{E} = -\nabla^2\vec{E}$$

$$\because \nabla.\vec{D} = 0$$

$$then\_the\_wave\_equation \qquad (1.25)$$

$$\nabla^2\vec{E} - n^2(\omega)\frac{\omega^2}{c^2}\vec{E} = 0$$

1.4.3 BOUNDARY CONDITIONS

In the regions free of the sources, we have the following boundary conditions:

- Continuity of the magnetic field and the component of the electric field tangential to the interface, i.e.,

$$H^{(1)} = H^{(2)}, E_{\parallel}^{(1)} = E_{\parallel}^{(2)} \qquad (1.26)$$

- Continuity of the normal component of the displacement vector, i.e.,

$$D_{\perp}^{(1)} = D_{\perp}^{(2)} \quad \text{or} \quad n_1^2 E_{\perp}^{(1)} = n_{\perp}^{(2)} \qquad (1.27)$$

1.4.4 RECIPROCITY THEOREMS

1.4.4.1 General Reciprocity Theorem

From the above source-free Maxwell equations, we have for two optical media of dielectric constants ε_1 and ε_2:

$$\nabla \times (\nabla \times E) = \omega^2 \varepsilon \mu E \qquad (1.28)$$

$$\nabla \times (\nabla \times H) = \omega^2 \varepsilon \mu H \qquad (1.29)$$

Using the above equations and the identity $\nabla \cdot (A \times B) = B \cdot (\nabla \times A) - A \cdot (\nabla \times B)$, we have

$$\nabla \cdot (E_1 \times H_2 - E_2 \times H_1) = j\omega(\varepsilon_2 - \varepsilon_1)E_1 \cdot E_2 \tag{1.30}$$

and its integral equivalence

$$\frac{\partial}{\partial z} \iint_{A_\infty} (E_1 \times H_2 - E_2 \times H_1) \cdot \hat{z}\, dA = j\omega \iint_{A_\infty} [\varepsilon_2(x,y) - \varepsilon_1(x,y)]E_1 \cdot E_2\, dA \tag{1.31}$$

1.4.4.2 Conjugate Reciprocity Theorem

Conjugate reciprocity theorem can be obtained in a similar way as above, except using the conjugate form of field expressions. This is particularly convenient in constructing the formulation for the lossless waveguides or couplers, in particular the expression of the power conservation. Following some algebra, we have

$$\nabla \cdot (E_\mu \times E_\nu^* + E_\nu^* \times H_\mu) = -j\omega(\varepsilon_\mu - \varepsilon_\nu)E_\mu \cdot E_\nu^* \tag{1.32}$$

$$\frac{\partial}{\partial z} \iint_{A_\infty} (E_\mu \times H_\nu^* + E_\nu^* \times H_\mu) \cdot \hat{z}\, dA =$$
$$- j\omega \iint_{A_\infty} (\varepsilon_\mu - \varepsilon_\nu)E_\mu \cdot E_\nu^*\, dA \tag{1.33}$$

1.5 CURRENT CONTINUITY

Consider a volume of charge Q enclosed in a volume surface S, which is varying as a function of time. If the volume is reduced, then the only possibility is that the charges are flowing through the surface. This flow of charges under the conservation of charges or energy is the current that must be equal to the rate of change of the contained charge. The flowing current I can be expressed in terms of the differential charge and the current density as

$$I = \oint_S \underset{\sim}{J} \cdot d\underset{\sim}{S} = -\frac{\partial Q}{\partial t} \tag{1.34}$$

where $\underset{\sim}{J}$ is the vectorial current density and t is the time variable $d\underset{\sim}{S}$, whose normal direction is the direction of the surface vector in the infinitesimal surface element and S is the surface area of the volume.

Using the divergence theorem[1] we can write

$$\oint_S \underset{\sim}{J} \cdot d\underset{\sim}{S} = -\oiint_V (\nabla \cdot \underset{\sim}{J}) dV = \int_V (\nabla \cdot \underset{\sim}{J}) dV \tag{1.35}$$

and

$$-\frac{\partial Q}{\partial t} = -\frac{\partial}{\partial t} \oiiint_V \rho_V \, dV \tag{1.36}$$

in which ρ_V is the charge density enclosed within the surface under consideration. Thus, comparing (1.34) to (1.36) we obtain

$$\nabla \cdot \underset{\sim}{J} = -\frac{\partial \rho_V}{\partial t} \tag{1.37}$$

   This is the point form of the current continuity equation, which indicates that the divergence of the current density equates to that of the differential changes of the charges flowing out of or into the volume under consideration. In the steady-state condition, when there is no change in the charge density, the conservation of charges leads to the fact that the total currents flowing into a node equate to the total current flowing out of a node. This is Kirchoff's current law, which is valid for transient or phasor steady-state conditions.

1.6   LOSSLESS TEM WAVES

Consider an x-polarized wave propagating along the $+z$-direction in a medium characterized by the material constants μ, ε and lossless; that is, the attenuation constant is zero. The field can be represented by

$$\underset{\sim}{E}(z,t) = E_0 \cos(\omega t - \beta z)\underset{\sim}{a}_x \tag{1.38}$$

Then applying the Faraday's law given in Table 1.1, we have

$$\nabla \times \underset{\sim}{E} = -\frac{\partial \underset{\sim}{B}}{\partial t} = -\mu \frac{\partial \underset{\sim}{H}}{\partial t}$$

$$\begin{vmatrix} \underset{\sim}{a}_x & \underset{\sim}{a}_y & \underset{\sim}{a}_z \\ \dfrac{\partial}{\partial x} & \dfrac{\partial}{\partial y} & \dfrac{\partial}{\partial z} \\ E_0 \cos(\omega t - \beta z) & 0 & 0 \end{vmatrix} = \beta E_0 \sin(\omega t - \beta z)\underset{\sim}{a}_y = -\mu \frac{\partial \underset{\sim}{H}}{\partial t} \tag{1.39}$$

Then by taking the integral of the differential component of the magnetic field flux we arrive at

$$\int d\underset{\sim}{H} = \underset{\sim}{H} = \frac{\beta E_0}{\omega \mu} \cos(\omega t - \beta z)\underset{\sim}{a}_y + C_1 \tag{1.40}$$

$$\therefore \underset{\sim}{H} = \frac{\beta E_0}{\omega \mu} \cos(\omega t - \beta z)\underset{\sim}{a}_y \qquad \because C_1 = 0 \qquad initial\_condition$$

where C_1 is the integral constant that turns out to be nil by using the initial condition. It is from (1.40) and (1.38) that the H and $\underset{\sim}{E}$ fields are orthogonal to each other, and their magnitudes differ by a scaling factor due to the characteristic of the medium and the propagation constant and radial frequency of the EM waves.

Now we can use the other Maxwell's equations to find the relationship between the propagation constant and the velocity of light $c = 1/\sqrt{\mu_0 \varepsilon_0}$ and the medium characteristic μ, ε, the permeability and permittivity.

$$\nabla \times H = J_C + \frac{\partial D}{\partial t} = \varepsilon \frac{\partial E}{\partial t} \rightarrow$$

$$\begin{vmatrix} \underset{\sim}{a_x} & \underset{\sim}{a_y} & \underset{\sim}{a_z} \\ \dfrac{\partial}{\partial x} & \dfrac{\partial}{\partial y} & \dfrac{\partial}{\partial z} \\ 0 & \dfrac{\beta}{\omega\mu} E_0 \cos(\omega t - \beta z) & 0 \end{vmatrix} = -\frac{\beta E_0}{\omega\mu} \frac{\partial}{\partial z} \cos(\omega t - \beta z) a_x \qquad (1.41)$$

$$= -\frac{\beta^2 E_0}{\omega\mu} \sin(\omega t - \beta z) a_x$$

therefore leading to

$$\rightarrow \qquad \frac{\partial E}{\partial t} = -\frac{\beta^2 E_0}{\omega\mu} \sin(\omega t - \beta z) a_x$$

$$\rightarrow \qquad E = -\frac{\beta^2 E_0}{\omega^2 \mu} \sin(\omega t - \beta z) a_x \quad \because \frac{\partial}{\partial t} = \omega \qquad (1.42)$$

The additional parameters of the amplitude of the electric field must be equal to top unity in order to satisfy the field expression assumed from the beginning. Thus we have

$$\beta = \omega \sqrt{\mu \varepsilon} \qquad (1.43)$$

And it follows that the phase velocity of the EM wave is given by

$$v_p = \frac{\omega}{\beta} = \frac{1}{\sqrt{\mu\varepsilon}} = \frac{1}{\sqrt{\mu_0\mu_r\varepsilon_0\varepsilon_r}} = \frac{c}{\sqrt{\mu_r\varepsilon_r}} \qquad (1.44)$$

Or the phase velocity of an EM wave is that of the velocity of light reduced by a factor related to the relative permeability and permittivity of the medium. It is noted in (1.44) that the propagation constant β is not restricted by the guiding condition but assumed as a plane wave in a medium. Under the guiding condition this propagation constant would be restricted by the confinement of the electric and magnetic components by the boundary conditions of the guided medium. We will describe these guiding conditions and the propagation constant or relative refractive indices as seen by the EM waves in such guided media in later chapters on guided waveguides.

Example

A y-polarized plane wave propagating along the x-direction and the medium is air. The frequency of the wave is 10 GHz, with a magnitude of 1.0 V/m. Write the expression for the electric field of this wave.

SOLUTION

$$\beta = \omega / c = 2\pi \times 10^{10} / 3.10^8 = \frac{2\pi}{3} \, rad / s, \underline{E} = 1.0 \cos\left(2\pi \times 10^{10} t - \frac{2\pi}{3} x \right) \underline{a}_y \, V / m$$

Example

A wave is propagating in a nonmagnetic medium whose magnitude and time-oscillating characteristic are given by

$$\underline{E}(x,t) = 10 \cos\left(10^8 \pi t + \frac{\pi}{10} x + \frac{\pi}{3} \right) \underline{a}_y \, V / m$$

Identify the frequency of the wave, its phase velocity, and the permittivity and permeability of the medium. Then, sketch the fields $\underline{B}$ and the flux $\underline{H}$.

SOLUTION

Inspecting the expression of the wave we have

$$f = \frac{10^8 \pi}{2\pi} = 5 \times 10^7 Hz = 50 \text{ MHz}$$

The propagation constant and the phase velocity of the wave are related by

$$\beta = \frac{2\pi}{3} = \frac{\omega}{v_p} \rightarrow v_p = \frac{3}{2\pi} 2\pi f \rightarrow 3f = 3 \times 50 \cdot 10^6 = 1.5 \times 10^8 \text{ m/s}$$

Thus, for the nonmagnetic medium ($\mu_r = 1$), we have

$$v_p = \frac{c}{\sqrt{\varepsilon_r}} \rightarrow \varepsilon_r = \frac{c^2}{v_p} = \frac{3 \times 10^8}{1.5 \times 10^8} = 2$$

The magnetic field density is then given as

$$H(x,t) = \frac{\beta E_0}{\omega \mu_0 \mu_r} \cos\left(10^8 \pi t + \frac{\pi}{10}x + \frac{\pi}{3}\right) a_z$$

$$H(x,t) = \frac{\frac{2\pi}{3} 10}{2\pi 5 \times 10^7 \mu_0} \cos\left(10^8 \pi t + \frac{\pi}{10}x + \frac{\pi}{3}\right) a_z$$

$$= \frac{2}{3\mu_0} \cos\left(10^8 \pi t + \frac{\pi}{10}x + \frac{\pi}{3}\right) a_z \qquad A/m$$

(c) The field vectorial directions can be sketched. They are straightforward and spatially orthogonal to each other.

1.7 MAXWELL'S EQUATIONS IN TIME-HARMONIC AND PHASOR FORMS

Given that the time-varying field is sinusoidal, substituting $\partial/\partial t = j\omega$ we have Maxwell's equations in differential form in which the subscript s stands for the phasor representation:

Equations	Differential Form
Gauss's law	$\nabla \cdot D_s = \rho$
Gauss's magnetic law	$\nabla \cdot B_s = 0$
Faraday's law	$\nabla \times E_s = -j\omega B_s$
Ampere's circuital law	$\nabla \times H_s = J_c + j\omega D_s$

1.8 PLANE WAVES

EM waves can be radiated from a source. Far away from the source the waves resemble a uniform phase front; this is called plane waves. In the uniform wave field the electric and magnetic components of the waves are orthogonal to the propagation direction—thus the name transverse electric and magnetic (TEM) waves.

1.8.1 GENERAL WAVE EQUATIONS

In this section it is assumed that the simple medium is linear, isotropic, and time invariant. Then Maxwell's equations can be written for this medium as

$$\nabla \cdot E = 0 \tag{1.45}$$

$$\nabla \cdot H = 0 \tag{1.46}$$

$$\nabla \times \underset{\sim}{E} = -\mu \frac{\partial \underset{\sim}{H}}{\partial t} \tag{1.47}$$

$$\nabla \times H = \sigma E + \varepsilon \frac{\partial E}{\partial t} \tag{1.48}$$

Taking the curl of both sides of (1.47) and the property of an isotropic medium we have

$$\nabla \times (\nabla \times E) = -\mu \frac{\partial}{\partial t} (\nabla \times H) \tag{1.49}$$

Then using the identity

$$\nabla \times \nabla A = \nabla \cdot A - \nabla^2 A \tag{1.50}$$

and combining (1.48) and (1.49) we arrive at

$$\nabla \cdot E - \nabla^2 E = -\mu\sigma \frac{\partial E}{\partial t} - \mu\varepsilon \frac{\partial^2 E}{\partial t^2} \tag{1.51}$$

Under the condition of charge-free, the divergence of is zero and (1.51) becomes

$$\nabla^2 E = +\mu\sigma \frac{\partial E}{\partial t} + \mu\varepsilon \frac{\partial^2 E}{\partial t^2} \tag{1.52}$$

This is the Helmholtz wave equation for the electric field component E. The medium is operating in the linear region; that is, the permittivity does not vary with the strength of the intensity of the field. We will see this in later chapters dealing with the nonlinear medium, especially in the case of guided waves in optical fibers with high power of the guided modes.

A similar equation can be derived for the magnetic component H without much difficulty.

Example

Write the wave equation for an EM wave whose field is expressed by

$$E(z,t) = \begin{pmatrix} E_x(z,t) \\ 0 \\ 0 \end{pmatrix}$$

propagating in a charge-free medium.

ANSWER

$$\frac{\partial^2}{\partial t^2} E_x(z,t) = +\mu\sigma \frac{\partial E_x(z,t)}{\partial t} + \mu\varepsilon \frac{\partial^2 E_x(z,t)}{\partial t^2}$$

1.8.2 TIME-HARMONIC WAVE EQUATION

When the wave is time harmonic, the wave equation can be written in term of the phasor of the electric field component as

$$\nabla^2 E_s = (j\omega\mu\sigma - \omega^2\mu\varepsilon)E_s \qquad (1.53)$$

Or, alternatively, as

$$\nabla^2 E_s - \gamma^2 E_s = 0 \qquad (1.54)$$

with $\gamma^2 = j\omega\mu\sigma - \omega^2\mu\varepsilon \rightarrow \gamma = \alpha + j\beta$

For the magnetic field component, we can similarly obtain

$$\nabla^2 H_s - \gamma^2 H_s = 0 \qquad (1.55)$$

Equations (1.54) and (1.55) are the well-known Helmholtz equations for time-harmonic fields propagating either in free space or in a confined and guided structure in the linear regions. When the waves propagate under the nonlinear regime of the medium, these equations would be modified to include a number of terms on the right-hand side (RHS). These equations would be termed as the nonlinear Schrodinger equations (NLSEs), which are described in later chapters on optical waveguiding in integrated and circular structures. These equations would then be subject to the boundary conditions, so that eigenvalue equations can be obtained, and thence the eigenvalues can be obtained. These values correspond to the propagation constant or wave vector that determines the speed of propagating in such guided structures.

Let us assume an x-polarized plane wave traveling in the z-direction whose electric field is given as

$$E_s(z) = E_{xs}(z)a_x \qquad (1.56)$$

The wave amplitude is solely dependent on the z-direction only. Then substituting this into the Helmholtz equation (1.54) we have

$$\frac{\partial^2 E_{xs}}{\partial z^2} - \gamma^2 E_{xs} = 0 \qquad (1.57)$$

This equation is a linear second-order and homogeneous differential equation. A possible solution for this equation is

$$E_{xs} = Ae^{\lambda z} \tag{1.58}$$

where A and λ are arbitrary constants. It is straightforward to prove that the differential equation can now be written as

$$\frac{\partial^2 E_{xs}}{\partial z^2} = \lambda^2 A E_{xs} e^{\lambda x} \tag{1.59}$$

Thence,

$$\lambda^2 - \gamma^2 = 0 \tag{1.60}$$

The only physical value of γ must be negative so that the wave does not grow to infinitive when the propagation distance becomes very large—thus with the time-dependent factor and the frequency component $e^{j\omega t}$. Both waves propagating in the forward and backward directions are possible. For the forward propagating waves with $A = E_0^-$ we have

$$E_{xsF} = Ae^{-\gamma z} \rightarrow E_{xsF} = E_0^- \cos(\omega t - \beta z) \tag{1.61}$$

Thus, we have

$$E_{xsF} = E_0^- e^{-\alpha z} \cos(\omega t + \beta z) \tag{1.62}$$

The waves in the backward direction take the form

$$E_{xsB} = E_0^+ e^{\alpha z} \cos(\omega t + \beta z) \tag{1.63}$$

Thence, the superposition of the two waves in the forward and backward directions arrives at

$$E_{xs} = E_0^- e^{-\alpha z} \cos(\omega t - \beta z) + E_0^+ e^{\alpha z} \cos(\omega t + \beta z) \tag{1.64}$$

By applying Faraday's law to the electric field the magnetic field can be found as

$$\nabla \times E_s = -j\omega\mu H_s = (-\gamma E_0^- e^{-\gamma z} + \gamma E_0^+ e^{\gamma z})a_y \tag{1.65}$$

Thus, we can obtain the magnetic field vector by expanding the curl of Equation (1.65) to give

$$H_s = \left(\frac{\gamma E_0^+}{j\omega\mu} e^{-\gamma z} - \frac{\gamma E_0^-}{j\omega\mu} e^{+\gamma z} \right) \underline{a}_y = (H_0^+ e^{-\gamma z} - H_0^- e^{+\gamma z}) a_y \qquad (1.66)$$

Hence, we can define the intrinsic impedance as

$$\eta = \frac{E_0^+}{H_0^+} = \frac{j\omega\mu}{\gamma} = -\frac{E_0^-}{H_0^-} \qquad (1.67)$$

with

$$\eta = \sqrt{\frac{j\omega\mu}{\sigma + j\omega\varepsilon}}$$

when the expression of γ given by (1.54) is used.

REFERENCE

1. E. Kreyszig, *Advanced Engineering Mathematics*, 8th ed., John Wiley & Sons, New York, 2008, Chapters 9 and 10.

2 Electrical Transmission Lines

In this chapter the transmission of waves at high frequency through transmission lines is studied. Transmission and reflection coefficients along transmission lines are important in the excitation of wire lines in transmission and receiving circuitry of radio frequency (RF) waves.

2.1 MODEL OF TIME-HARMONIC WAVES ON TRANSMISSION LINES

A transmission line is critical for waves to propagate from the source to the radiation devices, or from the receiving devices to the receiver. When the wavelength of the waves, the carrier, is compatible with the size of the transmission line, then lumped circuit treatment can no longer be valid. Both transmission and reflection of the waves can occur. This section gives an introduction to transmission lines and the characteristics related to the consideration of waves propagating through them.

2.1.1 DISTRIBUTED MODEL OF TRANSMISSION LINES

Consider a transmission whose dimension is compatible with the wavelength of the waves propagating through it, as shown in Figure 2.1. A differential distance Δz can be represented with its distributed parameters consisting of an inductor in series with a resistance and a shunt capacitance in parallel with a resistance. The inductor represents the effects of the induced magnetic field by the current flowing through the line, and its serial resistance represents the resistance of the line. The shunt admittance results from the charges, and hence potential, distributed on the positive- and negative-going lines of the transmission line. A capacitance is established due to the differential charges and the property of the insulator placed between them.

By applying Kirchoff's voltage law to the loop, we can write the relation between the voltages and current dropped and through the differential element and the currents and voltages at the terminals of the equivalent circuit of Figure 2.1 as follows:

$$v(z,t) - v(z + \Delta z, t) = i(z,t) R_d \Delta z + L_d \Delta z \frac{\partial i(z,t)}{\partial t} \qquad (2.1)$$

where the distributed capacitance, conductance, and inductance of the differential element of a coaxial transmission line (see Figure 2.1(c)) are given as

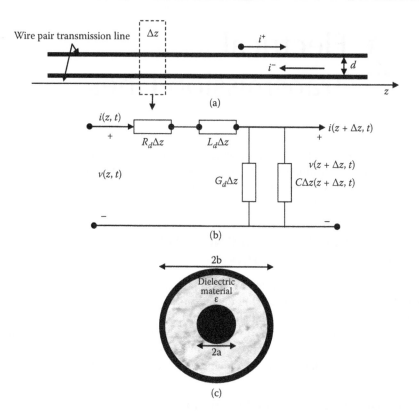

FIGURE 2.1 (a) Differential model of a distance Δz of a two-wire line, (b) its equivalent distributed parameters, and (c) geometrical cross section of the transmission line of a coaxial line.

$$C_d = \frac{2\pi\varepsilon}{\ln\dfrac{b}{a}} \; F/m; \qquad G_d = \frac{2\pi\sigma}{\ln\dfrac{b}{a}} \; S/m; \qquad L_d = \frac{\mu}{2\pi}\ln\frac{b}{a} \; H/m \qquad (2.2)$$

Then by reducing the distance $\Delta z \to 0$, we obtain the differential equation

$$\frac{\partial v(z,t)}{\partial z} = -i(z,t)R_d - L_d\frac{\partial i(z,t)}{\partial t} \qquad (2.3)$$

Similarly, by applying Kirchoff's current law at the node of the output voltage we have

$$\frac{\partial i(z,t)}{\partial z} = -v(z,t)G_d - C_d\frac{\partial v(z,t)}{\partial t} \qquad (2.4)$$

Equations (2.3) and (2.4) are commonly known as the *transmission line equations*, representing the relationship between the current and voltages and the resistance, inductance, and capacitance of a differential element of the lines.

2.1.2   TIME-HARMONIC WAVES ON TRANSMISSION LINES

The voltage applied to the transmission line is normally a data envelope modulating a carrier, e.g., in telecommunication microwave systems; thus we can represent the voltage as

$$v(z,t) = V(z)\cos(\omega t + \phi) \tag{2.5}$$

Or in terms of phasors:

$$v(z,t) = \text{Re}[V(z)e^{j\omega t}]$$

$$V(z) = V_{RMS}(z)e^{j\phi} \because V_{RMS} \triangleq \frac{V(z)}{\sqrt{2}} \tag{2.6}$$

where V_{RMS} is the root mean square value of the voltage $V(z)$. This means that the signal envelope can be represented by a vector with the root mean square (RMS) voltage amplitude and a phase ϕ, which is rotating with an angular frequency ω.

Thus the transmission line equations can be written in phasor forms as

$$\frac{\partial \tilde{V}(z,t)}{\partial z} = -\underset{\sim}{I}(z,t)R_d - L_d \frac{\partial \tilde{I}(z,t)}{\partial t} = -(R_d + j\omega L_d)\tilde{I}(z,t) \tag{2.7}$$

$$\frac{\partial \tilde{I}(z,t)}{\partial z} = -\tilde{V}(z,t)G_d - C_d \frac{\partial \tilde{V}(z,t)}{\partial t} = -(G_d + j\omega C_d)\tilde{V}(z,t) \tag{2.8}$$

Instead of solving these two equations, let us save some time by comparing them with the equations that arose from Maxwell's equations, $\nabla \times \underset{\sim}{E} = -j\omega\mu\underset{\sim}{H}$ and $\nabla \times \underset{\sim}{H} = (\sigma + j\omega\mu)\underset{\sim}{E}$, and assuming that only components in the z-direction exist in the electric field and magnetic field density. These equations result in exact expression, as given in Equations (2.7) and (2.8).

2.1.2.1   Characteristic Impedance

The characteristic impedance of a transmission line is defined as the ratio of the amplitudes of positive traveling voltage over that of the current of the waves, or

$$Z_0 = \frac{V_0^+}{I_0^+} = -\frac{V_0^-}{I_0^-} \tag{2.9}$$

With the voltage relation with the propagation parameters expressed in terms of the distributed parameters as

$$V(z,t) = V_0^+ e^{-\gamma z} + V_0^- e^{+\gamma z}$$

$$I(z,t) = I_0^+ e^{-\gamma z} + I_0^- e^{+\gamma z} \tag{2.10}$$

$$\gamma = \sqrt{j\omega\mu(\sigma + j\omega\varepsilon)} = \sqrt{(R_d + j\omega L_d)(G_d + j\omega C_d)}$$

we have

$$\frac{d}{dz}(V_0^+ e^{-\gamma z} + V_0^- e^{+\gamma z}) = -(R_d + j\omega L_d)(I_0^+ e^{-\gamma z} + I_0^- e^{+\gamma z}) \tag{2.11}$$

Then by substituting this differentiation with respect to z, we obtain

$$Z_0 = \frac{V_0^+}{I_0^+} = \sqrt{\frac{R_d + j\omega L_d}{G_d + j\omega C_d}} \tag{2.12}$$

Practical transmission lines are made of highly conducting materials, such as copper; thence the resistance becomes negligible. The insulating layer is also made of good dielectric materials such as Teflon, so that the conductance G is also very small. Thus we could see that the propagation constant is

$$\gamma = \alpha + j\beta = j\beta$$

$$\beta = \omega\sqrt{L_d C_d} \tag{2.13}$$

The propagation group velocity delay is

$$v_p = \frac{\omega}{\beta} = \frac{1}{\sqrt{L_d C_d}} \tag{2.14}$$

The characteristic impedance becomes

$$Z_0 = \frac{V_0^+}{I_0^+} = \sqrt{\frac{L_d}{C_d}} \tag{2.15}$$

The phase velocity is also commonly known for a transmission line or plane wave propagating in a medium with permittivity ε and permeability μ as

$$v_p = \frac{1}{\sqrt{\mu\varepsilon}}$$

(2.16)

Thus for a nonmagnetic medium we have μ_r, and thence

$$v_p = \frac{c}{\sqrt{\varepsilon_r}}$$

(2.17)

Then the characteristic impedance in (2.15) for a nonmagnetic medium is given as

$$Z_0 = \frac{V_0^+}{I_0^+} = \sqrt{\frac{L_d}{C_d}} = \frac{1}{(2\pi)^2}\frac{\mu}{\varepsilon}\ln\left(\frac{2b}{a}\right) = \frac{60}{\varepsilon^{1/2}}\ln\left(\frac{2b}{a}\right)$$

(2.18)

Example

A 1.0 mm diameter copper wire coaxial line has 1.0 mm thick Teflon and is jacketed by copper mesh wire. Assuming that this coaxial line is lossless, find (a) the propagation phase velocity and (b) the characteristic impedance of the line. (c) If the characteristic impedance is not equal to 50 Ω, what is the diameter of the Teflon to make it equal to 50 Ω?

SOLUTION

(a) For Teflon ε_r = 2.1; thus the propagation phase velocity is

$$v_p = \frac{c}{\sqrt{\varepsilon_r}} = \frac{3.10^8}{\sqrt{2.1}} = 2.1\times10^8 \text{ m/s}$$

(b) The characteristic impedance is

$$Z_0 = \frac{60}{\varepsilon^{1/2}}\ln\left(\frac{2b}{a}\right) = \frac{60}{2.1^{1/2}}\ln\left(\frac{2\times1.5}{0.5}\right) = 46\Omega$$

For (c) we can use these equations, but with known characteristic impedance of 50 Ω, the radius a and thence diameter can be found.

2.2   TERMINATED TRANSMISSION LINES

2.2.1   TERMINATED LINE

In practice transmission lines are terminated, connected to loads; such terminated transmission lines operating under the propagation of waves would create

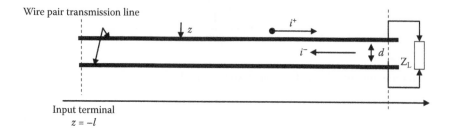

Wire pair transmission line

Input terminal
$z = -l$

FIGURE 2.2 Transmission line loaded with the load Z_L. Note the origin is located at the load.

transmitting and reflecting waves similar to any other waves of propagation phenomena, such as sounds in acoustic media, as shown in Figure 2.2.

At the load we have

$$Z_L = \frac{V_s(z=0)}{I_s(z=0)} = \frac{V_0^+ e^{-\gamma z} + V_0^- e^{+\gamma z}}{I_0^+ e^{-\gamma z} + I_0^- e^{+\gamma z}} \tag{2.19}$$

or

$$Z_L = \frac{V_0^+ + V_0^-}{I_0^+ + I_0^-} \tag{2.20}$$

Thence using the definition of the characteristic impedance Z_0, we arrive at

$$Z_L = Z_0 \frac{V_0^+ + V_0^-}{V_0^+ - V_0^-} \tag{2.21}$$

which can be rearranged as

$$V_0^- = Z_0 \frac{Z_L - Z_0}{Z_L + Z_0} V_0^+ \tag{2.22}$$

2.2.2 Reflection Coefficient

It is useful if one can define the reflection and transmission coefficient of EM waves propagating at the load. The reflection coefficient, Γ, is a complex quantity, including both amplitude and phase, and is given as

$$\Gamma = \frac{V_0^-}{V_0^+} = \frac{Z_L - Z_0}{Z_L + Z_0} \tag{2.23}$$

This equation is directly derived from (2.22). In general the reflection at any point z along the transmission line can be written as

$$\Gamma(z) = \frac{V_0^- e^{+\gamma z}}{V_0^+ e^{-\gamma z}} = \Gamma_L e^{+2\gamma z} \tag{2.24}$$

The reflection coefficient varies from 0 to 1. The reflected wave can be superimposed on the incident wave, and thus form a standing wave whose voltage standing wave ratio (VSWR) is defined as

$$VSWR = \frac{1+|\Gamma|}{1-|\Gamma|} \tag{2.25}$$

VSWR varies from 1 to infinity.

2.2.3 INPUT LINE IMPEDANCE

At any point along the transmission line one can find the input impedance looking into the line by finding the ratio of the total voltage to the total current at the point considered. The input impedance at the input, i.e., at $z = -l$, as indicated in Figure 2.2, is

$$Z_{in} = \frac{V_{z=-l}}{I_{z=-l}} = Z_0 \frac{V_0^+ e^{+\lambda l} + V_0^- e^{-\lambda l}}{I_0^+ e^{-\lambda l} - I_0^- e^{+\lambda l}} \tag{2.26}$$

Alternatively, we can manipulate the expression (2.26) to arrive at

$$Z_{in} = \frac{V_{z=-l}}{I_{z=-l}} = Z_0 \frac{Z_L + Z_0 \tanh(\gamma l)}{Z_0 + Z_L \tanh(\gamma l)} \tag{2.27}$$

If the line is lossless, we have

$$Z_{in} = \frac{V_{z=-l}}{I_{z=-l}} = Z_0 \frac{Z_L + Z_0 \tanh(\beta l)}{Z_0 + Z_L \tanh(\beta l)} \tag{2.28}$$

The use of this equation is that we can replace the line and the load with an equivalent input impedance looking into the line, or at a particular distance from the load we can replace the line and the load with equivalent impedance. This is particularly important when the load is not equal to the standard impedance of 50 Ω; then one can move away from this load a distance z such that the input impedance at this point reaches the standard level.

Examples

Consider a lossless transmission line with a characteristic impedance of 50 Ω. A load of 100 Ω is connected to the line and a voltage source is connected to the input line, the phasor of which is represented by $V_s = 10e^{j(\pi/3)}$ V. (a) Represent the circuit diagram of the source, transmission line, and load. (b) Find the equivalent impedance of the line and the load. (c) Hence find the voltage dropped across the input of the line.

SOLUTION

Considering the equivalent circuit given in Figure 2.3, we can represent the full circuit by an equivalent circuit with the characteristic impedance, the load, and the generator source. Thus,

$$V_{in} = V_s \frac{Z_{in}}{Z_s + Z_{in}} = V_{s(z=-l)} \tag{2.29}$$

Then at the input $z = -l$ we have

$$V_s(z) = V_0^+ e^{-\gamma z} + V_0^- e^{+\gamma z} = V_0^+(e^{-\gamma z} + \Gamma_L e^{+\gamma z})$$

$$at\ z = -l(\text{input}) \rightarrow V_0^+ = \frac{V_{in}}{e^{-\gamma l} + \Gamma_L e^{+\gamma l}} \tag{2.30}$$

At $z = 0$ (at the load end) we have $V_s(0) = V_0^+(1 + \Gamma_L)$,

$$\Gamma_L = \frac{Z_L - Z_0}{Z_L + Z_0} = \frac{100 - 50}{100 + 50} = 1/3 \tag{2.31}$$

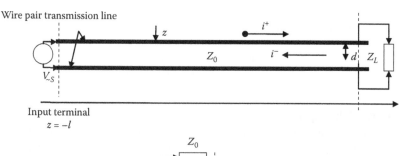

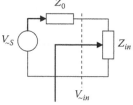

FIGURE 2.3 Source, transmission line, and load for a full circuit (top) and equivalent circuit (bottom) with Z_{in} as the equivalent load and line impedance.

Thus, the input impedance is

$$Z_{in} = \frac{V_{z=-l}}{I_{z=-l}} = Z_0 \frac{Z_L + Z_0 \tanh(\beta l)}{Z_0 + Z_L \tanh(\beta l)}$$

(2.32)

$$\beta l = \frac{2\pi}{\lambda} \frac{\lambda}{4} = \pi/2 \rightarrow \tanh(\beta l) = \infty \rightarrow Z_{in} = \frac{Z_0^2}{Z_L} = 25\Omega$$

Thus, we have

$$V_{in} = V_s \frac{Z_{in}}{Z_s + Z_{in}} = V_{s(z=-l)} = \frac{25}{25 + 25} = 1/2V_s = 5e^{j\pi/3}$$

$$V_0^+ = \frac{V_{in}}{e^{-\gamma l} + \Gamma_L e^{+\gamma l}} = \frac{5e^{j\frac{\pi}{3}}}{e^{-j\pi/2} + \frac{1}{3}e^{+j\pi/2}} = 7.5e^{-j\pi/3}$$

(2.33)

$$\therefore V_L = V_0^+(1 + \Gamma_L) = 7.5e^{-j\pi/3}(1 + 1/3) = 10e^{-j\pi/3}$$

$$\rightarrow v_L = 10\cos\left(\omega t - \frac{\pi}{3}\right) volts$$

2.3   SMITH CHART

The Smith chart is a graphical tool for use with transmission lines of RF and microwave components and circuits. It has been used for many years as a display on microwave measurement equipment, charts for student calculations, etc.

For use of the Smith chart, a number of steps must be conducted:

a. The line impedance must be normalized with respect to the characteristic impedance of the line.
b. Admittance can be used, but all the notations indicated in Figure 2.4 must be corrected (reverse).
c. The line impedance can be located on the chart by finding the intersection between the circles representing the normalized resistance and susceptance of the line impedance.
d. To find the input impedance to the line at a distance L, one must calculate the length of the line in terms of the wavelength of RF waves propagating in the line. Then the impedance is rotated by a distance equivalent in the Smith chart and the given impedance is read at the intersection point after the rotation. This is the normalized impedance looking into the line at a distance L from the load. One should make sure that the rotation direction is correct. Clockwise direction is for rotating toward the load, and anticlockwise is for rotating toward the generator.
e. Moving a full circle is equivalent to a half-wavelength distance or a quarter-wavelength for half-circle rotating.

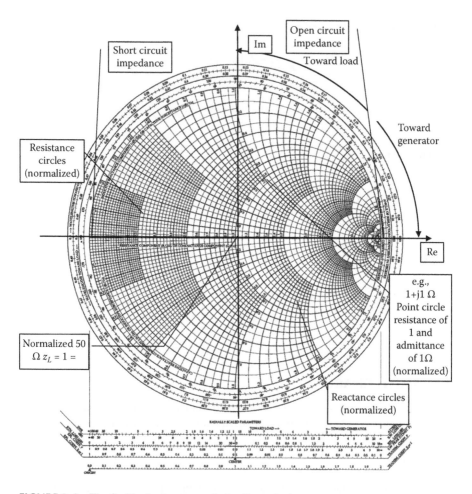

FIGURE 2.4 The Smith chart or transmission line calculator.

The circles of the Smith chart are defined by the following equations.[1] The reflection coefficients at any point along the transmission line given in (2.27) or (2.28) consist of a real part and imaginary parts. If normalized with respect to the characteristic impedance, this reflection coefficient at the load can be represented as

$$\Gamma = \frac{V_0^-}{V_0^+} = \frac{\dfrac{Z_L}{Z_0} - 1}{\dfrac{Z_L}{Z_0} + 1} = \frac{z_L - 1}{z_L + 1} = \frac{|V_{0\,max}|}{|V_{0\,min}|} \tag{2.34}$$

The reflection coefficient can also be given as the ratio between the maximum and minimum voltage levels of the standing wave formed by the forward and reflected waves along the transmission line. This is why the name VSWR is used to indicate the standing ratio.

In general, any point $z = l$ along the line can be written as

$$\Gamma_{z=l} = \Gamma_{z=0}e^{2\beta l} = \frac{z-1}{z+1} \tag{2.35}$$

where

$$z = \frac{Z_{z=l}}{Z_0} = r + jx$$

and the factor of $2\beta l$ indicates the total phase of the waves transmitting and reflecting from the load.

Thus, we can display the loci of the normalized impedances along the transmission lines by a number of circles for pure resistance and reactance as follows.

For resistance circles,

$$\left(\Gamma_{Re} - \frac{1}{2}\right)^2 + \Gamma_{Im}^2 = \frac{1}{4} \tag{2.36}$$

with Γ_{Re} and Γ_{Im}^2 being real and imaginary parts of the reflection coefficients.

For the reactance circles,

$$(\Gamma_{Re} - 1)^2 + (\Gamma_{Im} - 1)^2 = 1 \tag{2.37}$$

The illustration of these reactance and resistance curves on the Smith chart is given in Figure 2.5, also identified as the point of the normalized impedance $1 + j1 \ \Omega$.

Example

A transmission line of a characteristic impedance of 50 Ω is loaded with a load whose impedance is 50 + j100 Ω.

  a. Sketch the load and the transmission line with normalized values. Using the Smith chart, find the reflection coefficient at the load in terms of magnitude and phase.
  b. Find the location on the transmission line from the load where the input impedance is a pure resistance. You can also extend the line to achieve this at the load location.
  c. If moving from the load to a distance equivalent to ¼ wavelength, what is the impedance looking into the line at this point?

2.4   IMPEDANCE MATCHING

In circuit theory, for maximum power transfer and according to Thevenin's theorem on load matching, the load must be a complex conjugate of the source equivalent

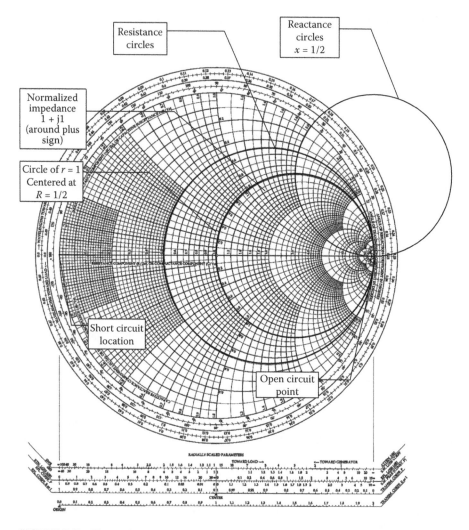

FIGURE 2.5 Illustration of reactance and resistance circles of the Smith chart. Also, the normalized impedance $1 + j1$ is identified as the intersection of two circles.

impedance. In transmission lines we can consider that the source impedance includes the source and the line, Z_S, Z_l, respectively, as shown in Figure 2.6. The impedance matching is a uniform generation of resistivity for alternating current; the quantity is a complex measure in term of ohms and is a function of frequency. A transmission line in circuit theory is dependent on the length and self-impedance, i.e., the per unit length resistance and capacitance or inductance. The transmission can be tuned so that the line impedance, together with the source, can be matched to that of the load impedance, which is normally varied with respect to frequency. The line impedance can be tuned by a parallel or series connection, shown as Z_{line} in Figure 2.6. The line impedance is adjusted so that the equivalent source and line impedance are the conjugate of the load impedance. Once the impedance is determined, the length of

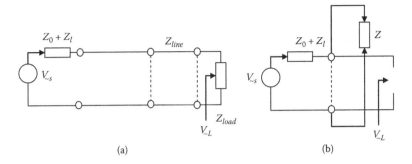

FIGURE 2.6 Equivalent load and internal plus line impedance for maximum power transfer or matching: (a) serial tuning line and (b) parallel tuning line.

the line can be found by using the Smith chart as described in the previous section, depending on whether serial or parallel lines are employed, as shown in Figure 2.6.

In integrated optical devices, transmission lines (see Chapter 5 on 3D optical waveguides and optical modulation) are commonly used in order to guide the electrical microwave signals along the electrodes so that a traveling wave electric field can be generated, and hence the interaction with the guided lightwaves, modulation of its phase, and then the interference or electroabsorption effects to change the intensity or phase based on some code, such as Gray code. The impedance matching is essential in order to avoid any reflection of the electrical waves, hence minimizing the interfering effects to the signal sources, especially when the microwave is amplified to a significant magnitude for maximum modulation level.

The traveling wave electrodes are sometimes called the microstrip lines and play a major role in the modulation of the lightwaves and the quality of the optical signals, depending on the matching of the microwave signals, especially when the frequency and bandwidth operation are very wide, reaching the order of several gigahertz, e.g., 30 GSymbols/s.

The reflection coefficient can be calculated using

$$\Gamma = \frac{Z_2/Z_1 - 1}{Z_2/Z_1 + 1} \tag{2.38}$$

where Z_1, Z_2 are the impedances of the first and second elements in cascade connection, respectively. An impedance mismatching would create a reflected wave and form a standing wave with the forward wave. Then a VSWR can be derived to indicate the degree of mismatch of the transmission line by the ratio between the maximum and minimum levels of the voltage levels of the standing wave along the transmission line. The higher the reflection, the larger the difference between the maximum and minimum levels, and thus the VSWR, defined as

$$VSWR = \frac{V_{max}}{V_{min}} = \frac{1+|\Gamma|}{1-|\Gamma|} \tag{2.39}$$

Example 1: A 50 Ω transmission line is terminated with a load of 30 Ω. Estimate the reflection coefficient and thence the VSWR.

Example 2: Continuing from Example 1, if the load is $30 + j30$ Ω, then estimate the VSWR and the magnitude and phase of the reflection coefficient using an analytical or Smith chart. Hint: Use Equations (2.38) and (2.39).

When the series impedance $Z_{line}^2 = Z_o Z_L$, then the length of the line for matching must be equal to a quarter of the wavelength of the RF wave. This follows from Equation (2.35) with the phase coefficient $\beta_l = \pi/4$.

Example

For a single-section quarter-wavelength transformer, determine the impedance of the line that would be placed in series with the line so that it could match to a load of 200 Ω at 100 MHz and a VSWR on the line remain less than 1.5. Given that the velocity of wave traveling in the RF cable is about 2×10^8 m/s, state the condition for matching and then the length of the cable required for no reflection.

   Hint: The quarter-wavelength section has impedance $Z_{line}^2 = Z_o Z_L = 50.200 \rightarrow Z_{line} = 100$ Ω. Using Equation (2.39) for VSWR gives a value of less than 1.5. Using the condition that the product of the propagation constant and the length of the RF cable equals $\pi/4$ at 100 MHz, one can find the length of the cable without much difficulty.

2.5 EQUIPMENT

This section gives a brief experimental setup for the study of transmission lines using different loads. The equipment is simple, using an RF signal generator, a stub as an RF transmission line, loads of different values, including open and short load impedance, and a spectrum analyzer for measuring the power of monitored RF points.

2.5.1 APPARATUS

The elements of the apparatus are as follows:

- A DC power supply and unit oscillator (or an RF synthesizer) capable of generating a sinusoidal output with a variable frequency range in the region of 1000 MHz.
- An attenuating pad to improve the matching between the synthesizer and the slotted line.
- A slotted line equipped with an RF diode detector and a tuning stub.
- A millivolt meter to measure the rectified output from the diode detector.
- A high-frequency terminator, which at different stages of the experiment will be a matched 50 Ω termination, a short circuit, or a load under test (LUT).
- A quarter-wave λ/4 transformer.

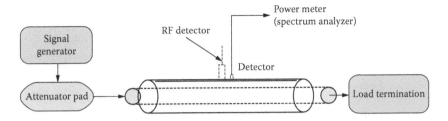

FIGURE 2.7 Experimental setup for transmission line study.

2.5.2 Experimental Setup

It is the responsibility of experimenters to note the details of the setup, including the dimensions and other related parameters. Please note that since 2006 newer and more accurate slotted lines have been used. You must note the model number and types of slotted lines. (See Figure 2.7.)

2.5.3 Notes on the Slotted Lines

The slotted line is designed to measure accurately the voltage standing wave pattern produced by any load. Basically, it is a rigid coaxial cable whose outer conductor is split by a narrow longitudinal slot approximately 500 mm long. The characteristic impedance of the line is approximately 50 Ω in the frequency region of interest.

A small shielded probe extends through the slot into the space between the inner and outer conductors. Its purpose is to capacitively couple to the transmission line and produce a voltage proportional to that existing between the inner and outer conductors of the line at the probe position. The probe is mounted on a carriage that slides along the outer conductor to any desired position.

Furthermore, on the carriage is a detector tuning stub whose purpose is to increase the sensitivity of the detector. Why and how?

2.5.4 Experiment

- Tune the unit oscillator to 1250 MHz. With a 50 Ω termination element connected to the slotted line, adjust the detector tuning stub to maximize the millivolt meter reading or the reading on the spectrum analyzer.
  - Connect the *short-circuit termination* to the slotted line and adjust the probe until the millivolt meter indicates a maximum or the spectrum analyzer indicates at maximum.
  - Adjust the output level control on the unit oscillator so that the millivolt meter reading lies between 1.0 and 3.0 mV. Note: Below 3 mV the diode detector operates in its square-law region. Above 3 mV the detector law changes, leading to difficulty for performing the measurements quantitatively. (Why?) On the other hand, if the reading is too low, the accuracy with which the voltage minima can be measured is compromised. The output of the detector, V_{DC}, in the square-law region of operation is related to the transmission line voltage phasor $\underline{V}$ by $V_{DC} = const|\underline{V}|^2$,

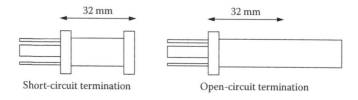

FIGURE 2.8   Typical short circuit and open loads with *N*-type connectors.

where *const* = the constant dependent on the location and phasor voltages and is unimportant if the ratios of the phasor voltages are used.

• To plot the voltage standing wave (VSW) patterns and calculate the VSWR it is only necessary to use $\sqrt{V_{DC}}$.

• With the *short-circuit termination* still connected, record the millivolt meter readings as a function of the probe position, starting from the load end. Continue the readings until at least two nulls are passed.

   • Plot the VSW pattern that is the transmission line voltage magnitude as a function of the probe position.

   • In your laboratory book you should note and compare your experimental results and those obtained from the theoretical expression as discussed in the above section.

• Determine the wavelength λ from the VSW pattern and compare it with that predicted from the unit oscillator frequency. By measuring multiples of a half-wavelength toward the short circuit from the voltage null positions on the slotted line scale, locate the electrical position of the short circuit.

   • Note: According to the manufacturer (General Electric), for the WN3 termination it lies 32 mm beyond the connector reference plane, as shown in Figure 2.8.

• Repeat the VSWR measurements with the WO3 open-circuit termination connected to the slotted line. Plot the results on the same graph, confirm the wavelength, and check that the electrical position of the open circuit is 32 mm beyond the connector reference plane as stated by the manufacturer.

• Fit the 25 Ω termination by connecting in parallel two 50 Ω termination elements to the branches of a 50 Ω T-piece. Determine the VSWR:

$$VSWR = \frac{\left|V\right|_{\sim max}}{\left|V\right|_{\sim min}}$$

• Compare your result with the theoretical value for resistive terminations:

$$VSWR = \frac{Z_0}{Z_L} \quad \text{or} \quad \frac{Z_L}{Z_0} \quad \text{whichever is } > 1$$

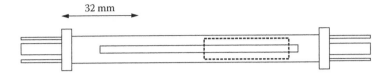

32 mm

FIGURE 2.9   A slotted line transmission piece.

- Again use the location of the voltage maxima and minima to find the electrical location of the 25 Ω load.
- A quarter-wavelength, λ/4, of transmission line of characteristic impedance terminated in the impedance Z_L presents at its input an impedance Z_l^2 / Z_L. Thus, a quarter-wave transformer of impedance 35.55 Ω will match the 25 Ω load to the 50 Ω impedance of the slotted line.
  - Now use the supplied λ/4 transformer as shown in Figure 2.9. For most of its length this device is a transmission line with a characteristic impedance of 50 Ω. Within a small adjustable section labeled λ/4, the outer diameter is reduced to produce a characteristic impedance close to 35 Ω.
  - Measure the length of the adjustable section and calculate the frequency at which it is λ/4. Adjust the unit oscillator to this frequency and repeat the measurements. That is, retune the detector and set the signal level again.
  - Connect the 25 Ω to the λ/4 transformer. Use the support provided to avoid straining the connector.
  - Adjust the sliding λ/4 section so that the edge nearest to the load is approximately λ/4 from the electrical location of the 25 Ω load. Observe the VSWR, and by carefully moving the sliding section back and forth slightly, determine the minimum VSWR that you can achieve.

2.5.5   TIME-DOMAIN REFLECTOMETRY

A time-domain reflectometer (TDR) is an electronic instrument used to characterize and locate faults in metallic cables (for example, twisted wire pairs, coaxial cables). It can also be used to locate discontinuities in a connector, printed circuit board, or any other electrical path. The equivalent device for optical fiber is an optical time-domain reflectometer. The operation principle is based on the propagation time and reflection of short pulses generated at the sources and detection of reflected pulses from any discontinuities or impedance mismatches along the transmission lines.

A typical setup of TDR is shown in Figure 2.10, consisting of a short pulse signal generator. The output of the generator is split into two parts. One is to be launched into the transmission line under test, and the other is fed to a signal processor, also triggering an oscilloscope. A number of different loads can be placed at the other end of the transmission line. We consider the following cases of loads.

Consider the case where the far end of the cable is shorted (that is, it is terminated into 0 Ω impedance). When the rising edge of the pulse is launched down the cable,

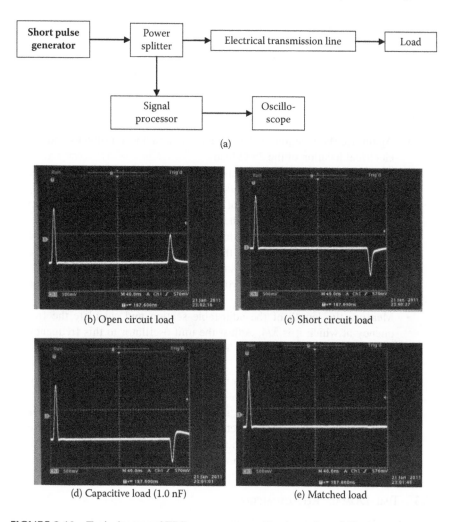

FIGURE 2.10 Typical setup of TDR measurement: (a) schematic and (b)–(e) typical oscilloscope traces under different loads *s* indicated.

the voltage at the launching point "steps up" to a given value instantly, and the pulse begins propagating down the cable toward the short. When the pulse hits the short, no energy is absorbed at the far end. Instead, an opposing pulse reflects back from the short toward the launching end. It is only when this opposing reflection finally reaches the launch point that the voltage at this launching point abruptly drops back to zero, signaling the fact that there is a short at the end of the cable. That is, the TDR has no indication that there is a short at the end of the cable until its emitted pulse can travel down the cable at roughly the speed of light and the echo can return back up the cable at the same speed. It is only after this round-trip delay that the short can be perceived by the TDR. Assuming that one knows the signal propagation speed in the particular cable under test, the distance to the short can be measured.

The propagation speed of electrical waves along coaxial cable is about $c/1.5$, with c the speed of light in vacuum. This is due to the fact that the electrical of the TEM mode traveling along the cable sees an effective permittivity of about two-thirds that of air.

A similar effect occurs if the far end of the cable is an open circuit (terminated into infinite impedance). In this case, though, the reflection from the far end is polarized identically with the original pulse and adds to it rather than cancelling it out. So after a round-trip delay, the voltage at the TDR abruptly jumps to twice the originally applied voltage.

Note that a theoretical perfect termination at the far end of the cable would entirely absorb the applied pulse without causing any reflection. In this case, it would be impossible to determine the actual length of the cable. Luckily, perfect terminations are very rare, and some small reflection is nearly always caused.

The magnitude of the reflection is referred to as the reflection coefficient Γ, as described in Section 2.2.2. The coefficient ranges from 1 (open circuit) to –1 (short circuit). The value of zero means that there is no reflection. The reflection coefficient can be calculated using Equation (2.34). Different loads would give different reflections of waves. The incident and reflected waves are superimposed on each other and standing waves are formed. However, due to the width of the pulses being short, there is no overlapping of the waves, and hence there are traces of incident and reflected pulses, as shown in Figure 2.10(c)–(e). It is obvious that when the load is matched to the transmission line, no reflection at all can be observed from (e). For a short-circuit load the reflection coefficient is negative, and thus the inverse of the pulse on the oscilloscope trace. Similarly for an open circuit, we see the positive reflected pulses. There are some differences for capacitive load compared to other passive load. Under the case of a capacitive load, it depends on how large the capacitance is. What happens here is that when the incident pulse has not arrived at the capacitive load, the capacitor is not yet charged, and thus it looks like an open circuit and has no trace at all. When the pulse arrives the capacitor is charging and it looks like a short circuit, hence the negative-going part of trace (d), until it is fully charged, when it again looks like an open-circuit load (the last part of the trace indicates the charging transient time).

The time distance between the incident and reflected pulses is determined by the length of the transmission line, the coaxial cable.

2.6 CONCLUDING REMARKS

This chapter has described the behavior of electromagnetic waves, the transverse electromagnetic waves when they are excited into electrical transmission lines whose cross section or length is shorter than the wavelength of the waves. This is the case when transmission lines are employed to connect different subsections of systems that operate at high-frequency regions, e.g., antenna transmission lines, traveling wave electrodes excited by ultra-broadband signals in integrated optical modulators. Various operation parameters of such transmission lines are given and should be sufficient background to allow readers to use in practice. The estimations and calculations can be done by hand or using the Smith chart.

2.7   PROBLEMS

2.7.1   PROBLEM ON TDR OPERATION ON TRANSMISSION AND REFLECTION

The TDR response measured for a 50 Ω characteristic impedance coaxial line terminated with a 50 Ω load is shown below.

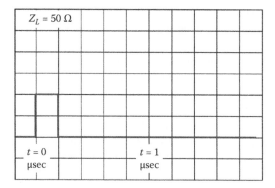

The line is 100 m long and is lossless. The velocity of propagation on the line is
200 m/μs.

a. What do the letters TDR stand for?
b. On each of the following accurately draw the TDR response that would be
   seen if the load was changed to those indicated on the plots.

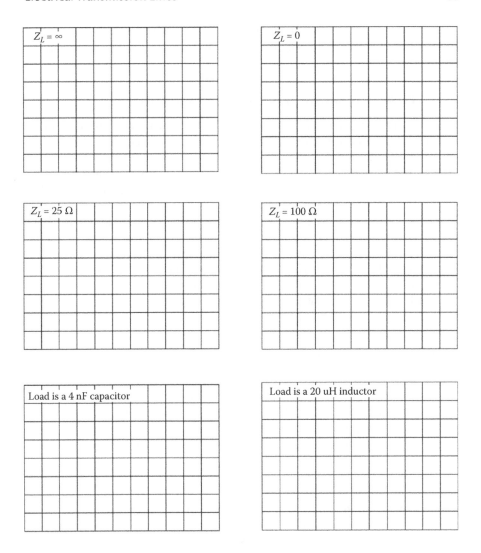

$Z_L = \infty$

$Z_L = 0$

$Z_L = 25\ \Omega$

$Z_L = 100\ \Omega$

Load is a 4 nF capacitor

Load is a 20 uH inductor

2.7.2 PROBLEM ON TRANSMISSION LINE

a. What is the relationship between the DC voltage measured by the detector probe and the RF voltage in the slotted line at the point where the probe is inserted?

b. If the detector measured a maximum voltage of 50 mV (DC) and a minimum voltage of 25 mV (DC) when measuring the voltages along the slotted line, what was the VSWR on the line?

c. When using the quarter-wave transformer to match the 25 Ω load to the 50 Ω slotted line, how far (in cm) did the closest end of the quarter-wave transformer need to be from the 25 Ω load?

d. Why could the quarter-wave transformer not be placed at the 25 Ω load?

e. Could we have matched the 25 Ω load by using a quarter-wave transformer of 12.5 Ω impedance at a different point on the line, and if so where should this be?

2.7.3   PROBLEM ON SLOTTED TRANSMISSION LINE EXPERIMENT

a. What is the relationship between the DC voltage measured by the detector probe and the RF voltage in the slotted line at the point where the probe is inserted?
b. If the detector measured a maximum voltage of 50 mV (DC) and a minimum voltage of 25 mV (DC) when measuring the voltages along the slotted line, what was the VSWR on the line?
c. When using the quarter-wave transformer to match the 25 Ω load to the 50 Ω slotted line, how far (in cm) did the closest end of the quarter-wave transformer need to be from the 25 Ω load?
d. Why could the quarter-wave transformer not be placed at the 25 Ω load?
e. Could we have matched the 25 Ω load by using a quarter-wave transformer of 12.5 Ω impedance at a different point on the line, and if so where should this be?

2.7.4   PROBLEMS ON TRANSMISSION LINES

1. Uniform TEM mode transmission lines:
   a. If in the experiment shown in Figure 2.7 the voltage probe measured a maximum voltage of 2 V and a minimum voltage of 1 V, what would be the VSWR in the slotted line?
   b. What are the maximum and minimum VSWRs you measured on the line?
   c. What load impedance gave you the maximum VSWR, and what load impedance gave you the minimum VSWR?
   d. What was the wavelength you measured in the slotted line?
   e. What was the free space wavelength?
   f. When the matching was successful, how far from the load reference position was the end of the quarter-wave transformer closest to the load?
   g. What were the impedances of the resistive load that was matched and the quarter-wave transformer that matched it?
   h. What is the function of the short-circuit stub, and why did it need to be tuned at each new frequency of measurement?
2. Sketch the schematic of an experimental setup that you may use for the study of slotted line transmission property with an operating frequency of 900 MHz. Describe the principles of operation of the setup. Indicate in your schematic the power level of the signals at the input of the slotted line, at the load, and at the detection system.
3. If the detector measured a maximum voltage of 50 mV (DC) and a minimum voltage of 25 mV (DC) when measuring the voltages along the slotted line, what was the VSWR on the line?

4. When using the quarter-wave transformer to match the 25 Ω load to the 50 Ω slotted line, how far (in mm) did the closest end of the quarter-wave transformer need to be from the 25 Ω load?
5. Why could the quarter-wave transformer not be placed at the 25 Ω load?

Refer to the above diagram to answer the following questions.

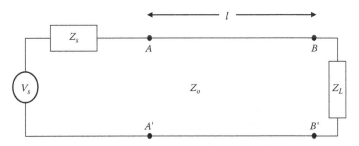

6. A lossless transmission line has a series inductance of 100 nH/m and a characteristic impedance of 75 Ω.
   a. What is the value of the line's shunt capacitance per meter?
   b. What is the propagation velocity of signals on the transmission line?
7. Write the formula for the voltage reflection coefficient in terms of characteristic impedance and load impedance.

   A transmission line with a source impedance of 50 Ω and characteristic impedance of 50 Ω is terminated in an impedance Z_L.

   What is the reflection coefficient at the load if:
   a. $Z_L = 10\ \Omega$
   b. $Z_L = 100\ \Omega$
   c. $Z_L = 50\ \Omega$
   d. $Z_L = 1\ M\Omega$
   e. $Z_L = 1\ m\Omega$
   f. $Z_L = 0\ \Omega$
8. A lossless transmission line with a source impedance of 50 Ω, a characteristic impedance of 50 Ω, and a length of 1500 m is terminated at an impedance of 50 Ω.

   The line has a forward traveling voltage wave $V^+(x - v_p t)$ on it such that

$$V^+(0, t) = 0\ \text{V}, \quad t \le 0$$

$$V^+(0, t) = 5\ \text{V}, \quad 0 < t < 2\ \mu s$$

$$V^+(0, t) = 0\ \text{V}, \quad t \ge 2\ \mu s$$

   $v_p = 200$ m/μs is the propagation velocity in the line

   a. Sketch the voltage on the line at times $t = 2, 4, 6, 8,$ and $10\ \mu s$.
   b. Sketch the voltage at the source of the line and at the load end of the line over the time period $0 \le t \le 20\ \mu s$.

   c.   Sketch the source voltage V_S as a function of time.
9. A lossless transmission line with a source impedance of 50 Ω, a characteristic impedance of 50 Ω, and a length of 1500 m is terminated in a 25 Ω
   impedance. The source impedance is 50 Ω.
   The line has a forward traveling voltage wave $V^+(x - v_p t)$ on it such that

$$V^+(0, t) = 0 \text{ V}, \quad t \le 0$$

$$V^+(0, t) = 5 \text{ V}, \quad 0 < t < 2 \text{ μs}$$

$$V^+(0, t) = 0 \text{ V}, \quad t \ge 2 \text{ μs}$$

$$v_p = 200 \text{ m/μs is the propagation velocity in the line}$$

   a.   Sketch the voltage on the line at times $t = 2, 4, 6, 8,$ and 10 μs.
   b.   Sketch the voltage at the source of the line and at the end of the line over
        the time period $0 \le t \le 20$ μs.
   c.   Sketch the source voltage V_s as a function of time
10. A transmission line with $Z_S = Z_0 = 50$ Ω has a length of 1000 mm. The
    velocity of propagation in the line is 200 m/μs.
    Its source voltage V_s is a step function $V_S(t) = 12u(t)$.
    Draw the voltage waveform seen at the terminals AA' and BB' if:
    a. $Z_L = 50$ Ω
    b. $Z_L = \infty$
    c. $Z_L = 0$
    d. $Z_L = 75$ Ω
    e. $Z_L = 25$ mΩ
11. Draw the current variation for the previous problem with time at the termi-
    nals AA' and BB'.
12. Write the expression for the input impedance seen by looking into a lossless
    line of length l that has a characteristic impedance Z_0 and is terminated in
    a load impedance Z_L.
13. What is the input impedance at 300 MHz of a lossless line of length l, which
    has a propagation velocity equal to that of free space, has a characteristic
    impedance $Z_0 = 50$ Ω, and is terminated in a load impedance Z_L if:
    a. $l = 750$ mm and $Z_L = 50$ Ω
    b. $l = 250$ mm and $Z_L = 50$ Ω
    c. $l = 650$ mm and $Z_L = 50$ Ω
    d. $l = 750$ mm and $Z_L = 100$ Ω
    e. $l = 1000$ mm and $Z_L = 100$ Ω
    f. $l = 500$ mm and $Z_L = 100$ Ω
    g. $l = 250$ mm and $Z_L = 100$ Ω
    h. $l = 600$ mm and $Z_L = 100$ Ω
    i. $l = 300$ mm and $Z_L = 100$ Ω
    j. $l = 1000$ mm and Z_L is a short circuit
    k. $l = 500$ mm and $Z_L =$ is a short circuit
    l. $l = 250$ mm and $Z_L =$ is a short circuit

m. $l = 350$ mm and $Z_L = $ is a short circuit
n. $l = 150$ mm and $Z_L = $ is a short circuit
o. $l = 1000$ mm and Z_L is an open circuit
p. $l = 500$ mm and $Z_L = $ is an open circuit
q. $l = 250$ mm and $Z_L = $ is an open circuit
r. $l = 350$ mm and $Z_L = $ is an open circuit
s. $l = 150$ mm and $Z_L = $ is an open circuit
   Note how long the line is in wavelengths in each case.

14. Given the results of question 8, what conclusions can you draw regarding
    a. The input impedance of a transmission line terminated with its own characteristic impedance.
    b. The input impedance of a half-wavelength long line.
    c. The input impedance of a wavelength long line.
    d. The input impedance of a quarter-wavelength long line.
    e. The input impedance of a quarter-wavelength long line terminated in a short circuit.
    f. The input impedance of a quarter-wavelength long line terminated in an open circuit.

15. A lossless line with a characteristic impedance of 100 Ω is 25 m long. It is fed from a 600 MHz sinusoidal source that has a source impedance of 100 Ω. The open-circuit voltage of the source is 60 mV. The line is terminated in an impedance of $(50 + j50)\Omega$.
    Assume a wavelength in the line is 67% of a wavelength in free space.
    a. What is the voltage reflection coefficient seen at the load?
    b. What is the magnitude of the forward traveling voltage wave?
    c. What is the magnitude of the reverse traveling voltage wave?
    d. What is the maximum voltage on the transmission line?
    e. What is the minimum voltage on the transmission line?
    f. What is the relationship between voltage maxima and voltage minima and the VSWR on the line?
    g. What is the relationship between forward traveling voltage magnitude and reverse traveling voltage magnitude and the VSWR on the line?
    h. What is the VSWR on the transmission line?
    i. What is the relationship between VSWR and the reflection coefficient?
    j. How much power (in dBm) could have been delivered to the load if the load was matched to the line characteristic impedance?
    k. What power (in dBm) will be delivered to the load?
    l. What is the relationship between the power in part (j), the power in part (k), and the reflection coefficient?

16. Give three common reasons an engineer may need to impedance match a load to a transmission line's characteristic impedance.

17. A load impedance of $(150 + j25)$ Ω terminates a transmission line that has a characteristic impedance of 75 Ω. The load is to be impedance matched to the line using a single shunt-connected stub. The stub is to comprise a short-circuited length of 75 Ω coaxial line. Determine:

    a. The position (in wavelengths relative to the load) at which to attach the stub.

    b. The length (in wavelengths) of the stub.

    c. How long should the stub be if it was terminated in an open circuit rather than a short circuit?

18. A load impedance of $(150 + j75)$ Ω terminates a transmission line with a characteristic impedance of 75 Ω.

    a. What is the reflection coefficient at the load?

    b. A 145 MHz signal is applied to the line. If the minimum voltage measured at any point on the line is 100 mV (RMS), what is the maximum voltage that would be measured on the line?

    c. What VSWR will occur on the line?

    d. Where (in terms of wavelengths from the load) will the voltage maxima occur?

    e. What will the impedance on the transmission line be at these points?

    f. What will the impedance be at points where there is a voltage minimum?

19. Give three common reasons why an engineer may need to impedance match a load to transmission line's characteristic impedance.

20. A load impedance of $(150 + j25)$ Ω terminates a transmission line that has a characteristic impedance of 75 Ω. The load is to be impedance matched to the line using a single shunt-connected stub. The stub is to comprise a short-circuited length of 75 Ω coaxial line. Determine

    a. The position (in wavelengths relative to the load) at which to attach the stub.

    b. The length (in wavelengths) of the stub.

    c. How long should the stub be if it was terminated in an open circuit rather than a short circuit?

21. A 900 MHz plane wave is propagating in the +x-direction of a rectangular Cartesian coordinate system. It is polarized in the y-direction. The electric field has an RMS magnitude of 50 mV/m.

    a. What is the wavelength of the plane wave?

    b. What is the magnitude of the magnetic field component of this electromagnetic wave?

    c. What is the power density carried by the wave?

    d. An antenna with an effective area of 0.5 m^2 is pointing in the negative x-direction and is polarized in the y-direction. How much power will it pick up from the plane wave?

    e. The plane wave is normally incident on an infinite, lossless wall that has one face in the yz-plane at $x = 0$. The other face of the wall is in the yz-plane at $x = $ d.

    The wall has a relative permeability of 4 and a relative permeability of 1. What percentage of the incident power is reflected back from the wall if

        i. $d = 83.25$ mm

       ii. $d = 41.625$ mm

22. A section of rectangular waveguide has internal dimensions of 12.5×25 mm.

    a. What modes will propagate in this guide at 9 GHz?

    b. What is the range of frequencies for which only one mode will propagate in the waveguide, and what is that mode?

    c. Sketch the electric field distribution of the TE10 mode in the waveguide.

    d. Sketch the magnetic field distribution of the TE10 mode in the waveguide.

    e. Sketch the electric field distribution of the TE21 mode in the waveguide

    f. What is the guide wavelength at 9 GHz?

REFERENCE

1. S. Wentworth, *Fundamentals of Electromagnetics with Engineering Applications*, John Wiley, New York, 2003, Chapter 6.

3 Antennae

3.1 INTRODUCTION

In wireless communications, the antenna is commonly known as an aerial, a transducer converting electrical energy into radiation energy and designed to transmit or receive electromagnetic (EM) waves. TV antennae or mobile antennae are designed specifically for reception of broadcast television signals or transmitting, propagating, and receiving of mobile signals that can be in either analog or digital form.

In practice the antennae need to be installed in a place that is sufficiently high for telecommunications and broadcasting, commonly known as antennae towers. There are a number of basic antenna structures, such as the dipole antenna, which is a simple structure constructed by two straight wires in opposite phase end-to-end, directional antennae. Alternatively, a beam antenna radiates greater energy/power directed toward some specific directions. A horn antenna is a type of directional antenna whose shape follows that of a horn. A meta-material antenna is a class of antenna incorporating meta-materials to enhance the performance of miniature antenna systems. An omnidirectional antenna is an antenna system radiating EM waves uniformly in all directions in one plane. A parabolic antenna has a radiating or reception area that follows the shape of a parabola in one or both planes. A typical parabolic antenna disk is shown in Figure 3.1. It is located in a remote area of Western Australia in order to avoid interference from human mobile communication signals so it can detect weak EM radiation from deep space or the outer boundary of the universe. All communications and data transfer works are implemented via fiber optical communication lines and systems.

Typically an antenna consists of an arrangement of metallic conducting elements, electrically connected through a transmission line to the receiver or transmitter. An oscillating electric current forced through the antenna by a transmitter will create an oscillating magnetic field around the antenna elements, while the flowing electric current also creates an oscillating electric field along the conducting elements. Hence the mutually orthogonal magnetic and electric fields are coupled with each other and propagate a long way. This is electromagnetic radiation.

There are a number of structures of antennae, depending on the geometrical arrangement, including linear loop antennae, monopole and dipole antennae, radiation aperture antennae, horn antennae, disk antennae, array antennae, etc. These antennae can act as reception or radiation systems. Figures 3.1, 3.3, 3.4, and 3.7 show the typical structures of these antennae. Figure 3.2 shows a typical pattern of radiating electric and magnetic field components, which are always orthogonal to each other. In this chapter, as an introduction to antennae for wireless applications, monopole and dipole antennae are treated as linear types for illustration of the principles of radiation and receiving. Further details of other types can be found in the literature, especially in reference books.

FIGURE 3.1   A parabolic antenna as an element of an antenna radio array, the square kilo-meter antenna (SKA), constructed in a remote region in Pilbarra, Western Australia, to avoid any wireless EM interference so that it can detect very weak EM radiation from stars located in deep outer space. (Artist image of SKA for radio astronomy. See http://www.skatelescope.org/the-technology.)

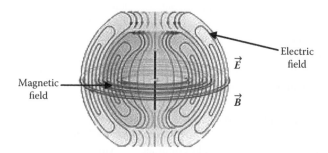

FIGURE 3.2   (a) A parabolic antenna as an element of an antenna radio array of (b) the square kilometer antenna (SKA) constructed in a remote region in Pilbarra, Western Australia, to avoid any wireless EM interference so that it can detect very weak EM radiation from stars located in deep outer space to determine the boundary of the universe. (Artist image of SKA for radio astronomy. See http://www.skatelescope.org/the-technology.)

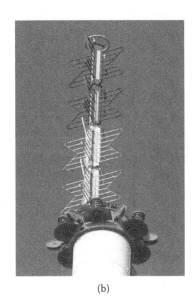

(a)   (b)

FIGURE 3.3   (a) Mobile antenna at a mobile base station. (b) VHF antenna tower.

An antennae tower is a tall structure to support antennae as aerials for telecommunication and broadcasting. The first tower was constructed in Munich Olympic Park for the television broadcasting purposes of the 1972 Olympics. A dipole antenna is a simple antenna usually constructed from two wires in opposite phases positioned end to end with respect to each other. A horn antenna is a type of directional antenna shaped like a horn. An omnidirectional antenna is an antenna system that radiates power uniformly in all directions in one plane. A parabolic antenna is an antenna shaped like a parabola in one or both planes.

This chapter describes a number of fundamental understandings of antenna theory and illustrates the technological aspects of antennae. A number of essential formulae for estimating antenna performance are given. They can be understood from the theory of electromagnetic waves, such as the field is an inverse proportional with distance. We will state these equations without proof in order to simplify the mathematics, but we will enhance the physical understanding of EM radiation and reception for engineering applications.

3.1.1   Differential Doublet and Dipole Antenna

Consider an elementary doublet consisting of two unconnected minute in-line elements located at the referenced center of the Cartesian coordinate system as depicted in Figure 3.4. The length of the doublet is δL, which is much shorter than the wavelength of the EM waves radiating from it. The doublet carries an alternating current I expressed as

$$I(t) = I_0 e^{j\omega t} \tag{3.1}$$

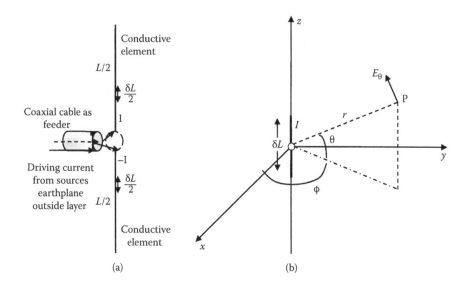

(a)   (b)

FIGURE 3.4 Coordinate and fields of a dipole antenna: (a) Exciting current $I(t)$ into the dipole antenna; (b) its differential antenna element.

where $\omega = 2\pi f$ is the angular frequency and I_0 is the excitation current at the origin. Thus I represents the phasor of the current I, that is, the magnitude of a wave rotating at an angular frequency. Normally this rotating frequency is too fast and can be considered "stationary" by vision. Any additional phase carried by the current I would be represented by an angle on a phasor plane equal to the phase radial angle.

This differential doublet is indeed considered an elementary element of a dipole antenna that consists of two wire lines coming from the center and connected to the positive and negative terminals, excited by an alternating current source as shown in Figure 3.4(b). Thus the total field considered at the point P can be found by integrating (3.1) over the entire length of the dipole antenna as shown in Figure 3.4(a).

3.1.2 FAR FIELD

If the point P is far away from the antenna, then the field is a far field. That is, the distance of consideration is much longer than the wavelength of the radiating EM waves. The imaginary part of this differential far field strength is given by

$$\delta E_\theta = -j\frac{I_0 \sin\theta}{2\varepsilon_0 cr}\frac{\delta L}{\lambda}e^{j(\omega t - kr)}a_\theta \tag{3.2}$$

This differential field strength is inversely dependent on the radial distance r, the angle θ, the velocity of light c, and the EM wave propagating velocity. ε_0 is the permittivity of the air medium. λ is the wavelength of the radiation EM waves; indeed, it is the wavelength of the excitation signal applied to the antenna. This field is projected onto the direction of the angular variable θ. The field is in the plane of the antenna and perpendicular to the radial direction. a_θ is the unit vector in the θ-direction. $k = 2\pi/\lambda$ is the EM propagation constant (or wavenumber) in free air.

3.1.3 NEAR FIELD

For completeness the near field of the E and H components in a spherical coordinate system is as indicated in Figure 3.4. Such fields of the radiation antenna, that is, the field in the distance less than one wavelength of the radiating wave divided by 2π, are given by

$$E_r = \frac{Z}{2\pi} I_0 \delta L \left(\frac{1}{r^2} - j \frac{\lambda}{2\pi r^3} \right) e^{j(\omega t - kr)} \cos\theta$$

$$E_\theta = j\frac{Z}{2\pi} I_0 \delta L \left(\frac{1}{r} - j \frac{\lambda}{2\pi r^2} - \frac{\lambda^2}{2\pi r^3} \right) e^{j(\omega t - kr)} \sin\theta \qquad (3.3)$$

$$H_\phi = j\frac{1}{2\lambda} I_0 \delta L \left(\frac{1}{r} - j \frac{\lambda}{2\pi r^2} \right) e^{j(\omega t - kr)} \sin\theta$$

where

$$Z_0 = \sqrt{\frac{\mu}{\varepsilon}} = \frac{1}{\varepsilon c} = \mu c = \mu_r \mu_0 c$$

is the characteristic impedance of the medium with permittivity $\varepsilon = \varepsilon_r \varepsilon_0$ and permeability $\mu = \mu_r \mu_0$. The characteristic impedance of free air is 120π or $377\ \Omega$.

Exercise

Confirm the characteristic impedance of free air or vacumm by substituting the permittivity of vacuum and the velocity of light in such a medium.

3.1.4 LINEAR ANTENNA CURRENT DISTRIBUTION

The radiation angular pattern of antennae is completely determined by the transverse component $F_\perp = \hat{\theta}F_\theta + \hat{\phi}F_\phi$ of the radiation vector; $\hat{F}, \hat{\theta}, \hat{\phi}$ are the unit vectors. A transverse plane of the radiation is determined as the plane perpendicular to a particular direction. In this case, if the antenna element is along the z-direction, then the transverse plane is in the xOy of a Cartesian coordinate system and $(\hat{\theta}, \hat{\phi})$ in a spherical coordinate. The current density J determines this field distribution. Here, we consider some examples of current densities describing various antenna types, such as linear antennae, loop antennae, and linear arrays.

For linear antennae, we may choose the z-axis to be along the direction of the antenna. Assuming an infinitestimally thin antenna, the current density *along the z-direction* will take the form

$$\tilde{J}(\tilde{r}) = \tilde{z}I(z)/\partial x \partial y \tag{3.4}$$

where $\partial x; \partial y$ are the differential elements in the x- and y-directions, respectively, $I(z)$ is the distribution of the excitation current along the z-direction of the antenna conducting elements, and is the unit vector in the z-direction. This current satisfies, approximately, the Helmholtz's equation:

$$\frac{d^2 I(z)}{dz^2} + kI(z) = 0 \tag{3.5}$$

where $k_0 = 2\pi/\lambda$ is the wavenumber of the radiating wave in free air. This second-order differential equation (3.5) represents the relationship of the current distribution along the conducting antenna element that may form by the resonanting of the currents to generate a standing wave—hence a maximum field induced to radiation when a resonance condition is satisfied. We will see later that this resonating phenomenon is the maximum condition for radiation EM fields.

The current $I(z)$ can be stated for different types of antenna elements with the excitation current at the origin of I_0 as a Hertzian dipole:

$$I(z) = I_0 l \partial z \tag{3.6}$$

A Hertzian antenna is made up of two small spherical disks connected by a wire, and the electric charge flows periodically back and forth between the spheres. The first antennae were built in 1888 by German physicist Heinrich Hertz in his pioneering experiments to prove the existence of electromagnetic waves as predicted by the theory of Maxwell (the Maxwell equations) at the time.

Uniform line element, monopole:

$$I(z) = I_0 \tag{3.7}$$

Small linear dipole:

$$I(z) = I_0(1 - 2|z|/l) \tag{3.8}$$

Standing wave antenna:

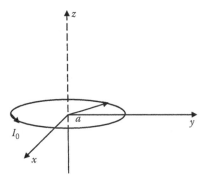

FIGURE 3.5 Loop antenna.

$$I(z) = I_0 \sin\left(k_0 \frac{l}{2} - |z| \right)$$ (3.9)

If the length of the antenna elements is equal to half the wavelength of the radiating frequency, we have a half-wave antenna. The electric oscillating current can be written as

$$I(z) = I_0 \cos(k_0 z) \text{ for half-wave antenna (with } l = \lambda/2)$$ (3.10)

Traveling wave antenna:

$$I(z) = I_0 e^{-j k_0 z}$$ (3.11)

where l is the length of the antenna element and the expressions are valid for $-l/2 \le z \le l/2$, and k_0 is the wavenumber of the radiating wave field. The term *traveling wave* is used when the length of the antenna is compatible with or longer than that of the wavelength of the radiating EM waves.

3.1.4.1 Loop Antenna

A loop antenna may be formed by a thin-wire circular loop that lies on the xy-plane and is centered at the origin. For a circular loop of radius a, the current flows azimuthally, as shown in Figure 3.5. The corresponding current density can be expressed in cylindrical coordinates $\mathbf{r} = (\rho, \varphi, z)$ as

$$\tilde{J}(\tilde{r}) = \hat{\varphi} I_0\, \partial(\rho - a)\partial z$$ (3.12)

3.1.4.2 Array Antenna

An array antenna is formed by a group of antenna elements, such as Hertzian or half-wave dipoles, that are arranged in particular geometrical configurations, such as along a particular direction. Some examples of antenna arrays composed of identical antenna elements whose current density can be written as a superimposition of each individual element referring to the same coordinate system follow:

$$\tilde{J}(\tilde{r}) = \hat{z}\sum_i^N a_i I(z)\partial(x - x_i)\partial y ... \text{array} \cdot \text{along} \cdot x\text{-direction}$$

$$\tilde{J}(\tilde{r}) = \hat{z}\sum_i^N a_i I(z)\partial(y - y_i)\partial x ... \text{array} \cdot \text{along} \cdot y\text{-direction} \tag{3.13}$$

$$\tilde{J}(\tilde{r}) = \hat{z}\sum_i^N \sum_j^M a_i I(z)\partial(x - x_i)\partial(y - y_j) ... \text{array} \cdot \text{along} \cdot z\text{-direction}$$

3.2 RADIATING FIELDS

The radiation angular pattern of antennae is completely determined by the transverse component, which is defined as the plane perpendicular to the antenna element of the radiation vector, $\tilde{F}$, which in turn is determined by the current density, $\tilde{J}$.

For an antenna element of length l excited by a current $I(z)$ the radiation field is given by

$$\tilde{F} = \hat{z}F_z = \int_{-l/2}^{l/2} I(z')e^{jk_z z}dz' \tag{3.14}$$

where k_z is the propagation constant of the radiation vector along the z-direction. The propagation vector can be represented in either spherical or Cartesian coordinate systems as

$$\tilde{k} = \hat{x}k_x + \hat{y}k_y + \hat{z}k_z = \hat{x}|k|\cos\phi\sin\theta + \hat{y}|k|\sin\phi\sin\theta + \hat{z}|k|\cos\theta \tag{3.15}$$

Thus (3.14) can be rewritten as

$$\tilde{F} = \hat{z}F_z = \int_{-l/2}^{l/2} I(z')e^{jkz'\cos\theta}dz' \tag{3.16}$$

This vectorial field can be resolved into the transverse plane as

$$\tilde{F} = \hat{z}F_z = \int_{-l/2}^{l/2} I(z')e^{j|k|z\cos\theta}dz' \tag{3.17}$$

which can be resolved into the (θ,ϕ) directions as

$$\tilde{F} = \hat{z}F_z = [\hat{r}\cos\theta - \hat{\theta}\sin\theta]F_z(\theta) = \hat{r}F_z(\theta)\cos\theta - \hat{\theta}F_z(\theta)\sin\theta \qquad (3.18)$$

Hence, the transverse component of the radiation field is the part involved, $\hat{\theta}$, and is written as

$$\tilde{F}_\perp = -\hat{\theta}F_z(\theta)\sin\theta \qquad (3.19)$$

Thus with an integration of the field component along the whole length l of the antenna element of the current I, we have the electric and magnetic fields $(\tilde{E},\tilde{H})$, which are orthogonal to each other, as

$$\tilde{E} = \hat{\theta}jkZ\frac{e^{-jkr}}{4\pi r}F_z(\theta)\sin\theta$$

$$\tilde{H} = \hat{\phi}H_\phi = \hat{\phi}jk\frac{e^{-jkr}}{4\pi r}F_z(\theta)\sin\theta \qquad (3.20)$$

The radiation field intensity of linear antennae would thus be given as

$$U(\theta) = \frac{Z^2 k^2}{32\pi^2}|F_z(\theta)|^2 \sin^2\theta \qquad (3.21)$$

By taking the real component of the cross-product of the electric and magnetic fields given in (3.20), we can define the real part of the Poynting vector, which is explained next.

The principal aims of constructing antennae and their uses are for transmitting energy and receiving radiated energy, namely, the transmitting antenna and receiving antenna, respectively. Thus in order to justify the performance of an antenna, the flux of the radiating field, then the radiating power, is very important. The radiating power is indeed the real part of the cross-product of the EM field components, $\tilde{E},\tilde{H}$, which are radiating from the excitation of the oscillating current to the antenna element and are mutually orthogonal. Indeed, wireless communication signals are the low-frequency data sequence represented by the envelope of the high frequency carriers to be carried or transported over an air medium by this transmitting field. The carrier frequency is the oscillating frequency of the exciting current applied to the antennae. Thus the radiating power, the real part of the Poynting vector, can be written as

$$P = \frac{1}{2}\text{Re}(\tilde{E}\times\tilde{H}) \qquad (3.22)$$

Thus (3.21) can be derived from substituting (3.20) into (3.22).

3.2.1 Radian Field of Hertzian Antenna

After the formation of Maxwell's equations involving time variants and radiation, Hertz of Kiel University in Nothern Germany initiated an experiment to create the generation of EM waves—thus the Hertzian antenna, which is a very small linear radiating element located at the origin of a coordinate system. The distribution of the excitation current is given as $I_0/\partial z$. By substituting into (3.16) and (3.21) and taking the integration over the length of the Hertzian antenna we obtain

$$F = \hat{z}F_z = \int_{-l/2}^{l/2} I(z')e^{j|k|z\cos\theta}dz' = I_0l \tag{3.23}$$

Then the radiation power can be found, by using (3.21), as

$$P_{rad}(\theta) = \frac{Z^2k^2}{32\pi^2}|I_0l|^2 \sin^2\theta \tag{3.24}$$

The gain of an antenna, $G(\theta)$, can be defined as

$$G(\theta) = \frac{P_{rad}(\theta)}{P_{rad,max}} = \sin^2\theta \tag{3.25}$$

where

$$P_{rad,max} = \frac{Z^2k^2}{32\pi^2}|I_0l|^2$$

This antenna gain and radiation pattern are plotted as shown in Figure 3.6(a) and (b). It is very obvious that this radiation pattern is maximum at an angle of $\pi/2$, that is, at the direction perpendicular to the Hertzian antenna element. The half-power point can be estimated to be $\pi/4$, i.e., at which the amplitude reaches the factor 0.7076 of the maximum of the field radiating pattern and 50% in the power distribution curve. The radiation pattern is thus fairly wide and not highly directional.

In order to achieve a more directional radiation pattern and, hence, control of the radiating beam to our target position, it is necessssary to design the arrangement of the antenna element. The solid beam angle can be derived by taking the integration over the entire beam distribution or estimating the half-power solid angle and multiplying by a scaling factor to get

$$\Delta\Omega = \frac{8\pi}{3} \tag{3.26}$$

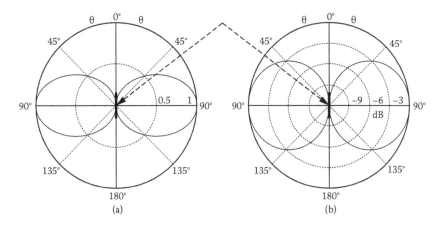

FIGURE 3.6 (a) Radiation and (b) gain in dB patterns of Hertzian antenna in polar coordinate system.

Having found the electric field one can compute the emitted power and thence the resistivce part of the series impedance of the antenna due to this radiating field, which is known as the radiation resistance, given as

$$P_{rad} = \frac{1}{2} R_{rad}^2 I_0 \rightarrow R_{rad} = \left(\frac{2 P_{rad}}{I_0} \right)^{1/2} \tag{3.27}$$

Indeed, we can consider the resistance or impedance to be that of a free air medium into which the radiation is entering, or this free air is the load of the antenna radiation source.

3.2.2 STANDING WAVE ANTENNA: THE HALF-WAVE DIPOLE ANTENNA

The most popular antenna structure is the standing wave type, especially the half-wave structure. A half-wave antenna structure is shown in Figure 3.7 whose length is equal to half of the wavelength of the radiating electromagnetic wave, that is, the ratio between the velocity of light in vacuum and the operating frequency. The main reason for the value of a half-wavelength is that a quarter of the wave is distributed to half of the antenna and the half to the other side. Thus a standing wave can be formed by the superposition of the forward wave and backward wave from the ends of the antenna elements. So the EM radiation forms a standing wave pattern with "close" ends, as shown in Figure 3.7; maximum field radiation is achieved due to the resonance of the fields over the length of the antenna. The current distribution of the half-wave antenna is given in (3.9). Thus the field radiation vector is given as

$$\tilde{F} = \hat{z} F_z = \int_{-l/2}^{l/2} I_0 \sin \left(k \left[\frac{l}{2} - |z'| \right] \right) e^{j|k|z \cos\theta} dz' = \frac{2 I_0}{k} \frac{\cos(kh\cos\theta) - \cos(kh)}{\sin^2\theta} \tag{3.28}$$

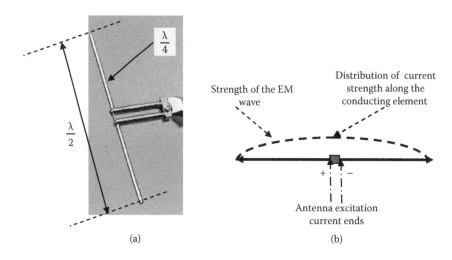

FIGURE 3.7 A half-wave dipole antenna consisting of two equal-length monopole wires: (a) structure image of 1–4 GHz operating frequency; (b) schematic.

By substituting (3.28) into (3.21) we obtain the normalized gain for the standing wave antenna as

$$G(\theta) = c_n \left[\frac{\cos(kh\cos\theta) - \cos(kh)}{\sin\theta} \right]^2 \tag{3.29}$$

where is chosen such that the maximum value of the gain is unity at the maximum value of the angle θ.

For a half-wave antenna the length l corresponds to half of the wavelength of the radiating EM wave, or $kl = \pi$; thus the normalized gain (3.29) becomes

$$G(\theta) = \frac{\cos^2\left(\dfrac{\pi}{2}\cos\theta\right)}{\sin^2\theta} \tag{3.30}$$

The plots of the distribution of the field and the normalized gain with respect to the angle θ are shown in Figure 3.8. We can see that the 3 dB point of the power radiation is at about 40° of arc. The radiating patterns of the resonance antennae having different lengths are shown in Figure 3.9. It is obvious that the radiation patterns would now exhibit multiple lobes, especially when the length is twice the wavelength.

3.2.3 MONOPOLE ANTENNA

The monopole antenna with length of a quarter-wave can be considered the version of a half-wave antenna whose second half is the image of the first. Thus the power radiating from the monopole antenna equals half of that of a dipole resonance antenna.

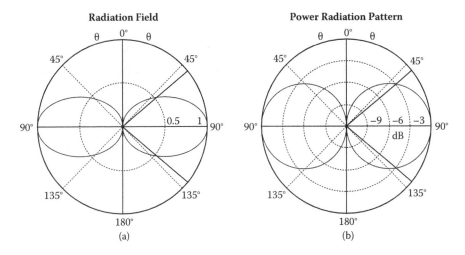

FIGURE 3.8 (a) Field pattern and (b) power distribution of the half-wave antenna.

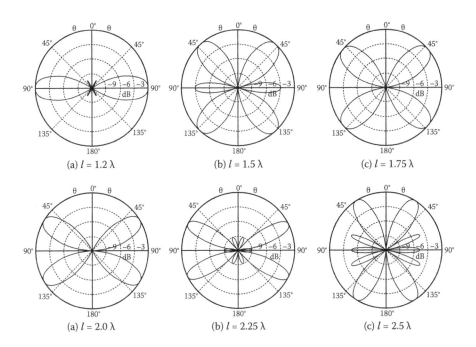

FIGURE 3.9 Field pattern and power distribution of resonance antenna with different total lengths, as indicated.

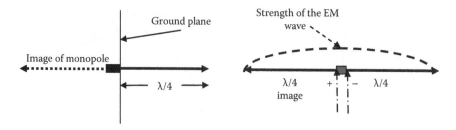

FIGURE 3.10 Structure of a monopole antenna and its equivalent image model using a half-wave antenna.

The simplicity of the monopole antenna and its "imaging" equivalence shown in Figure 3.10 gives tremendous advantage in modern mobile devices currently used in mobile networks. We have the radiating power given as

$$P_{monopole} = \frac{1}{2} P_{dipole} \tag{3.31}$$

3.2.4 Traveling Wave Antenna

The standing wave antenna considered in the previous section can be considered the superposition of forward and backward waves. If the end of the antenna has a load whose impedance matches that of the antenna, then all forward waves are matched, there is no reflection of waves, and there are zero backward waves. In this case the antenna is called a traveling antenna type. An example of this type of antenna is the traveling wave electrode integrated on the surface of an optical modulator substrate so that the electric fields generated across the optical waveguide can be established to modulate the change of the refractive index of the optical wave by the radiating field dropped across the guiding region via the electro-optic effects—hence the modulation of the phase of the lightwaves.

A schematic of a traveling wave is shown in Figure 3.11. The traveling wave looks like a wire excited from one end and matched with a load whose impedance equates that of the antenna—hence no reflection of the radiation wave and thus it is a traveling wave.

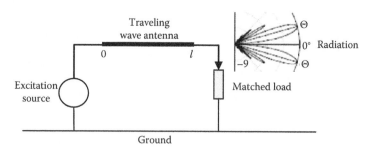

FIGURE 3.11 Schematic traveling wave antenna.

The current along the element of the antenna is given as

$$I(z) = I_0 e^{-jkz}, \qquad 0 \le z \le l \tag{3.32}$$

Thus, the radiation field can be found as

$$\tilde{F} = \hat{z} \int_0^l I_0 e^{-jkz'} e^{jk\cos\theta z'} \, dz' = \hat{z} \frac{1}{jk} \frac{1 - e^{-jkl(1-\cos\theta)}}{1 - \cos\theta} \tag{3.33}$$

The transverse component of the field in the θ-*plane* is given as

$$\tilde{F}_\theta = -\tilde{F}\sin(\theta) = -\frac{1}{jk}\sin\theta\frac{1 - e^{-jkl(1-\cos\theta)}}{1 - \cos\theta} \tag{3.34}$$

Similarly, the radiation intensity can be found as

$$P_{rad}(\theta) = \frac{Z_0|I|^2}{32\pi^2}|F(\theta)|^2 = \frac{Z_0|I|^2}{8\pi^2}\left|\frac{\sin\theta\sin\left(\pi\frac{l}{\lambda}(1-\cos\theta)\right)}{1-\cos\theta}\right|^2 \tag{3.35}$$

Hence, the antenna gain of the traveling wave type is given by

$$G(\theta) = C_n\left|\frac{\sin\theta\sin\left(\pi\frac{l}{\lambda}(1-\cos\theta)\right)}{1-\cos\theta}\right|^2 \tag{3.36}$$

where C_n is an arbitrary normalization constant. The radiation pattern plotted in Figure 3.12 represents the gain equation (3.36) for two typical lengths of 5 and 10 times the operating wavelength.

3.2.5 OMNIDIRECTIONAL ANTENNA

An omnidirectional antenna radiates radiowaves with power that is uniform over all directions, with an elevation angle above or below the plane and dropping to null at the antenna's axis. This radiation is often described as doughnut shaped, as shown in Figure 3.13(a) and (b). The radiated power is maximum in the horizontal plane, which is perpendicular to the antenna axis, and drops to zero at its axis. An omnidirectional antenna is commonly used in radio broadcasting and mobile devices that use radio waves, such as cell phones, FM radios, walkie-talkies, and cordless phones.

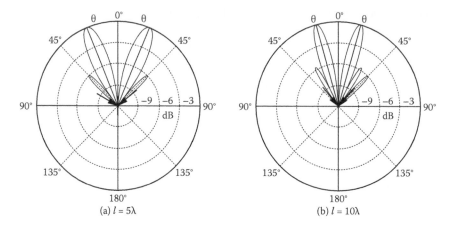

FIGURE 3.12 Radiation pattern plotted using the antenna gain equation of a traveling wave antenna with (a) $l = 5\lambda$ and (b) $l = 10\lambda$.

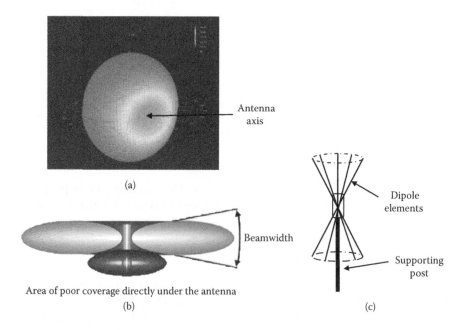

FIGURE 3.13 Computer-generated image of a doughnut-shaped radiation pattern of an omnidirectional antenna. (a) Radiation pattern. (b) Installed antenna on a ceiling and radiating pattern for indoor communication. (c) Omni antenna as a composite of dipole elements.

These omnidirectional antennae are useful for indoor communications, especially in underground transport networks for mobile transmitting and receiving signals.

Omnidirectional antennae can take the form of the simplest practical types of monopole and dipole structures, as shown in Figure 3.13(c).

3.2.6 HORN WAVEGUIDE ANTENNA

A horn antenna or microwave horn is an antenna that consists of a planar metal waveguide shaped like a horn, as shown in Figure 3.14. It consists of a metallic thin wall opening at the end and following a pyramidal shape with a narrow end where the signal is feeding through an *N*-type connector. The radiation coming out of the horn antenna follows a beam form. Horn antennae are commonly used as antennae at UHF frequencies above 300 MHz and microwave frequencies above 10 GHz. They are also commonly used as feeders to larger antennae, such as the parabolic antennae shown in Figure 3.1. The appendix at the end of this chapter gives a brief introduction to a metallic hollow waveguide, which can be used to feed the radiation waves at a distance to the radiating antenna system.

The horn antenna operates like an acoustic horn employed in music; electromagnetic waves are radiating within the walls of the horn, following a gradual transition to match the impedance of the waveguide to that of the free-space impedance (~377 Ω) so as to maximize the radiation. The horn acts as an impedance matcher that is gradual so that it can match to that of the free air and avoid reflection back to the waveguide and, thus, inefficiency. The length of the horn is longer than one wavelength; otherwise, the radiation is inefficient.

The EM waves traveling down a horn have a spherical wavefront whose origin is located at the apex of the horn. The pattern of the electric and magnetic fields at the aperture is a scaled image of the fields confined in the waveguide. However, due to the spherical wavefronts the phase increases gradually and smoothly from the center

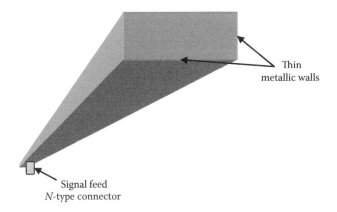

Thin
metallic walls

Signal feed
N-type connector

FIGURE 3.14 Structure of a pyramidal horn antenna with a typical bandwidth of 0.8 to 18 GHz, commonly used as a police antenna gun to measure the speed of a moving car by radiation and reflection from a metallic object (car).

of the aperture to its edges. The phase difference between the waves at the center of the aperture and those at its edges is the phase error, which increases with respect to the size of the horn aperture. Normally the radiation intensity is about −20 dB (or only 1% of that at the ecnter) at the edge compared to that at the center.

There are several common types of horn antennae. The flare angle of the horn and the gradient curve from the narrow end to the opening end determine their differences; that is, the shape of the opening end will allow either the electric component E or the magnetic component H radiating into some specific directions.

A pyramidal horn is a horn antenna whose shape follows that of a pyramid (see Figure 3.14). This type is very common, with the radiating frequency determined by the length of the rectangular metallic waveguide at the narrow end.

A sectorial horn is a pyramidal horn shape where only one side is a wide planar and the other is very thin, thus it looks like a fan. This leads to the cutting off the radiating of either an E- or H-field.

A conical horn has the shape of a cone, that is, a circular cross section of different sizes at the two ends and a gradual increase in diameter. Other horns include the corrugated horn, ridge horn, septum horn, and aperture-limited horn. Details on these horn structures can be found in numerous references.

Typical dimensions for an optimum pyramidal horn can be estimated as[1]

$$a_E = \sqrt{3\lambda L_E}$$
$$a_H = \sqrt{2\lambda L_H}$$

(3.37)

and likewise for a conical horn, the optimum condition is

$$d = \sqrt{3\lambda L}$$

(3.38)

where
    a_E = width of the aperture in the E-Field direction
    a_H = width of the aperture in the H-Field direction
    L_E = slant length of the side of the E-field direction
    L_H = slant length of the side of the E-field direction
    d = diameter of cylindrical horn aperture
    λ = wavelength of radiating EM waves

3.3 ANTENNA FIGURE OF MERIT

In electromagnetics, an antenna's power gain is a figure of merit (FoM) that combines the antenna directivity and electrical efficiency. As a transmitting antenna, this FoM describes how well the antenna converts input power into radiowaves radiated in a specified direction. On the contrary, as a receiving antenna, the FoM describes how well the antenna induces radiowaves arriving from a specified direction into electrical power. When no direction is specified, the gain is understood to refer to the peak value of the gain. A plot of the gain as a function of direction is called the radiation pattern.

Antenna gain is usually defined as the ratio of the power produced by the antenna from a far-field source on the antenna's beam axis to the power produced by a hypothetical lossless isotropic antenna, whose sensitivity to signals is uniform for waves from all directions. Usually this ratio is expressed in decibels, and these units are referred to as *decibels-isotropic* (dBi).

An alternate definition compares the antenna to the power received by a lossless half-wave dipole antenna, in which case the units are written as dBd. Since a lossless dipole antenna has a gain of 2.15 dBi, the relation between these units is the gain in dBd equals the gain in dBi minus 2.15 dB. For a given frequency the antenna's effective area is proportional to the power gain. An antenna effective length is proportional to the *square root* of the antenna gain for a particular inspecting frequency and radiation resistance. Due to reciprocity, the gain of any antenna when receiving is equal to its gain when transmitting.

Directive gain or directivity is a different measure that does *not* take an antenna's electrical efficiency into account. This term is sometimes more relevant in the case of a receiving antenna where one is concerned mainly with the ability of an antenna to receive signals from one direction while rejecting interfering signals coming from a different direction.

The antenna gain is thus given as

$$G(\theta,\phi) = \eta D(\theta,\phi)$$

(3.39)

where
η = conversion efficiency from current radiating power
$D(\theta, \phi)$ = gain directivity

With the azimuthal and elevation angles ϕ, θ, as indicated in the spherical coordinate systems of the antenna given in Figure 3.4, the directive gain, $D(\theta, \phi)$, signifies the ratio of radiated power in a specified direction relative to that of an isotropic radiator radiating the same total power as the antenna under test but uniformly in all directions, the isotropic property. It should be noted that a practical true isotropic radiator does not exist. On the other hand, the power gain defines the ratio of radiated power in a given direction relative to that of an isotropic radiator that is radiating the total amount of *electrical power* received by the antenna under test. This is in contrast to the directive gain, which ignores any reduction factor in the efficiency.

The radiation intensity can be found by integrating the radiation over the entire volume of the solid angle, and thus we can write

$$G(\theta,\phi) = \frac{\left(\dfrac{P_{ant}(\theta,\phi)}{S}\right)_{ant}}{\left(\dfrac{P_{iso}(\theta,\phi)}{S}\right)_{iso}}$$

(3.40)

where $P_{ant}(\theta, \phi)$; $P_{iso}(\theta, \phi)$ are the power radiation over the surface or intensity distribution dependent on the solid angle of the antenna under consideration and the isotropic antenna, respectively. $G(\theta, \phi)$ is the gain and relies on the intensity term and radiation pattern. Normally the intensity is calculated over the entire angle of the radiation pattern, from the peak to the half-power point (the full width at half maximum, FWHM).

Published figures for antenna gain are almost always expressed in decibels (dB), a logarithmic scale. From the gain factor G, one finds the gain in decibels as

$$G_{dBi} = 10 \log G \tag{3.41}$$

As an isotropic antenna cannot be constructed; a dipole antenna may be used, and hence this can be referred to as the referenced antenna, so the antenna gain can be rewritten as

$$G_{dBd} = 10 \log_{10} \frac{G}{1.64} \tag{3.42}$$

Example

Find the antenna gain factor in dBd with respect to that of a dipole antenna when its gain G is equal to 5.

ANSWER

Substituting the value of 5 as the gain into (3.42) gives the gain factor of 4.48 dBd.

In general the relationship between the gain of the isotropic antenna and that of a dipole is given by

$$G_{dBd} = G_{dBi} - 2.15 \text{ dB} \tag{3.43}$$

Note that dB is dimensionless, as it indicates the ratio of two same unit terms in a logarithmic scale. Partial gain of an antenna can also be used sometimes, especially when a certain plane is to be considered, for example, $G(\theta); G(\phi)$.

3.4 EXPERIMENT

3.4.1 BACKGROUND

Maxwell's equations predict the radiation of EM energy from time-varying current sources. Although radiation occurs at all frequencies, its relative magnitude is significant until the size of the source region is compatible to a wavelength, hence its corresponding frequency in a certain medium. A transmitting or receiving antenna is a structure designed to transfer EM energy from or to a transmission line or waveguide to or from free space or an essentially bounded medium. The characteristics of an antenna used for transmitting and the characteristics of the same antenna used for receiving are closely related, a fact that is expressed in the reciprocity theorem.[1] Some important antenna characteristics are the directive gain and effective area, polar radiation pattern, polarization, radiation efficiency, power gain, impedance, and frequency response.

In this experiment the complex admittance (reciprocal of the complex impedance) of a monopole antenna is measured over a band of frequencies. Such monopole antennae are widely used on vehicles for mobile communications, and generally consist of a quarter-wavelength ($\lambda/4$) vertical whip above a conducting ground plane that has large dimensions compared with a wavelength. At resonance their input resistance is around 37 Ω, so that the input conductance is around 27 mS at the base of the whip. Typically a monopole antenna is connected to a coaxial transmission line, and then to a source or a receiver. If a standard coaxial cable having a characteristic impedance of $Z_0 = 50\ \Omega$ (20 mS) is used, some mismatch will result, particularly at frequencies away from resonance, and there will be reflection loss.

A convenient measure of the mismatch is given by the voltage standing wave ratio (VSWR) in a transmission line defined by

$$VSWR = \frac{V_{max}}{V_{min}} = \frac{|V_+| + |V_-|}{|V_+| - |V_-|} = \frac{1 + \left|\frac{V_-}{V_+}\right|}{1 - \left|\frac{V_-}{V_+}\right|} = \frac{1 + |\rho_L|}{1 - |\rho_L|} \geq 1 \tag{3.44}$$

where the reflection coefficient of the load is related to the *normalized* load z_L and $y_L\ (= 1/z_L)$ as

$$\rho_L = \frac{V_-}{V_+} = \frac{z_L - 1}{z_L + 1} = \frac{1 - y_L}{1 + y_L} \tag{3.45}$$

and the normalized load impedance (or admittance) is the actual load impedance (admittance) divided by the characteristic impedance (admittance) of the transmission line, given as

$$z_L = \frac{Z_L}{Z_0}; \quad y_L = \frac{Y_L}{Y_0} \tag{3.46}$$

The normalized impedance, or its reciprocal, the normalized admittance, is conveniently plotted on a Smith chart, which can display all complex values within a circular area, the center representing $1 + j_0$ (a matched load). The VSWR of any point on the Smith chart is found by constructing a circle centered on $1 + j_0$ and passing through the point; the VSWR is then the intercept of this circle on the normalized resistance or conductance axis between 1.0 and ∞, while the radius of the circle compared with the full radius of the Smith chart is the magnitude of the reflection coefficient.

3.4.2 MEASUREMENT OF THE MONOPOLE ANTENNA ADMITTANCE

The monopole antenna to be measured has a finite 300×300 mm square aluminum ground plane and a whip length of 84.5 mm (not centered). It has been designed to operate over the mobile telephone band of 825–890 MHz, but is not claimed to be optimally tuned.

- Measure the antenna input admittance in steps of 10 MHz from 780 to 950 MHz using the radio frequency (RF) sensor included in an RF spectrum analyzer.
  - This instrument makes use of a null indication when comparing the unknown admittance with a standard conductance and susceptance (an adjustable stub), although it is not a true bridge.
  - For each frequency record the conductance and susceptance values indicated on the dials of the admittance meter as well as the multiplying factor used.
  - Normalize the complex admittance to 20 mS and plot it on a Smith chart.
- Display the antenna input admittance over a similar frequency band using HP Network Analyzer type 8410B (or equivalent Agilent or Rhode Schwartz or Anritsu products).
  - This instrument uses a sweep frequency oscillator and displays the results directly on a Smith chart.
  - Details of the operations of the setup will be explained during lab sessions.
  - Make a careful sketch of the displayed results on another hardcopy Smith chart.
- Compare the two sets of results.
  - For each set of results, what is the minimum VSWR and at what frequency does it occur?
  - What is the VSWR at 825 MHz and at 890 MHz?
  - From the admittance meter results plot the VSWR against the frequency on linear graph paper. Does the antenna seem to be correctly tuned to the mobile telephone band evenly?
  - If not, what whip length should have been chosen?
  - Assume that the frequency value of a particular point on the admittance versus frequency curve (or the VSWR versus frequency curve) scales inversely with the length of the monopole whip (that is, ignore any tuning effect of the finite ground plane, which should really be finite in extent).

Note and comment: One may experience some difficulty in obtaining consistent admittance readings because the antenna itself is radiating close to the experimental bench, other apparatus, and human bodies. This uncontrolled and changing environment causes small but detectable changes to the readings. Try to remain in a fixed position while adjusting the admittance meter to give a null indication.

3.5 CONCLUDING REMARKS

An antenna (or aerial) is an electrical device that converts electric currents into radio EM waves, and vice versa. It is usually used with a radio transmitter or radio receiver. In transmission, a radio transmitter applies an oscillating radio frequency electric current to the antenna's terminals, and the antenna radiates the energy from the current as electromagnetic waves (radiowaves). In reception, an antenna intercepts some of the power of an electromagnetic wave in order to produce a tiny voltage at its terminals, which is applied to a receiver to be amplified. An antenna can be used for both transmitting and receiving.

Antennae are essential components of all equipment that use radios. They are used in systems such as radio broadcasting, broadcast television, two-way radio, communications receivers, radar, cellphones, and satellite communications, as well as other devices, such as garage door openers, wireless microphones, Bluetooth-enabled devices, wireless computer networks, baby monitors, and radio frequency identification (RFID) tags on merchandise. For RFID, both transmitting and receiving antennae are integrated into one device that can receive interrogating RF signals and then transmit a response to integrating sources for identification. The design of such an RFID device requires transfer of power from the source to the receiving antenna, and uses this power for radiating at the transmitting antenna. These topics currently attract significant research for applications in modern-day complex supermarket systems and parcel transport and deliveries, as well as currency identification.

The treatment of antennae in this chapter is given a minimum level, which is necessary as an introduction for readers at the undergraduate level; further details can be found in several specialized textbooks on antennae.

3.6 APPENDIX: METALLIC WAVEGUIDE[2]

3.6.1 BRIEF CONCEPT

Microwave waveguides are used to couple the sources to the radiation antenna, for example, a large parabolic antenna with a feeder size larger than that of the waveguide. This appendix thus gives an introduction to the metallic waveguide, including the principles of operation and some resonance conditions, so that an introduction to such a waveguide can be appreciated. Depending on the frequency, waveguides can be constructed from either conductive or dielectric materials. Waveguides can be less than a millimeter in width. An example might be those that are used in extremely high-frequency (EHF) communications. The fundamental mode can be guided and estimated with a simple relationship between the depth b and the width a of a rectangular waveguide, as shown in Figure 3.15(a).

EM metallic waveguides can be analyzed by forming the wave equation using Maxwell's equations, or their reduced form, the electromagnetic wave equation. The solutions of such equations are subject to boundary conditions determined by the properties of the materials, and their interfaces give the conditions for physical solutions, the eigenvalue equations. The eigenvalue equations give the number of

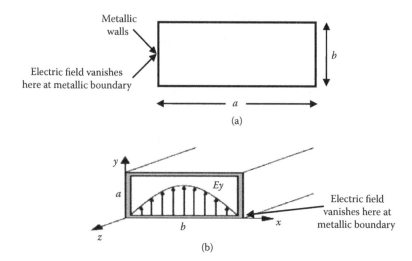

FIGURE 3.15 (a) Cross section of a rectangular metallic waveguide of dimensions a, b. (b) Coordinate and field distribution of the fundamental mode inside the rectangular waveguide.

solutions that are oscillating, thus waves, and eigenvalues that can be related to the effective propagation constant of the waves' guide in the waveguide. These equations have multiple solutions, or oscillating modes, that are the eigenfunctions of the equation system. Each mode is characterized by a cutoff frequency below which the mode cannot exist in the guide. Note that this cutoff condition is the reverse of the definition of the cutoff condition for a dielectric waveguide, optical fibers, or planar optical waveguides, which are described in later chapters. For the optical waveguide we say the cutoff condition involves the wavelength that is usually proportional to the inverse of the frequency.

Waveguide propagation modes depend on the operating wavelength and polarization, and the shape and size of the guide, as shown in Figure 3.15 for an air-filled rectangular metallic waveguide. The longitudinal mode of a waveguide is a particular standing wave pattern formed by waves confined in the cavity. The transverse modes are classified into different types: (a) TE (transverse electric) modes have no electric field in the direction of propagation, (b) TM (transverse magnetic) modes have no magnetic field in the direction of propagation, (c) TEM (transverse electromagnetic) modes have no electric or magnetic field in the direction of propagation, and (d) hybrid modes have both electric and magnetic field components in the direction of propagation.

In hollow waveguides (single conductor), TEM waves are not possible, since Maxwell's equations will show that the electric field must then have zero divergence and zero curl and vanish to zero at metallic boundaries, resulting in a zero field, or equivalently, $\nabla^2\phi = 0$, with boundary conditions guaranteeing only the trivial solution. However, TEM waves *can* propagate in coaxial cable because there are two conductors.

The mode with the lowest cutoff frequency is termed the fundamental or dominant mode of the guide, as shown in Figure 3.15(b). It is usual to choose the size of

TABLE 3.1

Characteristics of Some Standard IR-Filled Rectangular Waveguides

Type	a (inches)	b (inches)	Cutoff Frequency, f_c (GHz)	f_{min} (GHz)	f_{min} (GHz)	Band	Power MW	Attenuation Coefficient dB/m
WR-510	5.1	2.55	1.16	1.45	2.2	L	9	0.007
WR-284	2.84	1.34	2.08	2.6	3.95	S	2.7	0.019
WR-159	1.59	0.795	3.71	4.64	7.05	C	0.9	0.043
WR-90	0.90	0.40	6.56	8.2	12.5	X	0.25	0.110
WR-62	0.622	0.311	9.49	11.90	18.60	Ku	0.140	0.176
WR-42	0.42	0.17	14.05	17.6	26.7	K	0.027	0.583
WR-28	0.28	0.074	21.08	26.4	40	Ka	0.027	0.583
WR-15	0.148	0.074	39.87	49.80	75.80	V	0.0075	1.52
WR-10	0.10	0.05	59.01	73.80	112.00	W	0.0035	2.74

the guide such that only this one mode can exist in the frequency band of operation, with no interferences due to higher-order modes and this mode can occur, hence radiating pure EM waves. In rectangular and circular (hollow pipe) waveguides, the dominant modes are designated the $TE_{1,0}$ and $TE_{1,1}$ modes, respectively.

In the microwave region of the EM spectrum, a waveguide normally consists of a hollow metallic conductor. These waveguides can take the form of single conductors with or without a dielectric coating. Hollow waveguides must be *one-half wavelength* or more in diameter in order to support one or more transverse wave modes at the operating frequency band. This is similar to the half-wave antenna condition. Table 3.1 gives the size and type number of a number of air-filled rectangular waveguides that can operate from the gigahertz range (L-band) to the millimeter wave range (V- and W-bands). The frequency passband of the waveguide depends on the relative ratio between the depth and the width of the waveguide, as shown in Figure 3.16. The guide mode can be illustrated by the zigzag ray model, as shown in Figure 3.17.

Waveguides may be filled with pressurized gas to inhibit arcing and prevent multipaction, allowing higher-power transmission. Conversely, waveguides may be required to be evacuated as part of evacuated systems, for example, the electron beam systems. A slotted waveguide is generally used for radar and other similar applications. The waveguide structure has the capability of confining and supporting

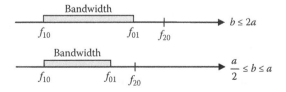

FIGURE 3.16 Operating frequency range or bandwidth of metallic rectangular waveguide is dependent on the relative dimension of the depth and width of the waveguide.

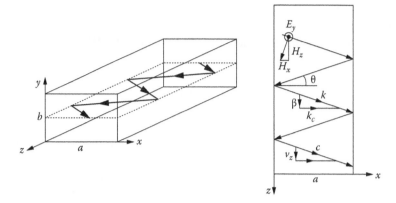

FIGURE 3.17   Equivalent ray model of waves propagating through the metallic rectangular waveguide.

the energy of an electromagnetic wave to a specific relatively narrow and controllable path.

A close waveguide is an electromagnetic waveguide (a) that is tubular, usually with a circular or rectangular cross section, (b) that has electrically conducting walls, (c) that may be hollow or filled with a dielectric material, (d) that can support a large number of discrete propagating/guided modes, though only a few may be practical, (e) in which each discrete mode defines the propagation constant for that mode, (f) in which the field at any point can be described in terms of the supported modes, (g) in which there is no radiation field, and (h) in which discontinuities and bends cause mode conversion but not radiation.

The dimensions of a hollow metallic waveguide (see Figure 3.15) determine which wavelengths it can support, and in which modes. Typically the waveguide is operated so that only a single mode is present. The lowest-order mode possible is generally selected. Frequencies below the guide's cutoff frequency will not propagate. It is possible to operate waveguides at higher-order modes, or with multiple modes present, but this is usually impractical. Waveguides are almost exclusively made of metal and mostly rigid structures. There are certain types of "corrugated" waveguides that have the ability to flex and bend, but they are only used where essential because they degrade propagation properties. Due to propagation of energy in mostly air or space within the waveguide, it is one of the lowest loss transmission line types and is highly preferred for high-frequency applications where most other types of transmission structures introduce large losses. Due to the skin effect at high frequencies, electric current along the walls penetrates typically only a few micrometers into the metal of the inner surface. Since this is where most of the resistive loss occurs, it is important that the conductivity of the interior surface be kept as high as possible. For this reason, most waveguide interior surfaces are plated with copper, silver, or gold.

The VSWR can be measured to determine that a waveguide is contiguous and no leaks or sharp bends exist. If such bends or holes in the waveguide surface are present, this may diminish the performance of both transmitter and receiver equipment connected at either end. Poor transmission through the waveguide may also occur as a result of moisture buildup, which corrodes and degrades conductivity of the inner surfaces and is crucial for low loss propagation. For this reason, waveguides are nominally fitted with microwave windows at the outer end that will not interfere with propagation but will keep out the elements. Moisture can cause fungus buildup or arcing in high-power systems such as radio or radar transmitters. Voltage standing waves occur when impedance mismatches in the waveguide cause energy to reflect back in the opposite direction of propagation. In addition to limiting the effective transfer of energy, these reflections can cause higher voltages in the waveguide and damage equipment. In practice, waveguides act as the equivalent of cables for super-high-frequency (SHF) systems. For such applications, it is desired to operate waveguides with only one mode propagating through the waveguide.

With rectangular waveguides, it is possible to design the waveguide such that the frequency band over which only one mode propagates is as high as 2:1, that is, the ratio of the upper band edge to lower band edge is 2. The relationship between the longest wavelengths that will propagate through a rectangular waveguide is a simple one. Given that the width, b, as shown in Figure 3.15, is the greater of its two dimensions, and lambda is the wavelength, then lambda equals twice the width b of the waveguide section, that is, the resonance condition with the two ends vanishing the EM waves. For circular metallic waveguides, the highest possible bandwidth allowing only a single mode to propagate is only 1.3601:1.

Because rectangular waveguides have a much larger bandwidth over which only a single mode can propagate, standards exist for rectangular waveguides, but not for circular waveguides, as both the $\tilde{E}^-$ and $\tilde{H}^-$ fields can be guided and no distinct polarization direction can be determined. In general, but not always, it is essential that standard waveguides are designed such that (1) one band starts where another band ends, with another band that overlaps the two bands; (2) the lower edge of the band is approximately 30% higher than the waveguide's cutoff frequency; (3) the upper edge of the band is approximately 5% lower than the cutoff frequency of the next higher-order mode; and (4) the waveguide height is half the waveguide width, $a = 2b$.

For the resonance of the oscillating TE waves along both sides of a rectangular waveguide the propagation wavenumbers $k_x; k_y$ in these directions must satisfy the following conditions:

$$k_x a = n\pi; \quad k_y b = m\pi \rightarrow k_x = \frac{n\pi}{a}; \quad k_y = \frac{m\pi}{b} \tag{3.47}$$

with n, m taking integer values, hence the order of the transverse electric-guided modes is abbreviated TE_{nm}. The cutoff wavenumber and hence the cutoff frequencies and wavelength of the modes take on quantized values given as

$$k_c = \sqrt{\left(\frac{n\pi}{a}\right)^2 + \left(\frac{m\pi}{b}\right)^2} \qquad \text{for } TE_{nm} \tag{3.48}$$

$$f_{nm} = c\sqrt{\left(\frac{n}{2a}\right)^2 + \left(\frac{m}{2b}\right)^2}$$

$$\lambda_{nm} = \frac{1}{\sqrt{\left(\frac{n}{2a}\right)^2 + \left(\frac{m}{2b}\right)^2}} \qquad (3.49)$$

3.6.2 EXPERIMENT ON WAVEGUIDE

3.6.2.1 Introduction

Propagation of electromagnetic (EM) waves may be either guided or unguided. Coaxial lines and waveguides are examples of transmission lines that guide waves from the source to the destination along a clearly defined path.

Free-space propagation of EM waves from a broadcast transmitter may serve as an example of unguided propagation resulting from radiation. The two cases are not mutually exclusive, and in some instances propagation along a clearly defined path may be accompanied by significant radiation in many directions.

The term *waveguide* is normally reserved for hollow metallic pipes, and in practice EM waves are wholly confined to their interior. Uniform rectangular and circular waveguides having the same cross section throughout their length are used most frequently (a pyramidal horn radiator being an example of a nonuniform waveguide).

All uniform waveguides are structured with no inner conductors and the same cross section shape, along the propagation direction of the guided electromagnetic waves. The phenomenon of cutoff is valid, and the waveguide behaves as a high-pass filter. Accordingly, the waveguide walls may have very small, and for many purposes negligible, losses below a certain frequency (known as the cutoff frequency). Thus the EM waves launched into the waveguide are subject to very high attenuation, whereas above that frequency their attenuation is normally exceedingly small. The attenuation below cutoff is essentially reflective and unrelated to wall losses, while the small attenuation above cutoff depends primarily on wall conductivity.

While the wavelength in free space is normally only a function of the velocity of the propagation (v_p) and frequency $f(\lambda = v_p/f; v_p = $ phase velocity), it can be shown that the wavelength inside the waveguide is additionally a function of the cross section geometry as well as the field pattern of the wave—the mode order.

The relationship between the cutoff wavelength λ_c, the wavelength in free space λ, and the guide wavelength takes the form

$$\frac{1}{\lambda^2} = \frac{1}{\lambda_g^2} + \frac{1}{\lambda_c^2} \qquad (3.50)$$

where $\lambda = v_p/f$, $\lambda_c = v_p/f_c$, and f and f_c are the excitation (variable) frequency and fixed cutoff frequency, respectively. We note that at cutoff the guide wavelength is infinite.

Equation (3.10) suggests a straight-line relationship between $1/\lambda^2$ and $1/\lambda_g^2$, making it possible to determine $1/\lambda_c^2$ and hence the cutoff wavelength λ_c from the

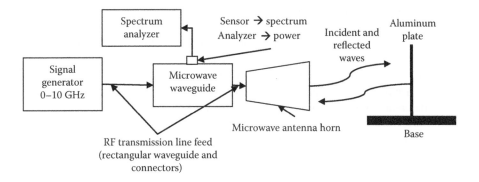

FIGURE 3.18   Experimental setup for the study of a microwave waveguide.

appropriate intercept on the $1/\lambda^2$ axis. The origin must be included. Use the same scale for both axes with units of m^{-2}. Perform a full error analysis.

With reference to Figure 3.18, λ_g represents twice the distance between the two minima of the field distribution displayed by the detector of the waveguide slotted line. On the other hand, the wavelength in free space λ can be deducted as twice the distance (outside the waveguide) between the two minima of the standing wave pattern when the slotted line probe is kept stationary and the reflecting sheet moved.

3.6.2.2   Experiment

An RF signal generator can be used as the source of microwave power, and in this experiment it is capable of delivering up to 15 dBm of power. At each frequency the reflector voltage must be adjusted for a maximum power output, and hence stable operation within each mode (there are several such modes, separately by a range of reflector voltages, at which oscillations are not possible).

Referring to Figure 3.18, note that as the waves leave the horn, they are reflected from the metallic target sheet and produce a standing wave. As the reflected waves reenter the waveguide, a standing wave will be produced as well, although the wavelengths of the two standing waves are not equal. The distance between the horn and the metallic reflecting sheet should not be less than about 100 mm.

a. Adjust the frequency to its upper limit, which is attained when the two bowed springs adjusting the screw come together.
b. Leaving the slotted line probe fixed in one position, move the reflecting sheet until the detector reading is at minimum. Make sure that the axis of the waveguide is at right angles to the plane of the sheet.
c. Make a record of the sheet position and then move it along a straight line coincident with the axis of the waveguide until the detector reading passes through another minimum. Repeat, noting that the average distance between the two minima represents half the wavelength in free space.
d. Leave the metal sheet fixed in a suitable position and obtain a few successive detector reading minima by moving the slotted probe instead. The

average between the two minima is now equal to half a waveguide wavelength λ_g.

e. Make a record of λ and the corresponding λ_g. Repeat for other frequencies
up to the lower limit. Typically a frequency range of 8500–9500 MHz can
be covered, the exact range varying somewhat from one unit to another.

f. Hence deduce with full error analysis of your measurements. Note that λ_c
theoretically should equal twice the broad a dimension of the rectangular
waveguide measured internally, that is, $\lambda_c = 2a$.

g. Measure the a and determine if your measured range of λ_c includes the
theoretical λ_c.

3.7   PROBLEMS

3.7.1   Waveguide Measurements

a. What waveguide fundamental mode can be supported in an air-filled metallic waveguide with a width of 60 mm and height of 30 mm?

b. If the waveguide has been changed to 50×25 mm internal dimensions,
what would the cutoff frequency for this mode be?

c. What was the cutoff frequency for the waveguide you used in this experiment?

d. Draw a schematic diagram of the equipment arrangement used to measure
the cutoff frequency of the metallic waveguide above.

3.7.2   Antenna Admittance

Draw a schematic diagram of the equipment you might use to measure the input
impedance of a horn antenna with waveguide excitation at 800 to 900 MHz. If you
were measuring the antenna admittance at 900 MHz, show on the schematic diagram the frequency of the signal being carried by each coaxial cable interconnecting
the subsystem equipment.

3.7.3   Waveguide

A section of rectangular waveguide has internal dimensions of 12.5×25 mm.

1. What modes will propagate in this guide at 9 GHz?
2. What is the range of frequencies for which only one mode will propagate in
the waveguide, and what is that mode?
3. Sketch the electric field distribution of the TE_{10} mode in the waveguide.
4. Sketch the magnetic field distribution of the TE_{10} mode in the waveguide.
5. Sketch the electric field distribution of the TE_{21} mode in the waveguide.
6. What is the guide wavelength at 9 GHz?

REFERENCES

1. T. Teshirogi and T. Yoneyama, *Modern Millimeter-Wave Technologies*, Wave Summit Course, Ohmsha, Ltd., Tokyo, 2001.
2. Institute of Electrical and Electronics Engineers, *The IEEE Standard Dictionary of Electrical and Electronics Terms*, 6th ed., Institute of Electrical and Electronics Engineers, New York, 1997.

4 Planar Optical Waveguides

4.1 INTRODUCTION

The term *integrated optics* was first coined by Miller in 1969,[1] as an analogy of light-wave to the electronic integrated circuits. Indeed, the present term used in industry is planar lightwave circuits (PLCs). Since that term was proposed, tremendous progress has been made. Various integrated optical devices have been researched, developed, and deployed in practical optical transmission systems and networks. Extensive surveys have been given over the years,[2,3] and even defined and described on the Internet.[4,5] Both linear and nonlinear integrated optics[6] have been exploited.

Recent experimental demonstrations have pushed the information transmission bit rate per channel to 100 Gb/s,[7–9] with the multiplexing of several wavelength channels reaching to tens of terabits/s.[10–12] At such speed there are needs of high speed modulations, switching, routing, etc. Thus the demands for optical modulators, switches, and optical preprocessors employing nonlinear integrated optical devices and the compensation devices for equalization of dispersion effects in single-mode optical fibers. Furthermore, integrated optic amplifiers are required for equalization of the propagation losses in fibers, the amplification of lightwave signals in the optical domain. Prior to the availability of erbium-doped fiber amplifiers (EDFAs), attempts were made to increase the span distance between repeaters of amplitude-modulated single-fiber transmission systems by employing coherent techniques,[13,14] in which a narrow linewidth and high power are used to mix with the received signals to improve the receiver sensitivity. This linewidth requirement on the local oscillator laser limits the deployment of such coherent transmission in real practice. Not until recently have coherent optical communications attracted much interest, once again as a possible technique to further increase the transmission spans.[15,16] Naturally the modulation formats, such as amplitude shift keying, phase shift keying, and frequency shift keying, are employed to reduce the effective signal bandwidth in order to minimize the dispersion effects in single-mode optical fibers. The detection of ultra-high-speed optical signals can thus be direct or coherent. This attracts the employment of integrated optical devices in the optical transmitters, receivers, and online components.

The basic component of these integrated photonic devices is the optical wave-guide formed by a thin film or diffused waveguiding layer structure on some substrate. From the mode propagation point of view, design optimization requires accurate estimation of the propagation constant, and thence dispersion characteristics, mode size, and group velocities, depending on the types of applications. These requirements led us to develop simple, accurate, and efficient methods of analysis of single-mode waveguides. This is the motivation of the first chapter of the book.

Most optical waveguides with graded index profile, especially lithium niobate types, are fabricated by a diffusion process that is commonly formed by diffusion of impurities into various substrate ferroelectric materials, such as lithium niobate and lithium tantalate. On the other hand, the proton exchange process can also be used so that the Li ions can be exchanged with the hydrogen ion. In both processes, diffusion and ion exchange, a crystal stress is established and a change of the refractive index is created, resulting in a graded index profile distribution. Complementary error function is usually used to represent the distribution of the impurity from the surface of the substrate to the depth of the substrate[17] and the time of diffusion of the metallic impurities. A Gaussian profile is expected when the diffusion time is sufficiently long[18] and is used to fit the experimental values of Se into CdS crystal. In addition, various profiles can be used to form optical waveguides using molecular beam epitaxy (MBE) and metallic organic chemical vapor deposition (MOCVD) techniques, as in the waveguiding structures for laser diodes of separate confinement of the heterojunction.[19]

Exact analytical solutions are available for the step,[20] exponential,[21] hyberbolic secant,[22] clad linear,[23] clad parabolic,[24] and Fermi[25] profiles. In general, approximate analytical or exact numerical methods are required to analyze general classes of profiles. For some practical profiles, universal charts describing the mechanism of waveguiding have already been presented by several authors. These have been obtained by variational analysis,[26] Wentzel–Kramers–Brilluoin (WKB),[27] and multilayer staircase.[28] However, these curves are only accurate for multimode waveguides, with the exception of the last two references. Single-mode planar optical waveguides are very important for integrated optic circuits for applications in advanced ultrahigh-speed optical communications; see, for example, the pioneering works by Korotky et al.[29] However, methods of analysis are confined predominantly to those originally used in multimode waveguides. In the single-mode regime, the variational and WKB methods are expected to perform poorly. In the former, the solutions are strongly affected by the choice of trial fields. In the latter, more accurate prediction of the phase changes at the turning point is required. In the Runge–Kutta method outward integration methods, instability is caused by the solution and error increasing at large x.[17] This problem is resolved by approximating the fields at sufficient depth in the waveguide by an evanescent field.[30] However, this requires the knowledge of the location of solution matching. Thus this chapter recognizes that any general technique of analysis must be numerical in nature due to stringent accuracy requirements for single-mode waveguides. If the single-mode waveguide is used as a nonlinear interaction medium, the phase matching is very important, and thus an accurate estimation of the dispersion curves plays a very important part in the conversion efficiency.[6,31]

This chapter is organized as follows: Section 4.2 gives the formation of the problems for solving single-mode optical waveguides in which all parameters are expressed in terms of normalized quantities. A novel relationship between a newly defined mode spot size and the normalized waveguide parameter is described. This is very much close to the definition of the mode spot size as defined for single-mode optical fibers commonly specified. Section 4.3 describes the simplest method for calculating the propagation constant and the model field. The stepping function recurrence method of integration originally developed by Killingbeck[32,33] can be used. The method is thence applied for asymmetrical optical waveguide structures. For diffused waveguides or graded index profile distribution of the refractive index from the surface of the waveguide to the deepest position, two widely used methods for waveguide modal analysis are given.

The variational method with a simple Hermite–Gaussian field was first introduced by Korotky et al.[29] to calculate the mode spot size in a diffused channel waveguide. In Section 4.4 we show that the method to estimate the modal characteristics of all diffused waveguide profiles is inaccurate and computationally intensive for the calculation of the dispersion characteristics, but a very close estimation of the mode spot size.

The equations required for the derivation of the wave equations, as well as an exact analysis of an asymmetric waveguide structure, are given in Appendix A.

4.2 FORMATION OF PLANAR SINGLE-MODE WAVEGUIDE PROBLEMS

A planar dielectric waveguide with the geometry shown in Figure 4.1 can support modes with two polarizations: TE and TM field-guided modes. In practice, either polarized mode can be excited. Provided certain boundary conditions are met, these modes are bounded and propagate along the z-axis, each with a unique effective phase velocity. If we allow for the uniformity of the refractive index and geometrical dimension in the propagation direction, the phase velocity is only a function of the transverse index profile. In this section we consider the simplest configuration with an index variation only in the $-x$-direction. Note that the notation of the coordinate system follows the right-hand side (RHS).

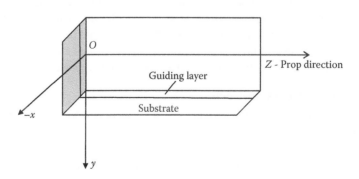

FIGURE 4.1 Schematic structure of a planar dielectric waveguide.

4.2.1   TE/TM WAVE EQUATION

The wave equation for the guided modes stems from the Maxwell equations given in the Chapter 1, together with the associated constituent relations. We consider the steady-state solutions in the dielectric medium free from any sources and losses. By omitting the common factor $e^{j(\omega t - \beta z)}$ from the equations, we can write, for a medium characterized with the refractive index $n(x)$, the well-known wave equations:

$$\frac{d^2 E_y}{dx^2} + [k_0^2 n^2(x) - \beta^2] E_y = 0$$

and   (4.1)

$$\frac{d^2 H_y}{dx^2} + [k_0^2 n^2(x) - \beta^2] H_y = \frac{1}{n^2(x)} \frac{d^2 n^2(x)}{dx^2} \frac{dH_y}{dx}$$

where k_0 = wavenumber in free space, β = propagation constant of the wave along the z-axis, $E_y(x)$ = transverse electric field, and H_y = transverse magnetic field.

4.2.1.1   Continuity Requirements and Boundary Conditions

In practical waveguides, it is common to expect that at least one region of dielectric continuity is encountered by the optical fields. At the location of the discontinuity, the wave equations are not valid. However, the identity of the modes is preserved by matching the fields and their derivatives on either side of the dielectric discontinuity. For the TE modes these boundary conditions impose the continuity $E_y; dE_y/d_y$ across the interface. For the TM modes we require the continuity $H_y; 1/n^2(x) \times dH_y/dy$. In addition, the bound modes satisfy the conditions that E_y, H_y vanishes at $x = \infty$. Together they give rise to the eigenvalue equation from which the propagation constant can be calculated.

Note that the principal object of the eigenvalue equation is to estimate the maximum value of the propagation constant so that the dependence of this propagation parameter on the optical frequency/wavelength is minimum and there is minimum dispersion of the waves at different wavelengths, which are normally found for the other spectral components of a modulated lightwave channel in an optical communication system. A maximum value of the propagation constant along the z-direction means that the direction of the wave vector is close to the propagation axis.

4.2.1.2   Index Profile Construction

For the purpose of computation and analysis it is customary to write the refractive index profile in the general form as

$$n^2(x) = \begin{cases} n_s^2 \left[1 + 2\Delta S\left(\dfrac{x}{d}\right) \right] & x \geq 0 \\ \\ n_c^2 & x < 0 \end{cases}$$

(4.2)

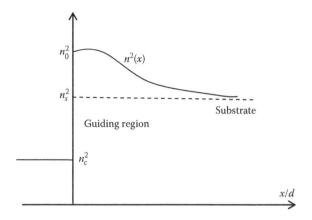

FIGURE 4.2 Square of the refractive index distribution profile for an asymmetrical waveguide.

where

$$\Delta = \frac{n_0^2 - n_s^2}{2n_s^2}$$

Δ is the profile height and n_c, n_0, and n_s are the refractive indices of the cover, the guiding layer, and the substrate, respectively. $S(x)$ is the profile shape function and d is the diffusion depth of a graded index distribution. Figure 4.2 shows a typical representation of the graded index profile. It turns out that further normalization of the shape profile can be represented as

$$S\left(X = \frac{x}{d} \right) = \frac{n^2(x) - n_s^2}{n_0^2 - n_s^2} \tag{4.3}$$

This definition of the profile shape is unaffected by the symmetry of the waveguide. In a symmetric waveguide structure with an axis of symmetry at $x = 0$, these equations are equally valid.

4.2.1.3 Normalization and Simplification

The presence of the nonzero term in the RHS of Equation (4.1) complicates the analysis. It is identically zero for a step index profile. The exact solutions of this equation are available for the exponential, hyperbolic secant, and an inverted x profile.[34] However, for smooth profiles normally encountered in practice, several authors[35] found by perturbation analysis that the RHS of the equation can be neglected. However, the fundamental mode of an infinite parabolic profile faces a 44% error in the group velocity of the guided mode.

Following Kolgenik and Ramaswamy[38] we can introduce the normalized parameter for the waveguide as follows:

$$V = \frac{dk_0}{n_0^2 - n_s^2}$$

$$A = \frac{n_s^2 - n_c^2}{n_0^2 - n_s^2} \tag{4.4}$$

$$b = \frac{n_e^2 - n_s^2}{n_0^2 - n_s^2}$$

where A is defined as the asymmetry factor, b is the normalized propagation constant, $n_e = \beta/k_0$ is the effective refractive index of the guided mode along the propagation axis, and V is the normalized frequency.

Guided mode requires

$$n_s < n_e < n_0 \tag{4.5}$$

Thus the normalized propagation constant must satisfy $B < 1$. The real advantages of the normalization are the analysis and design optimization points of view. Substituting the normalized parameters into the wave equation (4.3) we obtain

$$\frac{d^2\varphi}{dx^2} + V^2[S(X) - b]\varphi = 0 \tag{4.6}$$

where

$$\varphi(X) \equiv E_y; H_y$$

for the TE and TM modes, respectively. The propagation for the TM modes is accurately represented by that of the TE modes, except at cutoff or for a waveguide at large symmetry.[6] However, if extreme accuracy or mode splitting is required, then (4.6) can be modified to the changes involving only a slight modification of the boundary condition.

4.2.1.4 Modal Parameters of Planar Optical Waveguides

The solution of the wave equation (4.6) subject to the boundary conditions enables the determination of various optical parameters. The following are the most commonly used for the design of single optical guided wave devices.

4.2.1.4.1 Mode Size

The mode size Γ_a for an asymmetrical field is defined as the full width at half maximum (FWHM) power intensity. For a full description of the field, the peak position of intensity I_p and the field asymmetrical factor Γ_1/Γ_2, defined with respect to I_p, are required, as shown in Figure 4.3. The knowledge of the mode size is critical to match with that of the single-mode optical fiber for in-line integration with fiber

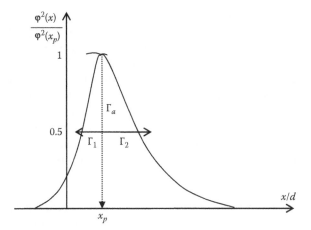

FIGURE 4.3 The minimum set of parameters required for characterization of a fundamental mode field in an asymmetrical planar waveguide.

transmission systems. These parameters are defined in terms of the optical power of the field due to practical reasons because the mode size is normally monitored using a CCD camera through which the intensity of the mode field is converted into the charge current and displayed or digitized for data processing.

4.2.1.4.2 Propagation Constant and Effective Refractive Index

The variation of the normalized propagation constant b as a function of the normalized frequency parameter V is normally required for the design and characterization of an optical waveguide. For example, this relationship specifies the diffusion depth required once the mode index at a specific operating wavelength is given. The purpose of this section is to present the relation of the modal field to the propagation constant.

If we integrate Equation (4.6) with respect to X then we have

$$\int_{-\infty}^{+\infty} \left\{ \frac{d^2\varphi}{dx^2} + V^2[S(X)-b]\varphi \right\} dX = 0 \quad \text{or} \quad \int_{-\infty}^{+\infty} \varphi'' dX + V^2 \int_{-\infty}^{+\infty} \{[S(X)-b]\varphi(X)\} dX \quad (4.7)$$

where the dashes denote the derivative with respect to X. If we further impose the condition that $\varphi(-\infty) = \varphi(+\infty) = 0$, the integral on the left becomes zero, and thus we obtain

$$b = \frac{\displaystyle\int_{-\infty}^{+\infty} S(X)\varphi(X)dX}{\displaystyle\int_{-\infty}^{+\infty} \varphi(X)dX} \quad (4.8)$$

This equation indicates that the dispersion characteristics for an arbitrary index profile must be a smooth curve. The rule of refractive index was used by Tien[37] to explain a host of new wave phenomena in integrated optical waveguides—that light tends to propagate in the region where the refractive index is largest. In the context of planar optical waveguides with arbitrary index profiles, this rule suggests that for a given profile at a given frequency, the mode field adjusts itself so that the maximum value of b is achieved. This corresponds to the minimum phase velocity allowed for the mode. This rule is indeed a direct statement of Fermat's law in ray optics and is a special case of a generalized rule in quantum mechanics formulated in the form of a well-known Feymann's path integral, of which Maxwell's equations are also satisfied.[38]

4.2.1.4.3 Waveguide Dispersion and Spot Size

A second and potentially useful relation between b and the modal field can be established from the stationary expression for b. This can be obtained from Equation (4.6) after multiplication by ϕ and taking the integration with respect to X from negative to positive infinity. After integrating the results by parts and imposing the boundary conditions $\varphi(-\infty)\varphi'(-\infty) = \varphi(+\infty)\varphi'(\infty) = 0$, we obtain the well-known stationary relation:

$$bV^2 = \frac{V^2 \int\limits_{-\infty}^{+\infty} S(X)\varphi^2(X)\,dX - \int\limits_{-\infty}^{+\infty} \varphi^2(X)\,dX}{\int\limits_{-\infty}^{+\infty} \varphi^2(X)\,dX} \tag{4.9}$$

This is the basic equation for the variational analysis. It possesses the unique property that for any trial fields that satisfy the boundary conditions above, the quotient remains stationary provided the mismatch between the trial and actual fields is small.[39]

Thus we can write

$$\frac{d(bV^2)}{dV^2} = \frac{\int\limits_{-\infty}^{+\infty} S(X)\varphi^2(X)\,dX}{\int\limits_{-\infty}^{+\infty} \varphi^2(X)\,dX} \tag{4.10}$$

or

$$\frac{1}{2}\left(bV + V\frac{d(bV)}{dV}\right) = V\frac{\int\limits_{-\infty}^{+\infty} S(X)\varphi^2(X)\,dX}{\int\limits_{-\infty}^{+\infty} \varphi^2(X)\,dX} \tag{4.11}$$

Taking the derivative a second time and using Equations (4.9) and (4.11), we obtain

$$\frac{1}{2}\left(V\frac{d^2(bV)}{dV^2}+2\frac{d(bV)}{dV}\right)=V\frac{d}{dV}\left(b+\frac{2}{V^2W_m}\right)+\frac{1}{2}\left(b+\frac{d(bV)}{dV}\right) \quad (4.12)$$

where we have to define a new spot size parameter as

$$W_m=\frac{2\int_{-\infty}^{+\infty}\varphi^2(X)\,dX}{\int_{-\infty}^{+\infty}[\varphi'(X)]^2\,dX} \quad (4.13)$$

Further algebraic manipulation leads to the simple relationship

$$V\frac{d^2(bV)}{dV^2}=4\frac{d}{dV}\left(\frac{2}{VW_m^2}\right) \quad (4.14)$$

This relation is analogous to the relation between Petermann's spot size and the waveguide dispersion in single-mode optical fibers.[40] The preceding analysis was first performed by Sansonetti,[41] which inspired Petermann to define a new spot size in the characterization of single-mode optical fibers from spot size measurement.

The relation (4.14) can be found for an optical waveguide with a profile following the Hermite–Gaussian variational field of

$$\varphi(X)=\begin{cases} A_0\alpha_0^{1/2}e^{-\alpha_0 X^2/2};X\geq 0 \\ 0;X<0 \end{cases} \quad (4.15)$$

$$W_m^2=\frac{4}{\alpha_0^2}=\Gamma_a^2 \quad (4.16)$$

where α_0 is the variational spot size parameter and A_0 is a constant. The RHS of (4.16) corresponds to Γ_a^2 as defined by Korotky et al.,[29] and has been defined above. Although α_0 is an approximate mode spot size, several experiments show excellent agreement between the theoretical and experimental values W_m and Γ_a.

4.3   APPROXIMATE ANALYTICAL METHODS OF SOLUTION

Despite the availability of direct numerical integration methods for the analysis of optical waveguides, approximate analytical solutions are still being used, improved upon, and sought after. We have to strike the balance between accuracy and simplification. Three well-known methods of analysis are described in this section. They are valid for single-mode planar optical waveguides.

The variational method[29] is applicable only to the fundamental mode of the asymmetrical waveguide due to the form of the trial field. The equivalent profile method is valid only for symmetrical waveguides because it requires the field to be monotonously decreasing. The WKB method can be used in both cases (see Appendix B). We group the methodological approaches into symmetry and asymmetry. In Section 4.3.1 the analytical formulae for a number of widely used profiles are obtained. We explore the improvements to the WKB method and limitations. In Section 4.3.2 we compare the accuracy of the equivalent profile moment methods using a step and a *cosh* reference profile.[42,43] The WKB method may not work at all for the analysis of the single-mode optical waveguides.

4.3.1 ASYMMETRICAL WAVEGUIDES

4.3.1.1 Variational Techniques

The variational method is based on the substitution of a TE_0 mode look-alike trial field into the stationary expression of the normalized propagation constant b given in Equation (4.8). The shape of the field is then adjusted to maximize b for all values of V (see Equation (4.9)). The mathematical procedure is given in Snyder and Love.[24]

Following Korotky et al.[29] and Riviere et al.[28] a trial solution can be proposed that closely fits the form of the TE_0 field:

$$\varphi(X) = \begin{cases} \sqrt{\alpha_0}\, e^{-\frac{\alpha_0 X^2}{2}} & X \geq 0 \\ 0 & X < 0 \end{cases} \qquad (4.17)$$

Note that one drawback of the form of this field is that it vanishes at $X = 0$. For single-mode optical waveguides, this condition is very nearly so only for guides with very large asymmetry. However, only a single parameter needs to be optimized; thus, the optimization scheme is simple. In the following sections it is shown that there exist closed-form formulae for several profiles.

4.3.1.1.1 Eigenvalue Equation

If we substitute (4.17) and the derivative of this trial field into (4.9) a simpler expression is obtained, after some tedious algebra:

$$b = I_1 - \frac{3\alpha_0}{2V^2} \qquad (4.18)$$

where

$$I_1 = 4\alpha_0 \left(\frac{\alpha_0}{\pi}\right)^{1/2} \int_0^\infty S(X) X^2 e^{-\alpha_0 X^2}\, dX$$

TABLE 4.1

Optimum Value of Selected Asymmetrical Clad-Diffused Waveguide

Profile/Distribution	$S(X)$	α_0
Gaussian	e^{-X^2}	$(\alpha_0+1)\left[\dfrac{(\alpha_0+1)}{\alpha_0}\right]^{1/2}-V=0$
Exponential	e^{-X}	$4V^2\sqrt{\alpha_0}-3\pi(\alpha_0+1)^2=0$
Complementary error function $erfc$	$erfc(X)$	$2(4\alpha_0+1)\sqrt{\dfrac{\alpha_0}{\pi}}-(1+6\alpha_0)\left[1-erfc\dfrac{1}{2}\sqrt{\alpha_0}\right]e^{\frac{1}{4\alpha_0}}-\dfrac{12\alpha_0^3}{V^2}$

is the only profile-dependent expression. The correct value of α_0 is obtained by noting that b must be stationary with respect to α_0; thus,

$$\frac{db}{d\alpha_0}=0=\frac{dI_1}{d\alpha_0}-\frac{3}{2V^2} \tag{4.19}$$

Substituting this α_0 into (4.18) we obtain the eigenvalues given in Table 4.1.

4.3.1.1.2   Fundamental Mode Cutoff Frequency

The lowest-order mode in an asymmetric optical waveguide has a nonzero cutoff frequency for $A \neq 0$. Thus we can set $b = 0$ and $V = V_c$ in (4.18) and (4.19) to obtain the desired cutoff frequencies. Since V_c appears in both equations, one can solve simultaneously for α_0 initially before substituting back to obtain the cutoff value for the V-parameter, V_c. It happens that the cutoff V_c for a Gaussian profile is given as $V_c = 1.9741$. Table 4.2 tabulates the analytical expressions for the cutoff frequency of the exponential and complementary error function profiles.

The computation of the propagation constant and the mode cutoff frequency for the TE_0 requires numerical integration that would be tedious. Fortunately, the commonly encountered graded profile waveguide shown in Table 4.1 involves only a root search for the estimation of α_0. There is no existing method to estimate the probable range of α_0 as a function of V. Thus, the consuming process in the computation is the correct estimation of the interval for a root search algorithm. Nevertheless, one would be interested in the instigation of the accuracy of the results over a selected range of V. The estimated values of the propagation constant b are calculated and tabulated for a number of profile distributions with a set of specific parameters: $n_c = 1.0$, $n_s = 2.177$, and $\Delta = 0.043$. This is a typical profile structure for an air cover diffused waveguide profile in LiNbO$_3$ or LiTaO$_3$ substrate. The trial field distribution is Hermite–Gaussian. The corresponding cutoff frequencies at the cutoff limit are given for TE_0 in Table 4.4.

The values of the propagation constants are as expected. The accuracy of the variational field fits to the actual field improves with increasing frequency. At large

TABLE 4.2

Optimum Value of Fundamental Cutoff Normalized Frequency Parameters of *exp* and *erfc* Profiles

Profile/Distribution	Parameter	Equation
Exponential	α_0	$\dfrac{(2\alpha_0^2 + 5\alpha_0 - 2)}{16\alpha_0(2\alpha_0 + 1)}\left[1 - erfc\dfrac{1}{2\sqrt{\alpha_0}}\right] + e^{\frac{1}{4\alpha_0}} + 1 = 0$
	V_c	$\left(\dfrac{3\alpha_0}{2} + \dfrac{1}{\sqrt{\alpha_0}}\right)\left[\left(1 + \dfrac{1}{2\alpha_0}\right)e^{\frac{1}{4\alpha_0}}\left(1 - erfc\dfrac{1}{2\sqrt{\alpha_0}}\right)\right] = V_c^2$
Complementary error function *erfc*	α_0	$\dfrac{(\alpha_0 + 1)}{\sqrt{\alpha_0}}\tan^{-1}\sqrt{\alpha_0} - \dfrac{\alpha_0}{(\alpha_0 + 1)} - 1 = 0$
	V_c	$V_c^2 = \dfrac{3\pi(\alpha_0 + 1)^2}{4\sqrt{\alpha_0}}$

V, the field in the cover decreases rapidly. Similarly, the evanescent field in the substrate follows a similar trend. Thus the mode field is confined within the guiding region, and its shape is accurately modeled by a Hermite–Gaussian function. This behavior was first observed experimentally by Kiel and Auracher.[44] This observation leads to the motivation of the Korotky et al. pioneering work in the use of this simple trial field. An earlier method by Taylor requires up to 21 terms in the variational field involving parabolic cylinder functions.[26] A simple relationship between and the mode spot size Γ_a can be obtained as[28]

$$\Gamma_a = \frac{1.555}{\sqrt{\alpha_0}} \tag{4.20}$$

Thus the mode size Γ_a can be obtained directly without using numerical computing. However, Korotky et al. found good agreement between experimental and theoretical results[29]; our analytical results given in Table 4.3 show that there are substantial discrepancies in the propagation constant for single-mode optical waveguides. This means that the mode spot size Γ_a may not be so dependent on the frequencies. Experimentally speaking, the variation of the wavelength of the guiding waveguide and detection is due to the sensitivity of the spot size image monitoring device. This must be taken into account for the measurements of the full width half mark of the image. To investigate this possibility, Γ_a is plotted versus the normalized frequency V-parameter shown in Figure 4.4. This step is also taken to examine the behavior of α_0 near the cutoff frequency of the fundamental mode. The difficulty in the calculation of the cutoff frequencies given in Table 4.4 can be observed. This is due to the volatility of the confinement of the mode near cutoff. This is a well-known phenomenon in optical fibers.[45] Figure 4.4 shows the variation with respect

TABLE 4.3

b-V Data for Selected Profiles Calculated with a Hermite–Gaussian Trial Field

	Exponential		Gaussian		Complementary Error Function	
V	b (var.)	b (exact)	b (var.)	b (exact)	b (var.)	b (exact)
2	0.066	0.105	—	—	—	—
3	0.193	0.299	0.216	0.275	0.015	0.068
4	0.289	0.321	0.370	0.413	0.121	0.169
5	0.362	0.390	0.476	0.510	0.213	0.255
10	0.560	0.578	0.719	0.732	0.477	0.497
100	—	0.897	0.970	0.971	0.883	0.885

Note: Variational method, exact = analytical expression.

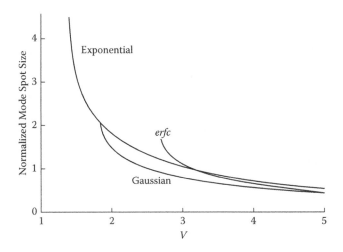

FIGURE 4.4 Mode spot size calculated using Hermite–Gaussian trial field for different profiles.

TABLE 4.4

Cutoff Frequencies of the TE_0 Mode

	Exponential		Gaussian		Complementary Error Function	
Profile	V_c (var.)	V_c (exact)	V_c (var.)	V_c (exact)	V_c (var.)	V_c (exact)
	1.563	1.087	1.974	1.433	2.839	2.085
α_0 at V_c		0.143		0.500		0.697

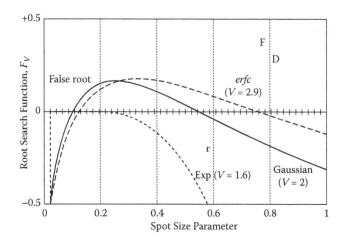

FIGURE 4.5 Multiple roots of the root search function of the variational method.

to V over a range of frequencies, including at $V = V_c$ for the profiles of exponential, Gaussian, and complementary error shapes. Two sets of data and curves are given so as to notice the method of using the root search algorithm to compute the optimum spot size parameters α_0. More than one root can exist in the search interval. The smooth set of curves given in Figure 4.3 is obtained by choosing only the negative-going crossover of the curves given in Figure 4.4. The kinks observed in Figure 4.5 are obtained when choosing smaller and incorrect root. The propagation constant computed from this false zero is much smaller and can be negative. Thus, it is preferred to operate the waveguide far from the cutoff region so that the mode spot size is not strongly dependent on the V-parameter, as seen in Figure 4.6. This scenario is important for the case when planar optical waveguides are used as an optical

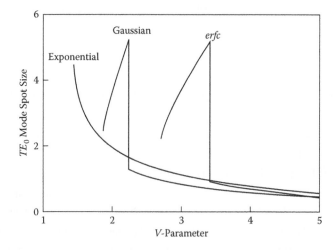

FIGURE 4.6 Mode spot size of TE_0 mode as a function of V for profiles of exponential, Gaussian, and complementary error functions estimated using the variational method.

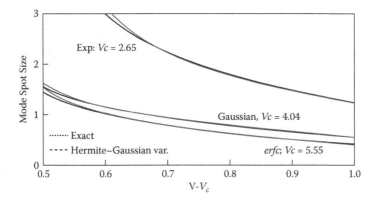

FIGURE 4.7 Spectral variation of mode spot size: accuracy of Hermite–Gaussian trial field fitting for single-mode diffused clad profiles. Single-mode diffused clad profiles with $n_c = 1.0$, $n_s = 2.177$, and $\Delta = 0.043$ ($A = 20$).

amplifier,[46–47] e.g., Er-doped LiNbO₃ waveguide. The wavelength of the pump beam is far from the operating wavelength region and may be close to the cutoff. This must be taken into account. If not, then the fluctuation of the mode spot size would alter the amplification gain of the amplifier. In practice, the refractive index profile could never be modeled by any form of analytical function and an equivalent profile may be used. The correct spot size behavior computed numerically using the numerical algorithm given here follows a trend similar to that shown in Figures 4.7 and 4.8. The variational spot size is superimposed on these curves for comparison; the agreement is remarkable. The wavy curves in Figure 4.8 are caused by numerical noises. The tolerance on each plot is 1%. Such accuracy is achieved due to the definition described here for the spot size that does not take into account the tails of the field. This is where serious agreement between the exact and Hermite-Gaussian occurs. This explains the discrepancies in the normalized propagation constant b as estimated by this method.

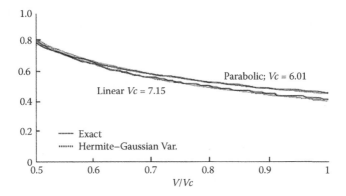

FIGURE 4.8 Accuracy of Hermite–Gaussian trial field fitting for single-mode clad power-law profiles. Single-mode clad power-law profiles with $n_c = 1.0$, $n_s = 2.177$, and $\Delta = 0.043$ ($A = 20$).

4.3.1.2 WKB Method

The WKB method was first developed by Jeffery[48] and applied to the calculation of energy eigenvalues in quantum mechanics by Wentzel,[49] Kramers,[50] and Brilluoin.[51] Due to the similarities between problems involving the quantum mechanical potential well and the refractive index profiles of optical waveguides,[37] the method can be easily adapted for use in guided wave optics. Marcuse first used the method for proving the eigenvalue equation of asymmetrical graded index optical waveguides.[23] A simplified derivation by Hocker and Burns[27] based on ray optics confirmed Marcuse's results. This is due to the equivalence of the WKB and ray optics formalism.[52]

The central problem involved in the techniques lies with the connection of oscillatory and evanescent fields at the turning point where the original WKB solutions are singular. Langer solved the problem by approximating the actual fields by Airy functions.[53] This is equivalent to replacing the actual profile locally by a linear segment. Its slope and position are implicitly related to the propagation constant in the eigenvalue equation. We have examined the turning point phenomena in detail (see Appendix A). We thus can state that the WKB method is not limited by the inaccurate phase prediction at the turning point. A more serious limitation is caused by the neglect of the cladding. Coupling effects between the turning point and cladding have been studied in detail by Arnold.[54] He found that the cladding effects can be isolated and built into the eigenvalue equation. However, the corrections involve a complicated nest of Airy functions, and the analytic simplicity of the method is lost.

We took a simpler and more practical approach to account for cladding effects and study the behavior of the WKB errors which are minute for asymmetrical waveguides provided that a simple correction is added. More discussions on the improvement of the method presented will be given in appropriate sections.

4.3.1.2.1   Derivation of WKB Eigenvalue Equation

An asymmetrical of the WKB eigenvalue equation is obtained by matching the field and its derivative at the dielectric interfaces. For the WKB method this is complicated by the fact that the solutions must be matched correctly at the turning point. Jeffery's solution can be referred to as the 0th-order WKB method, and Langer's method with the turning point correction as the first-order WKB method (see Figure 4.9). Following Gordon[55] and Marcuse,[23] one can write a graded index asymmetrical waveguide as

$$
\varphi(X) =
\begin{cases}
a_0 e^{V\sqrt{A+bX}}; X \le 0 \\[2mm]
p^{-1/2}(X)\cos[\phi(X)-\pi/4]; 0 \le X < X_t^- \\[2mm]
\left(\dfrac{2\pi\phi}{3p}\right)^{-1/2}[J_{1/3}(\phi)+J_{-1/3}(\phi)]; X = X_t^- \\[2mm]
\left(\dfrac{2\pi\phi}{3p}\right)^{+1/2}[I_{1/3}(\phi)+I_{-1/3}(\phi)]; X = X_t^+ \\[2mm]
\left(\dfrac{p(X)}{4}\right)^{-1/2} e^{-\phi(X)}; X_t^+ \le X < \infty
\end{cases}
\tag{4.21}
$$

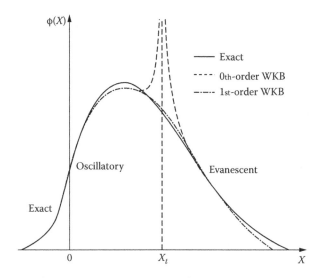

FIGURE 4.9 An illustration of regions of validity of the WKB solutions. The turning point is given by $S(Xt) = B$. The WKB eigenvalue equation is obtained by ensuring that the WKB solutions in the guide are matched to the exact field in the cover (superstrate).

where

$$\phi(X) = V \int_X^{X_t} \left|\sqrt{S(X) - b}\right| dX \quad \text{and} \quad p(X) = \left[\frac{S(X) - b}{1 - b}\right]^{1/2}$$

a_0 is a constant, and I and J are the Bessel functions' representation of the Airy solutions at the turning point, X_t.

The turning point is defined such that

$$S(X_t) = b \tag{4.22}$$

A general proof is given in the Appendix B, which shows that at a turning point, the approximation of the exact field by the Airy function is extremely good if

$$S'(X_t) \approx 0 \quad \text{and} \quad S''(X_t) \approx 0 \tag{4.23}$$

Furthermore, the oscillatory solution for $0 \le X < X_t^-$ and the evanescent field for $X > X_t^+$ are just asymptotic expansions of Bessel's solutions for $\phi(X) \gg 1$; i.e., $V \gg 1 >$. These are just the 0th-order WKB solutions (see Appendix B). They have to be used in these forms with the correct phase arguments to ensure uniformity of the WKB solutions in both the guide and the substrate, that they are already correctly matched.

The eigenvalue equation follows by ensuring the smooth matching of the WKB solution and the exact evanescent field at $X = 0$. The continuity of $\phi(0)$ and $\phi'(0)$ gives, after some algebraic manipulations,

$$V \int_0^{X_t} \left| \sqrt{S(X) - b} \right| dX = \left(m + \frac{1}{4} \right) \pi + \tan^{-1} \left(\frac{\sqrt{A+b}}{1-b} + \delta \right) \tag{4.24}$$

where

$$\delta = \frac{S'(0)}{4V\sqrt{1-b}}$$

If setting $d = 0$, then the WKB eigenvalue equation becomes

$$2V \int_0^{X_t} \left| \sqrt{S(X) - b} \right| dX - \frac{\pi}{2} - 2 \tan^{-1} \left(\frac{\sqrt{A+b}}{1-b} + \delta \right) = 2m\pi \tag{4.25}$$

as obtained by Marcuse.[23] For practical multimode waveguides, Hocker and Burn[27] claimed that

$$\tan^{-1} \left(\frac{\sqrt{A+b}}{1-b} \right) \approx \pi/2 \text{ since } A \gg 1$$

This led to the following relationship:

$$V \int_0^{X_t} \left| \sqrt{S(X) - b} \right| dX = \left(m + \frac{3}{4} \right) \pi \tag{4.26}$$

The third term of the RHS of (4.25) exists due to the phase shift undergone by the modal field by the discontinuity at $X = 0$. Indeed, Equation (4.25) is just the mathematical statement of the similar phase resonance condition of ray optics,[56] which states that the phase accumulated along the ray path over one period, including reflection as it traverses the guide from $X = 0$ to $X = X_t$, must be a multiple of 2π if constructive interference is to occur. Constructive interference is essential for maintaining a stable modal pattern. In fact, the phase changes at the turning point and the dielectric discontinuity are just

$$-\frac{\pi}{2} \quad \text{and} \quad -2 \tan^{-1} \left(\frac{\sqrt{A+b}}{1-b} \right)$$

Figure 4.10 illustrates the ray path of the process. To solve b for a given V, the turning point is tuned until the phase resonance condition is met. These results have been derived by Hocker and Burns without the use of WKB formalism.[27]

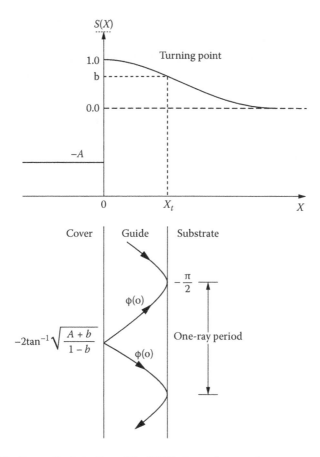

FIGURE 4.10   Ray optic derivation of the WKB eigenvalue equation.

4.3.1.2.2   Limitations of the WKB Method

Three sources of errors are inherent in the eigenvalue equation (4.25). It can be easily shown that Equation (4.24) can reduce to the eigenvalue equation for the TE modes of a step index profile waveguide provided the phase changes can be obtained correctly. For a graded index profile, this may not be the case, as described in the next section. Thus the phase accumulated in the guide as predicted by the WKB is only an approximation. This is due to the representation of the field by an equivalent cosine-like field.

Is the change at a dielectric interface dependent on the slope of the refractive index profile in the second medium at the interface? It is believed that in general $\delta \neq 0$ judging from the exact analysis of the linear clad profile. This factor was omitted from the results of Marcuse.[23] Furthermore the phase change at the turning point is estimated from (4.23) without resorting to the correctness of the evanescent field representation of the actual field beyond $X = X_t$. The question is whether one can lump together all sources of errors in the phase into a single-error parameter. Thus it is possible to propose, in general,

TABLE 4.5

Equations for Calculating b and V_c via the WKB Method

Profile	b	V_c
Clad linear	$1 - \dfrac{A_0}{V^{2/3}}$ $\qquad A_0 = \left[\dfrac{3\pi}{2}(m+\gamma) \right]^{2/3}$	$\dfrac{3\pi}{2}(m+\gamma)$
Clad parabolic	$1 - \dfrac{4(m+\gamma)}{V}$	$4(m+\gamma)$
Exponential	$b \colon \sqrt{1-b} - \sqrt{b}\, \tan^{-1}\!\left(\sqrt{\dfrac{1-b}{b}} \right) = \dfrac{\pi(m+\gamma)}{2V}$	$\dfrac{\pi}{2}(m+\gamma)$
Cosh^{-2}	$\left[1 - \dfrac{2}{V}(m+\gamma) \right]^2$	$2(m+\gamma)$
Step	$1 - \left[\dfrac{2(m+\gamma)}{V} \right]^2$	$(m+\gamma)\pi$
Gaussian	—	$(m+\gamma)\pi$

$$V \int_0^{X_t} \sqrt{S(X) - b} \cdot dX = (m+\gamma)\pi \tag{4.27}$$

where γ is the total accumulated phase change at the turning point; γ is normally equal to 3/4.

4.3.1.2.3 Profiles with Analytical WKB Solutions

The WKB integral given in Equation (4.26) is integrable for the step, clad linear, clad parabolic, exponential, and cosh graded index profiles. With the help of the integration formulae given in Gradshteyn and Ryzhik,[57] the results for the normalized propagation constant b and mode cutoff frequencies are presented in Table 4.5. For other profiles, the integral has to be integrated numerically. Although modern computing facility is done with ultra-high-speed processors, analytical solutions give us some insight into understanding the behavior of the wave solution. If numerical integration is conducted, then for each trial value of the normalized propagation constant b, the turning point change of b, especially when V becomes very small, approaches zero and the turning point value becomes very large. Thus the WKBs were not popular previously, but they are especially not now with modern and ultra-high-speed computing systems.

4.3.1.2.4 Ordinary WKB Results

We are now faced with two forms of the WKB eigenvalue equation, given by (4.24) and (4.27). This section studies the performance of this equation over a wide variety of profiles for representative values of V. The effect of the asymmetry factor on the dispersion was identified by Ramaswamy and Lagu,[17] who found that for $A > 10$, the error is negligible. However, their conclusion is only valid for multimode waveguides. In modern optical waveguides for advanced optical communication systems, single-mode optical waveguides are mainly the guided wave media for applications. The results obtained for single-mode optical waveguides prove otherwise when $A = 20$ is employed with the waveguide parameters $n_s = 2.177$ (lithium niobate as substrate), cover layer $n_c = 1.0$ (air), and $\Delta = 0.043$. Equation (4.26) reaches its asymptotic value when $A \rightarrow \infty$. The exact numerical results have been obtained with an integration step of 0.01. The profile truncation point is set at $X = 10$ for diffused profiles and $X = 1$ for the clad power-law profiles. The values of b and V for different profiles are calculated and tabulated in Tables 4.6 and 4.7. The improvement of the values of the normalized propagation constant and the V-parameter can be observed and is self-explaining. The value of V is set at a region close to that of the cutoff of the guided mode TE_0.

4.3.1.2.5   Enhanced WKB

There are serious drawbacks of both the variational and WKB methods in computing the dispersion characteristics of the diffused clad single-mode planar waveguides. The cosine-exponential trial field can substantially improve the accuracy of the variational analysis.[58] However, it requires optimization.

For a good field distribution and immunity to the bending of the waveguide, it is anticipated that the waveguide is operating in the region close to the cutoff of the TE_1 mode. Figures 4.11 and 4.12 show the range of applicability of each method for diffused clad as well as clad power-law profiles. Except for the Gaussian profile, all other profiles show that the enhanced WKB method is sufficiently accurate if the operating point lies in the range $0.75 < V/V_c < 1.0$. The discrepancies in the Gaussian profile are caused by the steepness of the profile. Errors caused by the presence of the uniform substrate index should taper off in the stated range of validity. The variational method with a Hermite–Gaussian field cannot be used for the calculation of b for any of these profiles due to the poor overlapping between the trial field and the actual field. The waveguide designs should use an appropriate method for a particular application.

4.3.2   SYMMETRICAL WAVEGUIDES

In this section we deal mainly with symmetrical optical waveguides and dwell mainly on the equivalent moment methods described above. The WKB method is also treated briefly. It involves only a slight modification of the previous equations and the entries given in Table 4.8. As presented, the WKB performs poorly in this kind of waveguide. On the other hand, the moment method (based on the $\cosh^{-2}$ profile) is accurate in the range of frequencies for single-mode operation.

4.3.2.1   WKB Eigenvalue Equation

For a symmetrical optical waveguide (4.27) becomes

TABLE 4.6
b-V for Clad Linear Profile, Clad Parabolic Profile, and Exponential Profile

	Clad Linear Profile			Clad Parabolic Profile			Exponential Profile		
	b (WKB)		b	b (WKB)		b	b (WKB)		b
V	Enhanced	Ordinary	(exact)	Enhanced	Ordinary	(exact)	Enhanced	Ordinary	(exact)
2	—	—	—	—	—	—	0.1086	0.0831	0.1050
3	—	—	0.0335	0.981	0.000	0.1577	0.23331	0.2054	0.2292
4	0.1333	0.0792	0.1479	0.3081	0.2500	0.3262	0.3249	0.2992	0.3212
5	0.2498	0.2045	0.2500	0.4417	0.4000	0.4475	0.3939	0.3705	0.3903
10	0.5218	0.5001	0.5182	0.7149	0.7000	0.7153	0.5809	0.5658	0.5781
100	0.8945	0.8923	0.8939	0.9705	0.9700	—	0.8974	0.8954	0.8968

TABLE 4.7

b-V for Gaussian Profile, erfc Profile, and Cosh^{-2} Profile

V	Gaussian Profile			erfc Profile			Cosh^{-2} Profile		
	b (WKB)		b	b (WKB)		b	b (WKB)		b
	Enhanced	Ordinary	(exact)	Enhanced	Ordinary	(exact)	Enhanced	Ordinary	(exact)
2	0.0452	0.0104	0.817	—	—	—	0.1001	0.0625	0.1231
3	0.2538	0.2071	0.2750	0.0575	0.0281	0.0677	0.2908	0.2500	0.3074
4	0.4008	0.3630	0.4133	0.1651	0.1293	0.1695	0.4244	0.3906	0.4357
5	0.3939	0.3705	0.3903	0.5013	0.4712	0.5095	0.2539	0.2198	0.2552
10	0.5809	0.5658	0.5781	0.7301	0.7173	0.7323	0.4991	0.4776	0.4971
100	0.9706	0.9702	0.9707	0.8959	0.8835	0.8852	0.9707	0.9702	0.9708

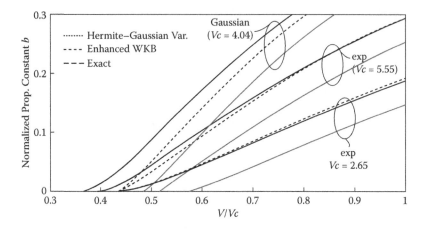

FIGURE 4.11 Dispersion characteristics: comparison of methods of analysis of single-mode clad power law with $n_s = 2.17$, $\Delta = 0.043$.

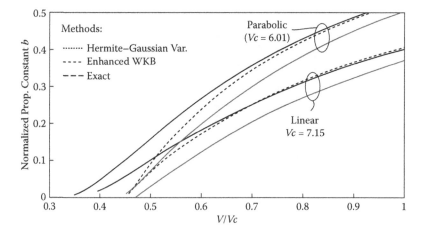

FIGURE 4.12 Dispersion characteristics: comparison of methods of analysis of single-mode clad power law with $n_s = 2.17$, $\Delta = 0.043$.

TABLE 4.8

WKB Calculated Cutoff Frequencies for the Two Lowest-Order Modes, TE_0 and TE_1

Profile	V_{c1}, V_{c2} (WKB)		V_{c1}, V_{c2} (exact)
	Enhanced	Ordinary	
Clad linear	3.2	3.53	2.46
	7.92	8.24	7.15
Clad parabolic	2.72	3.0	1.96
	6.72	7.0	6.01
Exponential	1.07	1.18	1.09
	2.64	2.75	2.65
Gaussian	2.14	2.56	1.43
	5.28	5.96	4.04
erfc	2.32	2.56	2.09
	5.72	5.96	5.55
$Cosh^{-2}$	1.36	1.50	1.24
	3.36	3.50	3.32

$$2V \int_0^{X_t} \sqrt{S(X)-b} \cdot dX = (m+\gamma_s)$$

$$(4.28)$$

$$\gamma_s = \frac{1}{2} \text{ diffused waveguides}$$

If the turning point coincides with a dielectric discontinuity, the correct phase shift formula is to be used. For buried modes the complicated expression is given in Appendix B. This section is limited to the profiles following diffused or clad power shape. The factor of 2 in (4.28) is accounted for by the WKB-defined effective guide width, now extended from $-X_t$ to X_t, the turning points on both sides of the guiding region. Thus the formulae in Table 4.5 for the estimation of b and V can be translated to a symmetrical optical waveguide by transforming $V \rightarrow 2V$ and $\gamma \rightarrow \gamma_s$.

4.3.2.2 Two-Parameter Profile Moment Method

The profile moment method is related to the variational formalism of optical wave-guide problems.[42] The trial field is derived from that of a reference profile where an exact analytical expression is available. It is known that the field distribution of the fundamental mode follows a bell-shape-like trend. Thus by adjusting the V-parameter, a close match to the modal field can be obtained. This condition can be satisfied by monotonously varying the variational parameters.

4.3.2.2.1 Theoretical Basis

The starting point is that for two symmetrical waveguides having the same substrate index, normalized mode propagation constants are related by[59]

$$\beta^2 - \beta_r^2 = \frac{k_0^2 \int\limits_{-\infty}^{\infty} [n^2(x) - n_r^2(x)] \varphi(x) \varphi_r(x) \, dx}{\int\limits_{-\infty}^{\infty} \varphi(x) \varphi_r(x) \, dx} \qquad (4.29)$$

where the subscript r indicates that the quantities belong to the reference waveguide. Then the condition for the two waveguides to be equivalent is $\beta = \beta_r$; thus, we have

$$\int\limits_{-\infty}^{\infty} [n^2(x) - n_r^2(x)] \varphi(x) \varphi_r(x) \, dx = 0 \qquad (4.30)$$

One can express the product of the field of this equation as a series:

$$\varphi(x) \varphi_r(x) = \sum_{l=0}^{\infty} c_l(k_0) x^{2l} \qquad (4.31)$$

where c_l are the frequency-dependent coefficients of the series. Thus (4.30) becomes

$$\sum_{l=0}^{\infty} c_l(k_0) \int\limits_{0}^{\infty} [n^2(x) - n_r^2(x)] x^{2l} \, dx = 0 \qquad (4.32)$$

Since we impose the condition that these waveguides have the same substrate index, we can write (4.32) in terms of the profile shape function $S(X)$, leading to

$$\sum_{l=0}^{\infty} c_l(k_0)[N_{2l} - N_{2lr}] = 0 \qquad (4.33)$$

in which N_{2l} can be identified by

$$N_{2l} = 2[n_0^2 - n_s^2] d^{2(l+1)} \Omega_{2l} = 0 \qquad (4.34)$$

where Ω_{2l} is defined as

$$\Omega_{2l} = \int\limits_{0}^{\infty} S(X) X^{2l} \, dX = 0 \qquad (4.35)$$

For profiles that are nearly identical, one can assume that for $\beta = \beta_r$ over the range of k_0, for which the fields are slowly varying, it is sufficient to retain only two terms in the series. Thus we have

$$N_0 = N_{0r}$$

$$N_2 = N_{2r}$$
(4.36)

Expanding these two terms we have

$$[n_0^2 - n_s^2]d\Omega_0 = [n_{0r}^2 - n_s^2]d_r\Omega_{0r}$$

$$[n_0^2 - n_s^2]d^3\Omega_2 = [n_{0r}^2 - n_s^2]d_r\Omega_{2r}$$
(4.37)

which can also be expressed in terms of the normalized frequency V-parameter as

$$\frac{V}{V_r} = \left\{ \frac{\Omega_0\Omega_2}{\Omega_{0r}\Omega_{2r}} \right\}^{1/4}$$
(4.38)

4.3.2.2.2 Estimation of Normalized Propagation Constant

The normalized propagation constant b can be expressed in terms of V and β as

$$b(V) = \left(\frac{d}{V}\right)^2 \frac{n_e^2 - n_s^2}{k_0^2}$$
(4.39)

where $n_e = \beta/k_0$ is the effective refractive index of the guided mode or the refractive index of the guided medium as seen by the mode along the z-direction. Since $n_e = n_{er}$, then

$$\frac{b(V)}{b(V_r)} = \frac{\Omega_0}{\Omega_{0r}} \left\{ \frac{\Omega_0\Omega_2}{\Omega_{0r}\Omega_{2r}} \right\}^{1/2}$$
(4.40)

This equation states that the propagation $b(V)$ of an arbitrary waveguide can be derived from that of a reference waveguide provided that the profile moments and the dispersion relation for $b_r(V_r)$ are known. Table 4.9 lists the three lowest moments of profiles having analytical forms of their shape functions. The profiles listed in this table, having step, clad linear, exponential, and cosh, have exact analytical solutions for their propagation constant. Thus any of these profiles can be employed as a reference profile.

We select two profiles, the step and cosh profiles, for two case studies that follow.

4.3.2.2.2.1 Step Reference Profile[20]

$$V_r\sqrt{1-b_r} = m\frac{\pi}{2} - \tan^{-1}\left(\sqrt{\frac{b_r}{1-b_r}}\right) \qquad m = 0,1,2\ldots$$
(4.41)

where

TABLE 4.9

Profile Moments of Selected Profile Shape

Profile	Ω_0	Ω_2	Ω_4	Ω_4/Ω_2	SDF (see Taylor[26])
Step	1	0.333	0.2	0.6	1.0
Clad linear	0.5	0.0833	0.0033	0.40	0.67
Clad parabolic	0.667	0.133	0.571	0.43	0.72
Exponential	1	2	24	12	10.08
Gaussian	0.866	0.443	2.659	6	5.04
$erfc(x)$	0.564	0.188	0.226	1.2	1.01
$Cosh^{-2}(x)$	1	0.693	0.823	1.19	1.00

$$V_r = [3\Omega_0\Omega_2]^{1/4} V$$

$$b = \left[\frac{\Omega_0^3}{3\Omega_2}\right]^{1/2} b_r \tag{4.42}$$

4.3.2.2.2   Cosh Reference Profile[59]

$$b_r = \left\{\left(1+\frac{1}{4V_r^2}\right)^{1/2} - \frac{1}{V_r}\left(m+\frac{1}{2}\right)\right\}^2 \qquad m = 0,1,2\ldots \tag{4.43}$$

where

$$V_r = \left[\frac{12\Omega_0\Omega_2}{\pi^2}\right]^{1/4} V$$

$$b = \frac{\pi}{2}\left[\frac{\Omega_0^3}{3\Omega_2}\right]^{1/2} b_r \tag{4.44}$$

These equations are required for the calculation to obtain the dispersion relation characteristics.

4.3.2.2.3   Estimation of TE₁ Mode Cutoff

The next higher-order mode is TE_1. The cutoff frequency of this mode is the upper limit of the single-mode operation. We can write the product of the guided waves of the reference waveguide and the one to be analyzed as

$$\varphi(x)\varphi_r(x) = \sum_{l=0}^{\infty} c_l(k_0)x^{2l+2} \tag{4.45}$$

and similarly

$$N_2 = N_{2r}$$
$$N_4 = N_{4r}$$

$$(4.46)$$

Thus, the relationship between the profile moments and the refractive index can be obtained as

$$[n_0^2 - n_s^2]d^3\Omega_2 = [n_{0r}^2 - n_s^2]d_r^3\Omega_2$$
$$[n_0^2 - n_s^2]d^5\Omega_4 = [n_{0r}^2 - n_s^2]d_r^5\Omega_4$$

$$(4.47)$$

The estimation of nonprofile moment terms using the definition of V for each guide gives the desired mode cutoff relation after setting $b = 0$.

4.3.2.2.3.1 *Step Reference Profile*[20]

$$V_c = \frac{\pi}{2}\left(\frac{5\Omega_4}{27\Omega_2}\right)$$

$$(4.48)$$

4.3.2.2.3.2 *Cosh Reference Profile*

$$V_c = \sqrt{2}\left(\frac{5\pi^2\Omega_4}{252\Omega_2^3}\right)^{1/4}$$

$$(4.49)$$

where the cutoffs for the reference profiles have been derived from (4.41) and (4.43).

The propagation constant and the cutoff frequency V_c are tabulated in Tables 4.10 and 4.11 for two typical profiles of the step and clad linear types, and Tables 4.12 to 4.15 are for clad parabolic, exponential, Gaussian, and $erfc(x)$ profiles, respectively. Note that the moments of the complementary error function are not listed in Table 4.9, as there are no closed-form solutions.

TABLE 4.10
***b-V* Data for Step Profile**

	b (moment)			
V	Step Reference	Cosh Reference	*b* (WKB)	*b* (exact)
0.5		0.192	<0	0.189
1.0		0.481	0.383	0.454
1.5	*b = b* (exact) for all *V*	0.697	0.726	0.628
2.0		0.848	0.846	0.725
3.0		>1	0.931	0.849
4.0		>1	0.961	0.902

TABLE 4.11
b-V Data for Clad Linear Profile

| V | b (moment) | | b (WKB) | b (exact) |
	Step Reference	Cosh Reference		
0.5	0.0560	0.0563	<0	0.0561
1.0	0.173	0.177	<0	0.174
1.5	0.286	0.300	0.149	0.290
2.0	0.375	0.404	0.297	0.384
3.0	0.491	0.558	0.464	0.515
4.0	0.558	0.660	0.557	0.579

TABLE 4.12
b-V Data for Clad Parabolic Profile

| V | b (moment) | | b (WKB) | b (exact) |
	Step Reference	Cosh Reference		
0.5	0.0951	0.0959	<0	0.0952
1.0	0.270	0.280	<0	0.272
1.5	0.419	0.448	0.333	0.423
2.0	0.525	0.580	0.500	0.535
3.0	0.653	0.762	0.667	0.673
4.0	0.721	0.877	0.750	0.751

4.3.2.2.4   Choice of Methods

There are some interesting insights as observed from Tables 4.10 to 4.15.

In the clad power-law profiles, the moment-ESI method consistently gives better results for both the propagation constant and the cutoff frequencies of the TE_0 mode. On the other hand, diffused waveguides characterized by nondecreasing

TABLE 4.13
b-V Data for Exponential Profile

| V | b (moment) | | b (WKB) | b (exact) |
	Step Reference	Cosh Reference		
0.5	0.142	0.148	0.0205	0.152
1.0	0.263	0.294	0.205	0.317
1.5	0.320	0.387	0.337	0.424
2.0	0.350	0.431	0.426	0.498
3.0	0.377	0.491	0.539	0.593
4.0	0.389	0.525	0.609	0.653

TABLE 4.14
_b-V_ Data for Gaussian Profile

	_b_ (moment)			
_V_	Step Reference	Cosh Reference	_b_ (WKB)	_b_ (exact)
0.5	0.146	0.148	<0	0.147
1.0	0.342	0.363	0.207	0.354
1.5	0.466	0.520	0.421	0.498
2.0	0.541	0.628	0.549	0.594
3.0	0.620	0.763	0.688	0.709
4.0	0.657	0.842	0.762	0.774

TABLE 4.15
_b-V_ Data for _erfc(x)_ Profile

	_b_ (moment)			
_V_	Step Reference	Cosh Reference	_b_ (WKB)	_b_ (exact)
0.5	0.0672	0.0679	<0	0.0678
1.0	0.187	0.194	0.0281	0.193
1.5	0.286	0.306	0.177	0.304
2.0	0.355	0.394	0.293	0.391
3.0	0.436	0.512	0.443	0.509
4.0	0.479	0.586	0.534	0.584

higher-order moments of Table 4.9 are more accurately modeled as y in the cosh profile. At low frequencies both approaches are asymptotically exact.

The eigenvalue equation (4.41) can be reduced to

$$b(V \to 0) = \left[\left(1 + \frac{1}{4V^2} \right)^{1/2} - \frac{1}{2V} \right]^2 \tag{4.50}$$

which is just the eigenvalue equation for the cosh profile. This comparison is valid only when the profiles have equal volume Ω_0.

For large values V their dispersion curve split and higher-order moment scaling factors in Equations (4.41) and (4.43) have to be used. However, due to different properties of the higher-order moments of the step and cosh profiles, neither one can be used to predict each other's dispersion characteristics accurately.

The WKB method gives consistently better results at large frequencies. To give an idea of the asymptotic range of applicability of the WKB and moment methods, the dispersion curves for the diffused as well as clad power-law profiles are plotted in Figures 4.13 to 4.16.

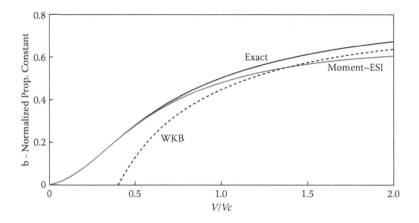

FIGURE 4.13 Dispersion characteristics: range of applicability of WKB method and moment method. Symmetric clad linear profile with $A = 20$; (mode) = 2.7995.

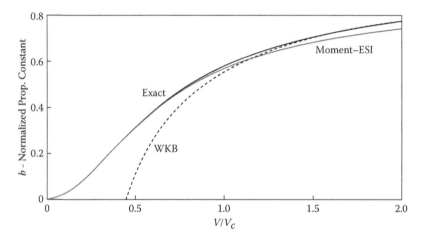

FIGURE 4.14 Dispersion characteristics: range of applicability of WKB method and moment method. Symmetric clad parabolic profile with $A = 20$; (mode) = 2.330, $n_c = 2.177$, $\Delta = 0.043$.

4.3.2.2.5 A New Method for Profile Classification

Table 4.15 and Figure 4.16 indicate that the dispersion characteristic of the complementary error function profile is well above the expected accuracy of the moment method as calculated from a cosh reference profile. Even at $V = 4.0$, a near-perfect agreement is obtained. To account for such observations a shape derivation factor (SDF) can be proposed as

$$SDF = \frac{\Omega_4 \big/ \Omega_2}{\Omega_{4r} \big/ \Omega_{2r}} \tag{4.51}$$

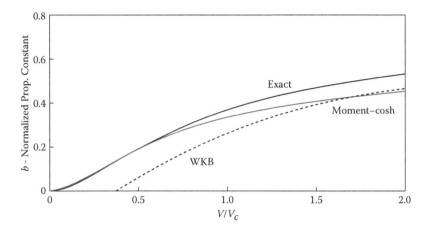

FIGURE 4.15 Dispersion characteristics: range of applicability of WKB method and moment–cosh method. Symmetric exponential profile with $A = 20$; (mode) = 1.2024, $n_c = 2.177$, $\Delta = 0.043$.

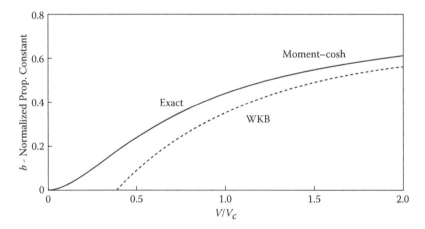

FIGURE 4.16 Dispersion characteristics: range of applicability of WKB method and moment–cosh method. Symmetric $erfc(x)$ profile with $A = 20$; (mode) = 2.3187, $n_c = 2.177$, $\Delta = 0.043$.

where the subscript r refers to the reference profile. For clad profiles one can choose the step profile as a reference, whereas for diffused waveguides the cosh profile offers a much better fit. This SDF parameter is thus entered Table 4.10. We could see the benefit of this factor for the $erfc(x)$ profile, which has an SDF factor of 1.01, compared to 10 for the exponential profile. Thus the former method offers better accuracy.

4.3.2.2.6 A New Equivalence Relation for Planar Optical Waveguides

By sketching the spatial distribution of the modal field of the TE_1 mode in a symmetrical waveguide and the TE_0 field in an asymmetrical waveguide, one can find out why the profile moment method does not work in both cases. However, there

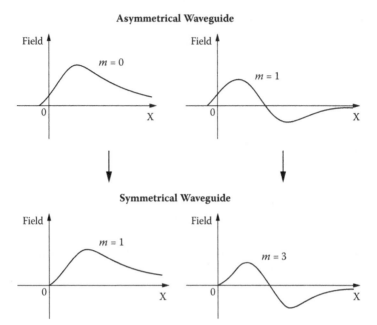

FIGURE 4.17 Correspondence between the modal fields of asymmetrical and symmetrical waveguides with the field distribution of odd modes.

is some surprise when $X \geq 0$: the distribution is similar if $A \to \infty$, as shown in Figure 4.17. For the same profile, if the field distributions are identical, then there exists a relation between the dispersion curves of both structures. Figure 4.18 illustrates the correspondence between the modes of both waveguide structures. One can postulate that

$$V_{ca1} = V_{cs2}$$
$$V_{ca2} = V_{cs4}$$
(4.52)

where the left-hand side denotes the cutoffs of the TE_0 and TE_1 modes in the asymmetrical waveguide. $V_{cs2} \simeq V_{cs4}$ are the cutoffs of the TE_1 and TE_3 modes of the corresponding symmetrical waveguide.

Furthermore, it is noted that the separation of the b-V characteristic curves at cutoff of the symmetrical waveguide is nearly uniform. For the step profile, this separation equals $\pi/2$, whereas for graded profiles, this is only approximately true. Thus we can write

$$V_{cs4} \simeq 3V_{cs2}$$
(4.53)

Combining (4.52) and (4.53) leads to

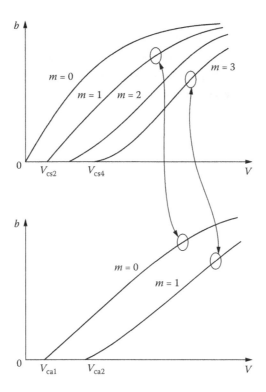

FIGURE 4.18 The m-mode dispersion curves of asymmetrical and symmetrical waveguides from the $(2m + 1)$th mode dispersion curve of the corresponding symmetrical waveguide.

$$\left.\frac{V_{ca2}}{V_{ca1}}\right|_{A\to\infty} \simeq 3 \qquad (4.54)$$

This equation allows us to conduct preliminary tests on the postulation above. Table 4.16 tabulates the ratio of the V-parameters at cutoff of the modes TE_0 and TE_1 for $A \sim 20$ and $A \to \infty$ for the listed profiles. It shows that (4.54) does not satisfy for step profile. This may be contributed by the error in the assumptions about the separation of the b-V curves at cutoff in Equation (4.53). To see if (4.52) can be satisfied one can compare the TE_0 mode cutoffs of the symmetrical waveguide as $A \to \infty$. The cutoffs are so close in the last two columns of the table, allowing them to be considered exactly equal. Thus this is the new corresponding relationship between the m-modes of an asymmetrical waveguide and the odd $(2m + 1)$th modes of the corresponding symmetrical waveguide. Table 4.17 tabulates the cut-off conditions for oddguided modes of planar waveguides.

4.3.2.2.7 Solution of a Simple Symmetric Waveguide

4.3.2.2.7.1 Structure A symmetric slab or planar optical waveguide consists of a slab (or core) of dielectric "transparent" material of refractive index n_1, embedded

TABLE 4.16

Ratio of TE_0 and TE_1 Mode Cutoff V-Parameter in Asymmetrical Waveguides

| Profile | $\dfrac{V_{ca2}}{V_{ca1}}$ | |
	$A \sim 20$	$A \to 20$
Step	3.33	3.0
Clad linear	2.91	2.67
Clad parabolic	3.07	2.78
Exponential	2.43	2.30
Gaussian	2.83	2.57
$erfc(x)$	2.65	2.47
$Cosh^{-2}$	2.68	2.45

Note: Obtained for profiles without analytical solutions by forward recurrence algorithm with 0.01 step size.

TABLE 4.17

Prediction of Odd Mode Cutoffs in Symmetrical Waveguides from the Mode Cutoffs of the Corresponding Waveguides

| Profile | V_{ca1} (Asymmetrical Waveguide TE_0) | | V_{ca4} (Symmetrical TE_1) |
	$A \sim 20$	$A \to 20$	
Step	1.35	1.57	1.57
Clad linear	2.46	2.80	2.80
Clad parabolic	1.96	2.26	2.25
Exponential	1.09	1.20	1.20
Gaussian	1.43	1.64	1.64
$erfc(x)$	2.09	2.35	2.37
$Cosh^{-2}$	1.24	1.41	1.41

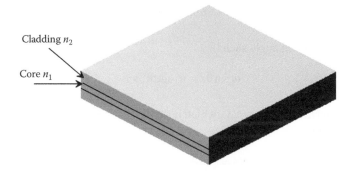

FIGURE 4.19 Cross section of a slab optical waveguide. The optical waveguide is assumed to be confined in the vertical direction x and extended infinitely in the lateral direction y. Lightwaves are guided and propagated along the z-direction.

between two layers of materials of index n_2 acting as the substrate and superstrate layers, as shown in Figure 4.19. The refractive index of the core is higher than those of the substrate and superstrate in order for lightwaves to be guided.

Assuming that the structure is extended to infinity in the y- and z-directions, and a guiding thickness of 2a, the materials are isotropic and lossless (i.e., the permittivities are real and scalar) and nonmagnetic.

4.3.2.2.7.2 Numerical Aperture If we assume at the moment that total internal reflections at the boundaries are required for guiding, what is the acceptance angle such that lightwaves can be launched? The ray path entering the optical fiber core (in axial plane) or a slab waveguide for total internal reflection is shown in Figure 4.20.

Applying Snell's law at the air-core and core-cladding boundaries of the dielectric waveguide, the total internal reflection can take place only if

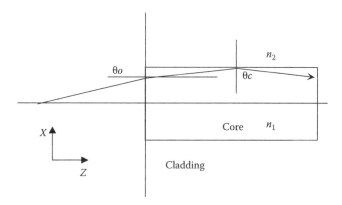

FIGURE 4.20 Numerical aperture of a dielectric waveguide. Lightwaves are approximated as light rays. This is true for the case when there are several lightwaves propagating in the waveguide. Light rays enter the waveguide interface and are refracted and then totally reflected at the core-cladding boundaries. The numerical aperture can be determined by calculating the maximum angle of the incident ray at the entry face of the fiber and air.

$$n_0 \sin \theta_0 \leq n_1 \cos \theta_c \qquad (4.55)$$

where θ_c is the critical angle such that

$$n_1 \sin \theta c = n_2 \sin 90 = n_2 \qquad (4.56)$$

Thus the numerical aperture (NA), which is defined as the maximum value of $\sin \theta_o$, is given by (4.55)

$$NA = (\sin \theta_0)_{max} = n_1 \cos \theta_c$$

Thus we have

$$NA = (n_1^2 - n_2^2)^{1/2} \qquad (4.57)$$

4.3.2.2.7.3   Modes of the Symmetric Dielectric Slab Waveguides   Consider a *monochromatic* (i.e., single ω or λ) wave propagating in the *z*-direction with its electric field:

$$E(x,y,z) = E(x)e^{j(\omega t - \beta z)} \qquad (4.58)$$

i.e., field dependent on *x*, and uniform along the *y*-direction. β is the propagation constant along the *z*-direction, then, in the absence of charges and currents. Maxwell's equations representing the fields of the optical magnetic waves of the optical guided modes following a wave equation are given as

$$\nabla^2 E + \frac{1}{c^2}\frac{\partial^2 E}{\partial t^2} = 0 \qquad (4.59a)$$

With the time dependent of the electric field, as in Equation (4.59a), we have

$$\frac{d^2}{dt^2} = -\omega^2 \quad \text{and} \quad \frac{d^2}{dz^2} = \beta^2$$

and substituting into Equation (4.59a) and using

$$\frac{\omega^2}{c^2} = n^2(\omega)k_0^2$$

we have the wave equation:

$$\nabla_t^2 E + (\beta^2 - n^2(\omega)k_0^2)E = 0 \qquad (4.59b)$$

where

$$\nabla_t^2 = \frac{\partial^2}{\partial x^2} + \frac{\partial^2}{\partial y^2} \tag{4.59c}$$

Therefore for a planar optical waveguide with an infinite extension in the y-direction we have d/dy, and for TE modes only E_y is significant; the wave equation becomes

$$\frac{d^2 E_y}{dx^2} + (\beta^2 - \omega^2 \mu \varepsilon) E_y = 0 \tag{4.60a}$$

where μ and ε are the permeability and permittivity of medium n_1 or n_2. [$\mu = \mu_0$ and $\varepsilon = \varepsilon_r \varepsilon_0$] are the nonmagnetic and (glass) dielectric materials. Similarly, a wave equation involving is given by

$$\frac{d^2 H_y}{dx^2} + (\beta^2 - \omega^2 \mu \varepsilon) H_y = 0 \tag{4.60b}$$

The relationship between the lightwave angular frequency and the velocity of light are given as

$$k = \frac{\omega}{c} \quad \text{and} \quad c = \frac{1}{\sqrt{\mu_0 \varepsilon_0}}$$

is the light velocity in vacuum and k is the wavenumber in vacuum) as

$$\frac{d^2 E_y}{dx^2} + (\beta^2 - k^2 n_j^2) E_y = 0 \tag{4.60c}$$

$$\frac{d^2 H_y}{dx^2} + (\beta^2 - k^2 n_j^2) H_y = 0 \tag{4.60d}$$

where $n_j = n_1, n_2$, depending on whether the equations are applied in the core or cladding regions; $n_1 = (\varepsilon_{r1})^{1/2}$, $n_2 = (\varepsilon_{r2})^{1/2}$ with ε_{r1} and ε_{r2} are the relative permittivities of the core and cladding, respectively. From (4.60a) and (4.60d) we observe the variation of the fields along Ox as:

- *Sinusoidal* behavior when $k^2 n_j^2 > \beta^2$ or guided waves inside the core
- *Exponential* (decay) behavior when $k^2 n_j^2 < \beta^2$, i.e., no radiation in the cladding region

In other words, for a properly designed optical waveguide the optical field is oscillating in regions where the longitudinal propagation constant is smaller than the plane-wave propagation constant and "evanescent" with an exponential-like behavior elsewhere. Thus the lightwaves are "trapped" in the core region and guided through the waveguide length. In the next section we analyze the wave equation so that conditions for guiding lightwaves are established.

4.3.2.2.7.4   Guided Modes
4.3.2.2.7.4.1   General Solutions and the Eigenvalue Equation   Optical waves are guided along the waveguide when their EM fields are oscillatory in the slab waveguide region and exponentially decay in the cladding region, that is,

$$kn_2 \leq \beta \leq kn_1 \tag{4.61}$$

We now define a transverse propagation constant u/a and transverse decay constant v/a as

$$\frac{u^2}{a^2} = k^2 n_1^2 - \beta^2 \tag{4.62a}$$

$$\frac{v^2}{a^2} = -k^2 n_2^2 + \beta^2 \tag{4.62b}$$

Adding Equations (4.62a) and (4.62b) gives

$$\frac{u^2}{a^2} + \frac{v^2}{a^2} = k^2 (n_1^2 - n_2^2) \tag{4.63}$$

or alternatively,

$$V^2 = u^2 + v^2 = k^2 a^2 (n_1^2 - n_2^2) \tag{4.64}$$

We observe that Equations (4.62a) and (4.62b) represent the propagation constant in the transverse direction, as illustrated in Figure 4.21.

In order for the lightwaves to be guided or effectively oscillating in the transverse direction, we can see that the transverse propagation constant u/a must be positive in the core region and negative in the cladding region.

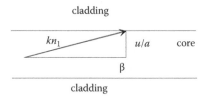

FIGURE 4.21   Representation of the propagation constant along the "ray" direction, the propagation direction z, and the transverse direction.

The parameter V is defined as the normalized frequency (V), which is dependent only on the guide thickness and light frequency (i.e., wavelength) and the refractive index difference between the core and cladding regions (the slab and superstrate or substrate regions for planar optical waveguides).

The fields E_y for TE modes and H_y for TM modes are a linear combination of $\cos(ux/a)$ and $\sin(ux/a)$ inside the core layer (i.e., when $|x| \leq a$), and exponential decay forms on the outside of the core or in the cladding layer (that is, when $|x| > a$) with $\exp(-vx/a)$ and $\exp(+vx/a)$ in the superstrate or substrate. We therefore have a continuum of optical guided modes, depending on whether the solution function follows a symmetrical or antisymmetric pattern (e.g., cosine or sine functions).

Mathematically the general solution of the wave equations given in (4.60c) and (4.60d) would be a combination of the sine and cosine or even and odd functions, respectively. In the following sections we split the solution into two parts, the even and odd modes corresponding with the even and odd functions. We can write a combination of these solutions, as we have seen done usually in the mathematics of differential equations.

4.3.2.2.7.4.2 Even TE Modes (for Modes with Solution Function Cosine) For $|x| \leq a$, that is, inside the core region, the only significant components for the TE mode are E_y and H_x:

$$E_y(x) = A\cos\frac{ux}{a} \tag{4.65a}$$

$$H_y \quad \text{and} \quad H_z = A\sin\frac{ux}{a} \tag{4.65b}$$

For $|x| > a$, that is, the field portion of the lightwaves in the cladding region,

$$E_y = Ce^{-\frac{v}{a}(x-a)} \tag{4.66}$$

The arbitrary constants A and C can be found by applying the boundary conditions as follows.

The value of E_y at $x = a^+$ and $x = a^-$ must be equal; using (4.65a) and (4.65b):

$$A\cos u = Ce^{-\frac{v}{a}(x-a)} \tag{4.67}$$

Evaluated at $x = a^+$ this equality becomes

$$C = A\cos u \tag{4.68}$$

The coefficient A (thus C) can then be found by using H_z and one of Maxwell's equation as H_z (at core $x = a^+$) = H_z (at cladding $x = a^-$). Then at $x = a^+$ in the core we have

$$H_z = \frac{1}{j\omega\mu_0}\frac{dE_y}{dz} = \frac{1}{j\omega\mu_0}\left(-\frac{u}{a}A\sin\frac{u}{a}x\right) \tag{4.69}$$

and in cladding at $x = a^-$:

$$H_z = \frac{1}{j\omega\mu_0}\left(-C\frac{v}{a}e^{-\frac{v}{a}(x-a)}\right) \tag{4.70}$$

Therefore, equating the boundary conditions in (4.69) and (4.70), we obtain

$$C = A\frac{u}{v}\sin u \tag{4.71}$$

The eigenvalue equation for the even modes can be achieved by equating (4.68) and (4.71):

$$v = u\tan u \tag{4.72}$$

This equation is called the eigenvalue equation and can be solved to find the propagation constant β along the z-direction. The number of guided modes that can be supported by the slab optical waveguide can then be easily determined. The number of possible values of β gives the number of guided even TE modes. Thus, whether the waveguide is single mode or multimode depends on the number of odd modes possibly supported by the waveguide. This is investigated in the next section.

4.3.2.2.7.4.3   Odd TE Modes   Similarly for odd TE modes, we have the solution function following a sine function. Writing the solution for E_y in the core region and the evanescent field in the cladding regions, and then applying the boundary conditions at the core-cladding interface, we obtain the eigenvalue equation for odd modes:

$$v = -\frac{u}{\tan u} \tag{4.73}$$

4.3.2.2.7.4.4   Graphical Solutions   Combining Equations (4.72) and (4.73) we observe that the waveguides can support only discrete modes, and the propagation constant β related to u- and v-parameters can be found by solving graphically the intersection between circles of V and curves representing Equations (4.64), (4.72), and (4.73). These solutions are illustrated in Figure 4.22.

4.3.2.2.7.4.5   Cutoff Properties   From Figure 4.22 we observe that

- $V = 0$: Zero optical frequency or λ is zero; that is, there exist no lightwaves. Thus, we observe that we always have at least one guided (even) mode, TE_o.
- $V < \pi/2$: There exists only one guided mode (fundamental even mode).

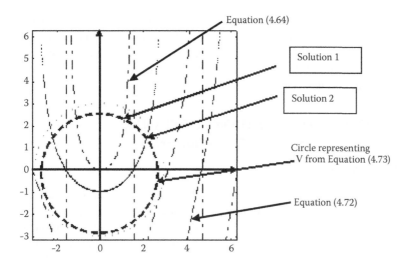

FIGURE 4.22   Graphical solution of (4.64), (4.72), or (4.73): _____ $v = u\tan(u)$, __ _ __ $v = -u/\tan(u)$, and— — — - $V^2 = u^2 + v^2$.

- $V > \pi/2$: An odd mode appears (second mode—fundamental odd mode).
- $V = \pi$: Third mode (TE_{e1}, first order even mode).

That is, each time V reaches a multiple integer of $\pi/2$, a new TE mode reaches its cutoff. The fundamental TE even mode is cut off only when $V = 0$, that is, when there is no waveguide.

4.3.3   CONCLUDING REMARKS

The variational method incorporating a Hermite–Gaussian trial field is inaccurate for calculating the b–V dispersion characteristics of single-mode optical waveguides. However, it is an accurate and convenient tool for mode spot size calculations provided that the mode is well confined in the single spectral range.

The WKB approach is numerically straightforward without any divergence. The enhanced WKB formulae with the correct phase connection at the dielectric interfaces yield very accurate results even for single-mode optical waveguides. This is in contrast with a number of published works.

The profile moment method is applicable only to symmetrical profiles. An SDF is defined to allow a decision on the choice of reference profile to obtain optimum performance of this method. For diffused profiles, the hyperbolic cosh reference profile is required to yield accurate results. For clad profiles, the step reference profile offers accurate results and is significantly better when SDF → 1.

However, the profile moment method does not cover the entire single-mode region for all profiles. At larger V its accuracy deteriorates significantly. Neither do the WKB formulae cover the single-mode region; a hybrid profile moment-WKB method has to be considered. A simple profile characterization factor is presented to assess the applicability of the profile moment method.

From the analysis of the mode field distribution and numerical computation, the m-modes of an asymmetrical waveguide and the $(2m + 1)$ odd modes of its corresponding asymmetrical waveguide are directly related. Computations of their mode cutoffs allow us to establish the exact correspondence for an asymmetrical waveguide with an infinite asymmetrical factor.

4.4 DESIGN AND SIMULATIONS OF PLANAR OPTICAL WAVEGUIDES: EXPERIMENTS

4.4.1 INTRODUCTION

As we have outlined above on planar optical waveguides, we have treated slab optical waveguides as an extension of an optical fiber (see also Chapter 6 on circular optical waveguide or optical fibers) where the structure is restricted to one dimension, and the other dimension of its cross section has been extended to infinity. The optical waveguides considered in this experiment are not a step index slab type, as considered in the theoretical section above, but rather they are graded indexed; i.e., the refractive index is gradually decreased from the core to the cladding region.

This section can be used as an introductory experiment of optical waveguides and aims to familiarize potential optical communications engineers with the structures and behavior of the optical field distribution in a number of optical guided wave structures, such as straight, bend, and Y-junction. The computer experiment is written in such a way that you can read and perform the preliminary work and experiment in stages. Software packages, including executable files and sources for different types of waveguides, are provided for downloading. The web address can be obtained from the publisher at CRCPress.com, or by contacting the author at lnbpbc69@gmail.com.

The objectives of this section are

- To design parameters of slab optical waveguides so that they can support a certain number of guided modes in the single- or multimode regions.
- To use the fundamental mode of the optical waveguide for observation and measurement of optical fields of several waveguide composite structures.
- To propagate the fundamental optical field through a number of optical waveguide structures, such as straight, bend, and Y-junction optical guided wave structures.

4.4.2 THEORETICAL BACKGROUND

4.4.2.1 Structures and Index Profiles

Optical waveguides are the fundamental elements in modern optical communications and photonic signal processing systems. Optical fibers are the guiding medium for optical signal transmission and are formed by a circular core inside a circular cladding region. The mathematics required to represent the electric and magnetic field components of the guided waves in optical fibers involve Bessel functions. These waveguides are treated in detail in most fourth-year courses of optical communications

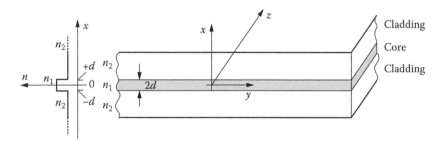

FIGURE 4.23 Schematic structure of the slab optical waveguide: the guiding layer is sandwiched between the superstrate (upper cladding region) and substrate (lower cladding region) of *identical* refractive indices.

engineering. A simplified version of optical fibers is the slab optical waveguide, whose structure is shown in Figure 4.23. The cladding regions are the superstrate and substrate and have identical constant refractive indices. The guiding region is a slab or thin-film layer sandwiched between the cladding regions. The refractive index of the slab region must be higher than that of the cladding, and its thickness must be sufficiently thick to support confined (bound) optical guided modes.

The refractive index profile of the step-slab region can be uniform, i.e., constant throughout, or graded where $n(x)$ decreases gradually from the center of the slab to the cladding. For the sake of simplicity to obtain an analytical solution to the wave equation representing the guided field, the index profile of our slab structures would have a "cosh^{-2}" distribution (graded index profile) given by

$$n^2(x) = n_s^2 + \frac{2n_s\Delta n}{\cosh^2\dfrac{2x}{h}}\tag{4.74}$$

or approximately

$$n(x) = n_s + \frac{\Delta n}{\cosh^2\dfrac{2x}{h}}\tag{4.75}$$

where n_s = cladding refractive index (for both the superstrate and substrate), Δn = refractive index difference between the cladding and slab regions, and h is the total thickness of the guiding layer (slab thickness). It is convenient to define a normalized parameter V as

$$V = kh\,(2\,ns\,\Delta n)^{1/2}\tag{4.76}$$

where $k = 2\pi/\lambda$, with λ being the operating wavelength of the optical waves in vacuum. It is noted that the expression of parameter V is identical with that of the circular optical fiber described in Chapter 6. However, in this experiment we are dealing with planar optical waveguide structures.

4.4.2.2 Optical Fields of the Guided TE Modes

Normally the optical fields in a slab waveguide would consist of two quasi-polarizations TE (transverse electric) and TM (transverse magnetic) where the nonzero electromagnetic field components are (E_y, H_x, H_z) and (E_x, E_z, H_y) for TE and TM modes, respectively. In this experiment we consider only the behavior of TE modes in slab optical waveguides.

The wave equation for TE modes can be derived from Maxwell's equations; in the case where the refractive index difference is small, the wave equation can be approximated to have a scalar form as

$$\frac{d^2 E_y}{dx^2} - (\beta^2 - n^2 k^2) E_y = 0 \tag{4.77}$$

When the refractive index distribution of the waveguide structure $n(x)$ has a $\cosh^{-2}$ profile in the slab region and constant in the cladding regions, the field solution of Equation (4.77) would have an analytical form of

$$E_y(x) = \frac{u_\nu(2x/h)}{\cosh^{-2}(2x/h)} \tag{4.78}$$

where $\nu = 0, 1, 2, 3, \ldots$. Equation (4.78) is subject to the boundary condition that the field must vanish at a distance very far from the slab-cladding interface. The function $u_\nu(x/h)$ would take the following forms.

4.4.2.2.1 For Even TE Modes, $\nu = 0, 2, 4, \ldots$

$$u_\nu(2x/h) = 1 - \frac{1}{2}\nu(2s - \nu)\frac{\sinh^2(2x/h)}{1.1!}$$
$$+ \frac{1}{4}\nu(\nu - 2)(2s - \nu)(2s - \nu - 2)\frac{\sinh^4(2x/h)}{(1.3.2!)} + \ldots \tag{4.79}$$

with $s = 1/2\{(1 + V^2)^{0.5} - 1\}$ being the *total number of guided even modes* that this kind of optical waveguide can support.

4.4.2.2.2 For Odd TE Modes, $\nu = 1, 3, 5, \ldots$

$$u_\nu(2x/h) = \sinh(2x/h) \cdot \left[\begin{array}{l} 1 - \dfrac{1}{2}(\nu - 1)(2s - \nu - 1)\dfrac{\sinh^2(2x/h)}{3.1!} \\[3mm] + \dfrac{1}{4}(\nu - 1)(\nu - 3)(2s - \nu - 3)\dfrac{\sinh^4(2x/h)}{(3.5.2!)} + \ldots \end{array} \right] \tag{4.80}$$

with $s = 0.5[(1 + V^2)^{1/2} - 1]$ being the *maximum number of guided odd modes*.
For lower-order modes, we have

$$u_0 = 1 \tag{4.81a}$$

$$u_1 = \sinh\frac{2x}{h} \tag{4.81b}$$

$$u_2 = 1 - 2(s-1)\sinh^2\frac{2x}{h} \tag{4.81c}$$

$$u_3 = \sinh\frac{2x}{h}\left[1 - \frac{2}{3}(s-2)\sinh^2\frac{2x}{h}\right] \tag{4.81d}$$

The propagation constant β_υ and the effective indices, $n_{eff} = \beta_\upsilon/k$, of the υth order modes are given by

$$\beta_\upsilon^2 = n_s^2 k^2 + 4(s-\upsilon)^2/h^2 \tag{4.82}$$

$$n_{eff}^2 = n_s^2 + (s-\upsilon)^2(\lambda/\beta_\upsilon)^2 \tag{4.83a}$$

and the normalized propagation constant b is defined as

$$b = \frac{n_{eff}^2 - n_s^2}{n^2 - n_s^2} \tag{4.83b}$$

where n is the refractive index at the center of the guide, or approximately

$$b = \frac{n_{eff}^2 - n_s^2}{2n_s\Delta n} \tag{4.83c}$$

Thus, the optical field of a slab optical waveguide can be found if we can specify the following parameters: the slab thickness h, the cladding refractive index, the refractive index difference Δn, and the operating wavelength.

4.4.2.3 Design of Optical Waveguide Parameters: Preliminary Work

Choose the parameters n_s, Δn, and h of your waveguide. Some typical refractive indices of certain transparent materials for superstrate and substrate of optical waveguides are

- $n_s = 1.447$ for silica glass at the operating wavelength of 1310 nm. This also is the base material for modern telecommunication optical fibers.
- $n_s = 3.6$ for GaAs semiconductor waveguide at 1300 nm. This is the base material for optical waveguides formed in the resonant cavity and waveguide of semiconductor lasers for optical fiber communications.

- $n_s = 2.2$ for lithium niobate crystal at a certain crystal axis, as seen by the TE waves. This is also the base material for optical modulators for optical communications.

Make sure that the chosen parameters would form an optical waveguide that would support no more than four guided modes at the operating wavelength of 1300 nm. Can you design an optical waveguide such that it supports only one guided mode, which is the only fundamental mode of the optical waveguide? In fact, we can plot the b–V curve for $\upsilon = 0, 1, 2, \ldots$, and from this diagram we can design monomode, two-mode, etc., optical waveguides. Notice that only TE modes are considered here.

4.4.3 Simulation of Optical Fields and Propagation in Slab Optical Waveguide Structures

A number of computer simulation programs have been written (available for download from crcpress.com; contact the publisher for details) to study the evolution of optical fields in slab optical waveguides that form the basic component for several optical waveguiding devices.

To numerically study the behavior of optical waves, particularly the fundamental mode field, in these structures, the whole waveguide region, including the slab and cladding regions, i.e., W and L, is sliced into several intervals along the propagation direction z as well as in the vertical direction for numerical calculations.

The field in the first plane, i.e., at $z = 0$, can be found by (4.80). This field would then be propagated to the next plane through a discredited equation by applying the finite difference method to the wave equation with the z-dependence in Maxwell's equation. This para-axial wave propagation equation is given by

$$2jkn_s \frac{\partial E_y}{\partial z} = \frac{\partial^2 E_y}{dx^2} + k^2[n^2(x,y) - n_s^2]E_y \tag{4.84}$$

which can be written using the center finite difference technique as

$$jkn_s \frac{E_{i,k+1} - E_{i,j}}{\Delta z} = \frac{E_{i-1} - 2E_i + E_{i+1}}{\Delta x^2} + k^2[n_i(x,y) - n_s^2]E_i \tag{4.85}$$

where $j = \sqrt{-1}$, and the subscripts i and j denote the variation of E with respect to the x- and z-directions, respectively. The cross section plane that is partitioned into several layers with order ith and the propagation steps along the z-direction are assigned with order jth. The obtained results of the field at location j would then be used as the field initial distribution for propagating through the structure to obtain the optical field of the next plane $j + 1$ and so on. Thus, we can employ an analytical method to obtain the field solution for the optical field in the transverse plane. A numerical method (the finite difference) is used to study the evolution of the optical field propagating along the optical waveguide structure.

In this section (or experiment) we are not going to study the finite difference method but the evolution of the optical field in optical waveguides. In the following parts we will examine the optical field behavior in the structures illustrated in Figure 4.23. An additional dimension, the z-axis, is now added to the optical waveguide devices. These structures are shown in Figures 4.24 to 4.27.

4.4.3.1 Lightwave Propagation in Guided Straight Structures

Do not hesitate to seek assistance from the author (email: lnbpbc69@gmail.com) for procedures in running the simulation package. Typical steps for simulation are run MATLAB®; go to directory Straight (see Figure 4.24); run FD1, the program for the beam propagation method; and choose parameters as prompted by the program, such as (1) waveguide region to be analyzed, (2) operating wavelength, (3) slab thickness, (4) the number of x intervals for the optical field and number of propagation steps in the z-direction, (5) the refractive indices of slab and cladding, i.e., cladding index and the index difference, and (6) propagation distance.

a. After successfully obtaining the guiding of optical waves, keep one or two parameters constant (such as the waveguide thickness) and vary stepwise other parameters, e.g., index difference, wavelength. Observe the evolution of the optical field and plot the 3D guided wave field profile and the field contour. *Note*: When specifying the number of planes to be plotted by MATLAB the product of the number of intervals in x- and z-directions must not exceed the MATLAB limit, which is 8188, depending on the available computer memory and MATLAB version.

b. Observe the field evolution with respect to the change in refractive index difference, waveguide thickness, etc.

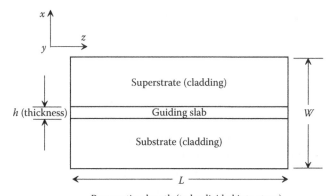

Propagation length (to be divided into steps)

FIGURE 4.24 Side view of a slab optical waveguide in a straight optical device structure. h is the waveguide thickness, W is the total width of the structure in the transverse plane to be specified for numerical simulation, and L is the total length of the device. W is to be divided into several equispacing layers for numerical simulation. The length L along the propagation direction is also split into several steps for propagation from one plane to the other, and so on.

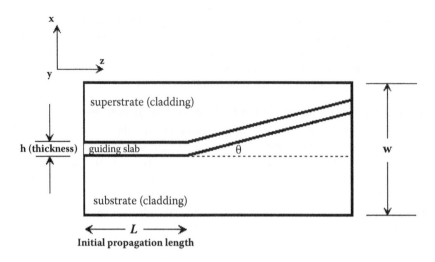

FIGURE 4.25 Side view of the optical waveguide device using a slab waveguide in bend structure. Notice the bending section. θ is the bend angle; *L* is the length of the straight section.

An example of the waveguide and propagation in a straight waveguide is shown in Figure 4.28.

Note: The graphical facilities are dependent on the computer networking at the time of the experiment.

4.4.3.2   Lightwave Propagation in Guided Bent Structures

Similar to the steps as above: (1) Choose the most suitable optical waveguide structures of the straight structure to enter into the bend structure parameters. (2) The Bend directory (Figure 4.25) is to be evoked. (3) Additional parameters for this structure are the bend angle θ and the length *l* of the straight section. Start with a bend angle of about 0.5° of arc. (4) Now vary the bend angle in steps of 0.5° or 1° of arc to about 10° of arc. (5) Observe the radiation of the guided field at the bend section and report the guided and radiated optical fields. (6) Vary the refractive index difference and run the program for the bend angle of about 2–4° and observe the confinement and radiation of the optical field at straight and bend sections.

4.4.3.3   Lightwave Propagation in Y-Junction (Splitter)
and Interferometric Structures

As illustrated in Figure 4.27, the Y-junction or optical splitter is considered a combination of two identical bend sections. Note that the section right of the Y-junction has a width that is wider than that of the straight section. Thus the number of guided modes would be higher for this very short section: (1) Choose the half-angle θ of the Y-junction from 0.5 to 5° of arc, run the appropriate program to observe the field evolution, and measure the field strength, i.e., intensity or optical power, distribution across the whole device. (2) You can vary the refractive index difference for a small half-angle at the Y-junction to observe the splitting effect. Report the field behavior at these Y-junctions. (3) If time permits, use the facility provided for simulation of the

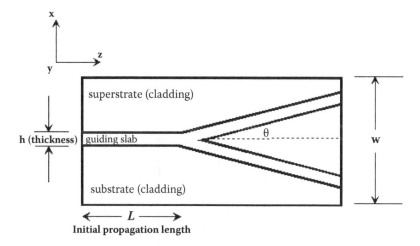

FIGURE 4.26 Side view of the Y-structure using slab optical waveguide. θ is the half Y-junction angle; L is the straight initial section before splitting.

propagation of lightwaves through an interferometric optical waveguide structure, the Mach–Zhender, in which the input lightguide is split into two paths and then combined into one output port. A typical evolution of the guided optical field in a planar straight waveguide is shown in Figure 4.28, with the following parameters: cladding refractive index of 2.2, refractive index difference of 0.02, propagation length of 100 μm, step size of 0.5 μm, and step size in the x-direction of 0.02 μm.

4.5 APPENDIX A: EXACT ANALYSIS OF CLAD LINEAR OPTICAL WAVEGUIDES

The exact analysis of TE modes guided in an optical waveguide whose refractive index profile follows linear shape is described in this section. The profile was first analyzed exactly by Marcuse,[23] then treated in full by Adams[60] and applied to the study of a low-threshold current laser diode. The results presented here are different from the published formulae, as they are expressed in terms of Bessel functions of real positive order. Starting with the eigenvalue equation we derive the propagation constant and the cutoffs of the waveguide. The treatments of symmetrical and asymmetrical profiles are given separately.

4.5.1 ASYMMETRICAL CLAD LINEAR PROFILE

4.5.1.1 Eigenvalue Equation

The eigenvalue equation is given by[23]

$$\frac{Ai'(\alpha_0)-V^{1/3}\sqrt{A+b}\,Ai'(-\alpha_0)}{Bi'(\alpha_0)-V^{1/3}\sqrt{A+b}\,Bi'(-\alpha_0)}=\frac{Ai'(\alpha_1)-V^{1/3}\sqrt{b}\,Ai(\alpha_1)}{Bi'(\alpha_1)-V^{1/3}\sqrt{A+b}\,Bi(-\alpha_1)} \qquad (4.86)$$

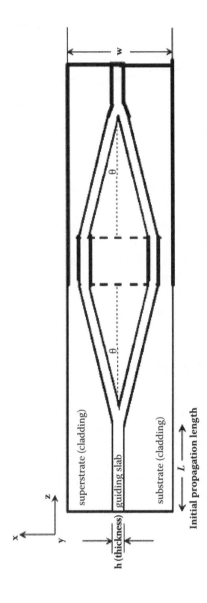

FIGURE 4.27 Side view of the interferometric structure using slab optical waveguide. θ is the half Y-splitting angle; l is the straight initial section before splitting.

3D View of the Field Intensity

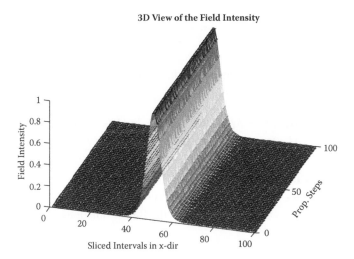

FIGURE 4.28 Typical evolution of the guided mode field propagating through a straight slab optical waveguide. Waveguide parameters: cladding refractive index = 2.2, refractive index change = 0.02, propagation length = 100 μm, step size = 0.5 μm, step size in x-direction = 0.02 μm.

where $\alpha_0 = (1-b)V^{2/3} \simeq \alpha_1 = bV^{2/3}$; Ai and Bi are Airy functions and the dash denotes the derivatives with respect to the argument.

Using the relations between the Airy and Bessel functions of Abramowitz and Stegun,[61] we can convert (4.86) into an immediate form:

$$\frac{J_{-2/3}(\gamma_0)-J_{2/3}(\gamma_0)+Q[J_{-1/3}(\gamma_0)+J_{1/3}(\gamma_0)]}{J_{-2/3}(\gamma_0)-J_{2/3}(\gamma_0)+Q[J_{-1/3}(\gamma_0)-J_{1/3}(\gamma_0)]} = $$
$$\frac{-I_{-2/3}(\gamma_1)+I_{2/3}(\gamma_1)+I_{-1/3}(\gamma_1)-I_{1/3}(\gamma_1)}{I_{-2/3}(\gamma_1)+I_{2/3}(\gamma_1)+I_{-1/3}(\gamma_1)+I_{1/3}(\gamma_1)}$$

(4.87)

where the arguments $\gamma_0; \gamma_1$ of the Bessel functions and Q are given as

$$\gamma_0 = \frac{2}{3}V(1-b)^{3/2}$$

$$\gamma_1 = \frac{2}{3}V(b)^{3/2}$$

(4.88)

$$Q = \left[\frac{A+b}{1-b}\right]^{3/2}$$

In arriving at Equation (4.87) we have also used[*]

---

[*] Note: Equation given in Abramowitz and Stegun (1972) does not have the negative sign (–), a vital error.

$$Ai'(\alpha_1) = -\frac{\alpha_1}{3}(I_{-2/3} - I_{2/3}) \tag{4.89}$$

Finally, using the recurrence relations for the J and I functions results in the required version of the eigenvalue equation:

$$\frac{(2+3\gamma_0 Q)J_{1/3}(\gamma_0) - 3\gamma_0 J_{4/3}(\gamma_0)}{(4Q+3\gamma_0)J_{2/3}(\gamma_0) - 3\gamma_0 J_{5/3}(\gamma_0)} = \frac{(2+3\gamma_1 Q)I_{1/3}(\gamma_0) - 3\gamma_1 I_{4/3}(\gamma_1)}{(4+3\gamma_1)I_{2/3}(\gamma_1) - 3\gamma_1 I_{5/3}(\gamma_1)} \tag{4.90}$$

4.5.1.2 Mode Cutoff

At cutoff we have $b = 0$ and $Q = [A]^{1/2}\gamma_0 = 2/3V$; $\gamma_1 = 0$. The RHS of (4.90) $\rightarrow \infty$ and a substitution of the asymptotic formula for the Bessel functions gives the mode cut off for modes with large V_c:

$$\frac{2}{3}V_c \sim \left(m + \frac{1}{12}\right)\pi - \tan^{-1}(\sqrt{A}) \qquad m = 0,1,2... \tag{4.91}$$

4.5.2 Symmetrical Waveguide

4.5.2.1 Eigenvalue Equation

The eigenvalue equations for the odd and even modes are given in Adams[60] and can be derived from (4.86). Only the LHS (left-hand side) is affected, and the eigenvalue equations for these modes are given as

$$\frac{Ai'(-\alpha_0)}{Bi'(-\alpha_0)} = \frac{Ai'(\alpha_1) - V^{1/3}\sqrt{b}\,Ai(\alpha_1)}{Bi'(\alpha_1) - V^{1/3}\sqrt{A+b}\,Bi(-\alpha_1)} \qquad \text{even mode}$$

$$\frac{Ai(-\alpha_0)}{Bi(-\alpha_0)} = \frac{Ai'(\alpha_1) - V^{1/3}\sqrt{b}\,Ai(\alpha_1)}{Bi'(\alpha_1) - V^{1/3}\sqrt{A+b}\,Bi(-\alpha_1)} \qquad \text{odd mode} \tag{4.92}$$

Similarly, using the relations between the Airy and Bessel functions of Abramowitz and Stegun,[61] we can convert (4.92) into an immediate form:

$$\frac{-2J_{-2/3}(\gamma_0) + 3\gamma_0 J_{4/3}(\gamma_0)}{3\gamma_0 J_{2/3}(\gamma_0)} = \frac{-I_{-2/3}(\gamma_1) + I_{2/3}(\gamma_1) + I_{-1/3}(\gamma_1) - I_{1/3}(\gamma_1)}{I_{-2/3}(\gamma_1) + I_{2/3}(\gamma_1) + I_{-1/3}(\gamma_1) + I_{1/3}(\gamma_1)}$$

$$\frac{-3\gamma_0 J_{1/3}(\gamma_0)}{3\gamma_0 J_{5/3}(\gamma_0) - 4J_{2/3}(\gamma_0)} = \frac{-I_{-2/3}(\gamma_1) + I_{2/3}(\gamma_1) + I_{-1/3}(\gamma_1) - I_{1/3}(\gamma_1)}{I_{-2/3}(\gamma_1) + I_{2/3}(\gamma_1) + I_{-1/3}(\gamma_1) + I_{1/3}(\gamma_1)} \tag{4.93}$$

4.5.2.2 Mode Cutoff

The RHS of (4.90) goes to infinity at cutoff, so we have

$$J_{2/3}\left(\frac{2V_C}{3}\right) = 0 \qquad\qquad m = 0,2\ldots \text{ even modes}$$

$$J_{2/3}\left(\frac{2V_C}{3}\right) - V_C J_{5/3}\left(\frac{2V_C}{3}\right) = 0 \quad m = 1,3\ldots \text{ odd modes}$$

(4.94)

Alternatively these equations can be obtained from (4.91).

4.6   APPENDIX B: WKB METHOD, TURNING POINTS, AND CONNECTION FORMULAE

4.6.1   INTRODUCTION

Consider the scalar wave equation:

$$\frac{d^2\varphi(x)}{dx^2} + K^2\varphi(x) = 0 \qquad\qquad (4.95)$$

where $K^2 = k_0^2 n^2(x) - \beta^2$ is the transverse propagation constant and $\varphi(x)$ is the modal field in the transverse plane. The characteristic mode factor $e^{j(\omega t - \beta z)}$ is omitted. This version of the wave equation is selected to present the turning points in the subsequent analysis. The turning point is defined by $x_t \rightarrow K(x_t) = 0$.

When the refractive index function $n(x)$ has certain simple forms, (4.95) can be solved explicitly for $\varphi(x)$, the good behavior of this function restricting the axial propagation constant β to discrete values. These are the characteristics of the bound modes. However, in most practical optical waveguides, explicit solutions of the fields are not available and approximation methods of solutions must be developed.

The WKB method is based on an asymptotic expansion in k_0^{-1}, the first term of which leads to geometrical optic results, or the 0th-order WKB solutions, and higher-order terms lead to exact modal solutions. The principal concern of this method lies in the transitional region, which connects the oscillatory fields and its evanescent neighbors. These are the turning points of the problem where the semiclassical approximation breaks down. The way in which the WKB solutions are valid on either side of the turning point connects remains the central problem of the method.

Before proceeding to derive the WKB solutions we assign the following symbols: $\phi(x)$, $\varphi(x)$, and $\Phi(x)$ = exact modal field, WKB solution, and approximate modal field valid at the turning point, respectively.

4.6.2   DERIVATION OF THE WKB APPROXIMATE SOLUTIONS

Following established tradition, we postulate a solution of (4.95) in the form of

$$\phi(x) = Ae^{jk_0 S(x)}$$

$$j = \sqrt{-1};\ A = \text{constant}$$

(4.96)

Thus Equation (4.95) can be transformed into the Riccati equation:

$$j\frac{1}{k_0}\frac{d^2S(x)}{dx^2} - \left(\frac{dS(x)}{dx}\right)^2 + (n^2(x) - n_e^2) = 0 \tag{4.97}$$

$$n_e \equiv \text{effective index of waveguide mode}$$

Now let $y = S'$ and assume that y admits of a formal series expansion of the form

$$y = \sum_{n=0}^{\infty} k_0^{-n} y_n \tag{4.98}$$

Thence

$$y' = S'' = \sum_{n=0}^{\infty} k_0^{-n} y_n' \tag{4.99}$$

and

$$y^2 = S^2 = \left[\sum_{n=0}^{\infty} k_0^{-n} y_n'\right]^2 = y_0^2 \left\{1 + k_0^{-1}\frac{2y_1}{y_0} + k_0^{-2}\left(\frac{2y_1}{y_0} + \frac{y_1^2}{y_0^2}\right) + \ldots\right\} \tag{4.100}$$

Therefore, substituting (4.99) and (4.100) into (4.97) and equating the coefficients of the like power of k_0 leads to the following recurrence relations:

$$y_0^2 = n^2 - n_e^2 = \frac{K^2}{k_0^2}$$

$$jy_0' = 2y_1 y_0 \tag{4.101}$$

$$jy_1' = 2y_2 y_0 + y_1^2$$

Thus, we can obtain

$$y_0' = \pm\frac{K}{k_0}$$

$$\pm j\frac{K'}{k_0} = \pm\frac{2K}{k_0} y_1 \tag{4.102}$$

which can be integrated into the form

$$\frac{j}{2k_0} \ln K + C = \frac{1}{k_0} \int_{x_0}^{x} y_1 dx \tag{4.103}$$

where C is the integration constant.

Hence,

$$\phi(x) = Ae^{jk_0 S(x)} = Ae^{jk_0 \int_{x_0}^{x} \sum_{n=0}^{\infty} k_0^{-n} y_n dx} = Ae^{jk_0 \left\{ \int_{x_0}^{x} y_0 dx + k_0^{-1} \int_{x_0}^{x} y_1 dx + k_0^{-2} \int_{x_0}^{x} y_2 dx + \dots \right\}}$$

$$= A_{\pm} e^{\pm j \int_{x_0}^{x} K dx + jk_0 C} \tag{4.104}$$

$$\rightarrow \varphi(x) = A_{\pm} K^{-1/2} e^{\pm j \int_{x_0}^{x} K dx}$$

where $A_{\pm}$ denotes the constants corresponding to the $\pm$ solutions, respectively.

For $K^2 > 0$ we have

$$\varphi_1(x) = DK^{-1/2} \cos\left(\pm j \int_{x_0}^{x} K \, dx + \delta \right) \tag{4.105}$$

$$D, \delta \equiv \text{arb. constants}$$

For $K^2 < 0$ we have

$$\varphi_2(x) = B_{\pm} |K|^{-1/2} e^{\left(\pm \int_{x_0}^{x} |K| dx \right)} \tag{4.106}$$

Clearly at the turning point defined by $K^2(x) = 0$ both the oscillatory and evanescent fields diverge. Hence neither form can be retained during the transition from one interval to the other, where K^2 changes sign.

Furthermore, a back substitution of the WKB solutions, for example, φ, into the original equation (4.95) produces an inhomogeneous equation:

$$\frac{d^2 \varphi(x)}{dx^2} + K^2 \varphi(x) = W(x)$$

$$W(x) = \frac{3}{4} \left(\frac{K'}{K} \right)^2 - \frac{1}{2} \frac{K''}{K} \tag{4.107}$$

For this equation, any point where $K^2(x)$ vanishes is singular. However, far from the singularity, a higher-order WKB solution can be obtained with (4.107) as the starting point and incorporating $XW(x)$ into K^2. This method contrasts with the straightforward idea using a high-order recurrence relation to obtain higher-order terms.[62]

4.6.3   TURNING POINT CORRECTIONS

4.6.3.1   Langer's Approximate Solution Valid at Turning Point

For a refractive index profile that is unbounded we can write, for an nth-order zero at $x_0 = 0$,

$$K^2(x) = (x - x_n)^n f(x) \tag{4.108}$$

where

$$f(x) = \sum_{i=0}^{\infty} C_i (x - x_i)^i$$

is a nonvanishing polynomial of x at $x_0 = 0$. For simplicity and without loss of generality, let $x_0 = 0$. Thus, for values of x close to zero we can write

$$K^2(x) = C_0(x)^n \tag{4.109}$$

In the vicinity of the turning point, we can represent the wave equation by an approximate differential equation:

$$\frac{d^2\phi(x)}{dx^2} + C_0 x^n \phi(x) = 0 \tag{4.110}$$

where

$$\phi \equiv \varphi \text{ at } x = 0$$

The solutions to (4.110) are Bessel functions. Thus, it would be necessary to transform the wave equation before setting the condition (4.109). We can now introduce the Liouville transform:

$$\xi = \int_0^x K(x)dx \quad \text{and} \quad \phi = K^{1/2}(x)v \tag{4.111}$$

After some algebra the equation becomes

$$\frac{d^2v}{d\xi^2} + \left\{\frac{3}{4}K^{-4}\left(\frac{d^2K}{dx^2}\right)^2 - \frac{1}{2}K^{-3}\frac{d^2K}{dx^2} + 1\right\}v = 0 \tag{4.112}$$

Now using the form of the approximate transverse propagation constant in Equation (4.109) and evaluating the terms in the bracket reduces Equation (4.112) to the required intermediate form

$$\frac{d^2v}{d\xi^2} + \left\{1 + C_0^{-1}x^{-(n+2)}\left(\frac{n^2+4n}{16}\right)\right\}v = 0 \tag{4.113}$$

Then using (4.111) we obtain

$$\frac{1}{\xi^2} = C_0^{-1}x^{-(n+2)}\left(\frac{n^2+2}{2}\right)^2 \tag{4.114}$$

Substituting into (4.113) we arrive at

$$\frac{d^2v}{d\xi^2} + \left\{1 + \frac{1}{\xi^2}\cdot\left(\frac{n(n+4)}{4(n+4)^2}\right)\right\}v = 0 \tag{4.115}$$

Now changing the variable $v \to \xi^{1/2}W$ leads to

$$\frac{d^2v}{d\xi^2} = \xi^{1/2}\frac{d^2W}{d\xi^2} + \xi^{-1/2}\frac{dW}{d\xi} - \frac{1}{4}\xi^{-3/2}W \tag{4.116}$$

Finally substituting (4.116) back into (4.115) and multiplying by $\xi^{3/2}$ throughout, the resultant equation is transformed to the desired canonical form:

$$\frac{d^2W}{d\xi^2} + \xi\frac{dW}{d\xi} + \left\{\xi^2 - \frac{1}{(n+2)^2}\right\}W = 0 \tag{4.117}$$

This is the Bessel equation of order $1/(n+2)$ with independent solutions denoted by plus and minus signs.

$$W = \left\{\frac{\pi^2}{4}A_\pm\right\}J_{\pm m}(\xi) \tag{4.118}$$

with $m = \dfrac{1}{n+2}$

Thus, from $v \rightarrow \xi^{1/2}W$ and (4.111) we can recover the solutions of the original form (4.110) at the turning point as

$$\phi = K^{-1/2}\xi^{1/2}W(\xi) = \left\{\frac{\pi^{1/2}}{2^{1/2}}A_{\pm}\right\}\left(\frac{\xi}{K}\right)^{1/2}J_{\pm m}(\xi) \tag{4.119}$$

where $A_{\pm}$ is an arbitrary constant and ξ is related to $K(x)$ via the Liouville transformation. The relationship between Equations (4.117) and (4.110) can be found in standard textbooks on Bessel's function.[57] We include it here for interested readers.

To study the behavior of the approximate solutions at the turning point, we obtain the differential equation (DE) satisfied by ϕ. To do this we represent (4.119) in the form

$$\phi = G(x)\xi^m J_{\pm m}(\xi)$$

$$\text{with } G(x) = \left\{\frac{\pi^{1/2}}{2^{1/2}}A_{\pm}\right\}K^{-1/2}(x)\xi^{1/2-m} \tag{4.120}$$

Differentiating (4.120) with respect to x we obtain

$$\frac{d\phi(x)}{dx} = \frac{\left\{\frac{\pi^{1/2}}{2^{1/2}}A_{\pm}\right\}\xi^{1/2-m}}{G(x)}\{\xi^m J'_{\pm m} + m\xi^m J'_{\pm m}\} + G'(x)\xi^m J_{\pm m} \tag{4.121}$$

and

$$\frac{d^2\phi(x)}{dx^2} = \xi^m J_{\pm m}G''(x) - \frac{\pi^2}{4}A_{\pm}^4 G^{-3}(x)\xi^{2-3m}J_{\pm m}$$

Using (4.120) we can write in more compact form

$$\frac{d^2\phi(x)}{dx^2} = \frac{G''(x)}{G(x)}\phi(x) - K^2\phi(x) \tag{4.122}$$

Thus, we have the relation

$$K(x) = \frac{d\xi(x)}{dx} = \frac{\pi}{2}A_{\pm}^2\xi^{1-2m}G^{-2}(x) \tag{4.123}$$

and the Bessel identity

$$\xi^2 J'' + \xi J' = (m^2 - \xi^2) J(\xi) \tag{4.124}$$

4.6.3.2 Behavior of the Turning Point

Equation (4.122) can be recast into

$$\frac{d^2\phi(x)}{dx^2} + \{K^2(x) - \theta(x)\}\phi(x) = 0 \tag{4.125}$$

where

$$\theta(x) = \frac{G''(x)}{G(x)}$$

This is the DE that satisfies the approximate Langer's solution ϕ at the turning point. To prove the validity of ϕ, we only have to show that $G(x)$ is bounded at $x = 0$, so that the coefficient of ϕ does not possess singularity. To do this we write

$$\frac{G(x)}{\left[\dfrac{\pi^{1/2}}{2^{1/2}} A_\pm\right]} = K^{-1/2}(x)\xi^{1/2-m} = x^{-n/4} f^{-1/2}(x)[I_1(x)]^{1/2-m} \tag{4.126}$$

Then, using the expression for K in (4.109),

$$I_1(x) = \int_0^x x^{n/2} f^{1/2}(x)\,dx \tag{4.127}$$

Thence integration by parts gives the expanded form

$$I_1(x) = \frac{x^{n/2+1} f^{1/2}}{n/2+1} - \left\{1 - \frac{\displaystyle\int_0^x x^{n/2+1} f'(x) f^{-1/2}(x)\,dx}{2x^{n/2+1} f^{1/2}(x)}\right\} \tag{4.128}$$

Substituting Equation (4.128) into (4.126) we obtain

$$\frac{G(x)}{\left[\dfrac{\pi^{1/2}}{2^{1/2}} A_\pm\right]} = \frac{f^{-m/2}}{(n/2+1)^{1/2-m}}[1 - I_2(x)]^{1/2-m} \tag{4.129}$$

where the intermediate integral is given by

$$I_2(x) = \frac{\displaystyle\int_0^x x^{n/2+1} f'(x) f^{-1/2}(x)\,dx}{x^{n/2+1} f^{1/2}(x)}$$ (4.130)

Applying the l'Hospital rule shows that

$$\theta(x) = \frac{G''(x)}{G(x)} = \frac{3}{4}\left(\frac{K'}{K}\right)^2 - \frac{1}{2}\left(\frac{K''}{K}\right) + (m-1/4)\frac{K^2}{\varsigma^{1/2}}$$ (4.131)

while we observe that $f(0) \neq 0$ by definition. Hence $G(x) \neq 0$ at the turning point, and the proposition has been proved.

4.6.3.3 Error Bound for φ Turning Point

To investigate the accuracy with which the DE for an error bound for φ represents the exact wave equation at the turning point, one only has to compute the value of $\theta(x)$ at $x = 0$. Thus, we find

$$\theta(x) = \frac{G''(x)}{G(x)} = \frac{3}{4}\left(\frac{K'}{K}\right)^2 - \frac{1}{2}\left(\frac{K''}{K}\right) + (m-1/4)\frac{K^2}{\varsigma^{1/2}}$$ (4.132)

Note that the first two terms of this equation are the function $W(x)$ defined earlier. For $m = 1/2$ (i.e., $n = 0$) we have the turning point of order zero and $\phi \rightarrow \varphi$. Therefore, it can be concluded that in a region removed from any turning point the WKB solutions are of the same form as Φ and vice versa. This observation is substantially proven by investigations of the asymptotic behavior of φ away from the turning point. The proof of this result paves the way for the important connection formulae, which is the eventual purpose of this exercise.

Now, using the form of $K^2(x)$ given in Equation (4.108), we can put $\theta(x)$ in terms of $f(x)$ as

$$\frac{3}{4}\left(\frac{K'}{K}\right)^2 = \frac{3}{16}\left\{\left(\frac{f'}{f}\right)^2 + \frac{2n}{x}\left(\frac{f'}{f}\right) + \frac{n^2}{x^2}\right\} - \frac{1}{2}\left(\frac{K''}{K}\right)$$

$$= \frac{1}{8}\left(\frac{f'}{f}\right)^2 - \frac{n}{4}\left(\frac{n}{2}-1\right)x^{-2} - \frac{n}{4}\left(\frac{f'}{f}\right)x^{-1} - \frac{1}{4}\left(\frac{f''}{f}\right)$$ (4.133)

and

$$\left(m^2 - \frac{1}{4}\right)K^2\xi^{-2} = \left(m^2 - \frac{1}{4}\right)x^n f(x)\frac{1}{I_1^2(x)} \tag{4.134}$$

Thus we can rewrite (4.128) in a more suitable form as

$$I_1(x) = \left\{x^{\pi/2+1}f^{1/2} - \frac{1}{2}\int_0^x x^{n/2+1}f'(x)f^{-1/2}(x)dx\right\}(n/2+1)^{-1} \tag{4.135}$$

Since the integral $\to 0$ in the limit of $x \to 0$, we can expand I_1^{-2} in a binomial series as

$$I_1^{-2} = \left(\frac{n}{2}+1\right)^2\left(x^{\frac{n}{2}+1}f^{1/2}\right)^{-2}\left\{1+\frac{I_3}{x^{\frac{n}{2}+1}f^{-2}}\right\}+\frac{3}{4}\left\{\frac{I_3^2}{\left(x^{\frac{n}{2}+1}f^{1/2}\right)^2}\right\}+... \tag{4.136}$$

with $I_3 - \int_0^x f^{1/2}f'x^{\frac{n}{2}+1}dx$

Integrating by parts again, we can expand into the form as

$$I_3 = \frac{1}{\left(\frac{n}{2}+2\right)}\left\{x^{\frac{n}{2}+2}f^{-1/2}f' - \int_0^x x^{\frac{n}{2}+2}f^{-1}\left(f^{1/2}f'' - \frac{1}{2}(f')^2f^{-1/2}\right)dx\right\} \tag{4.137}$$

$$\therefore I_3^2 = \left(\frac{n}{2}+2\right)^{-2}\left\{x^{\frac{n}{2}+4}f^{-1}(f')^2 - 2x^{\frac{n}{2}+2}f^{-1/2}f'I_4 + (I_4)^2 +...\right\}$$

where I_4 is the integral expression given in (4.136). Finally, substituting (4.137) and (4.136) into (4.133) we obtain

$$\left(m^2 - \frac{1}{4}\right)K^2\xi^{-2} = -\frac{n}{16x^2}\left(\frac{n+4}{2}\right) - \frac{2}{8}\frac{1}{x}\left(\frac{f}{f'}\right) - \frac{3}{16}\frac{n}{n+4}\left(\frac{f'}{f}\right)^2$$

$$+\left[\frac{n}{8}\frac{1}{x^2}\left(\frac{f}{f'}\right) + \frac{3}{8}\frac{n}{n+4}\frac{1}{x}\left(\frac{f'}{f}\right)^2\right]\frac{I_4(x)}{x^{n/2+1}f^{1/2}(x)}+... \tag{4.138}$$

Substituting (4.138) and (4.133) into (4.132) we arrive at

$$\theta(x) = \frac{3}{2}\frac{1}{n+6}\left(\frac{n+5}{n+4}\right)\left(\frac{f'}{f}\right)^2 - \left(\frac{f'}{f}\right) + \text{terms in } x \text{ and higher} \qquad (4.139)$$

Equation (4.139) shows that if $f(x)$ is a relatively slowly varying function at $x = 0$, then $\theta(0) \to 0$ and the Langer's approximate solution Φ is a good approximation of the actual mode field at the turning point. This is a general result, of which the expression obtained by Marcuse[23] is a special case.

4.6.4 CORRECTION FORMULAE

It is shown above that the DE satisfied by Langer's approximate solution Φ is non-singular. Therefore, unlike the WKB solution, Φ is single-valued. It is not restricted to yield representations of the solution of the wave equation (4.95) in the intervals on one to the other side of the turning point. Therefore, in problems involving a single tuning point no question of connection formulae arises in association with it. However, in actual waveguides, the simplest scalar wave equation contains two turning points at the minimum. Thus, a single function such as Langer's approximate solution, Φ, valid at a turning point, cannot possibly describe the modal fields since essentially there are two regions of evanescent field behavior connected by a region where the field is oscillatory.

As a first step to obtain a connection formulae, we introduce a new variable,

$$t = -\int_0^x |K|(x)dx = j\xi$$

$$j = \sqrt{-1}$$
\qquad (4.140)

so that we can express Φ as ϕ_1 or ϕ_2 to denote the solutions in regions in which $K^2 > 0$ and $K^2 < 0$, respectively. We now have

$$\phi_1 = \left\{\left(\frac{\pi}{2}\right)^{1/2} A_\pm \left[\frac{\xi}{K}\right]^{1/2}\right\} J_{\pm m}(\xi); \quad x \geq 0$$

$$\phi_2 = \left\{\left(\frac{\pi}{2}\right)^{1/2} B_\pm\right\}\left[\frac{t}{K}\right]^{1/2} I_{\pm m}(\xi); \quad x < 0$$
\qquad (4.141)

The asymptotic solutions of the Bessel functions of large argument are well known, and we can write[61]

$$\phi_1^{as} = \phi_1\big|_{\xi \to \infty} = A_{\pm} K^{-1/2} \cos\left(\xi \mp \frac{m\pi}{2} - \frac{\pi}{4}\right)$$

$$\phi_2^{as} = \phi_2\big|_{t \to \infty} = B_{\pm}|K|^{-1/2} e^{-t + j2\pi\left(\mp m - \frac{1}{4}\right)}$$

(4.142)

These relations confirm our earlier hypothesis that $\phi_1^{as}; \phi_2^{as}$ are just linear combinations of the WKB solutions $\phi_1 = A_{\pm} K^{-1/2} e^{\pm j\xi}$ and $\phi_2 = B_{\pm}|K|^{-1/2} e^{-t}$.

We consider the case in which the two turning points are sufficiently far apart so that we can use a turning point of order $n = 1$. Now the solutions for the linear turning point are just the Airy functions or Bessel functions of order 1/3. These are

$$\Phi_1 = \left(\frac{\pi}{2}\right)^{1/2} A_{\pm}\left(\frac{\xi}{K}\right)^{1/2} J_{\pm 1/3}(\xi); \quad x \geq 0$$

$$\Phi_2 = \left(\frac{\pi}{2}\right)^{1/2} B_{\pm}\left(\frac{t}{|K|}\right)^{1/2} I_{\pm 1/3}(t); \quad x < 0$$

(4.143)

These expressions are just alternative ways of writing the same solutions. They must be identical. Continuity requirements at $x = 0$ give

$$B_+ = -A_-$$

$$B_- = A_-$$

(4.144)

Thus, the asymptotic forms can be written as

$$\phi_1^{as} = A_{\pm} K^{-1/2} \cos\left(\xi \mp \frac{\pi}{6} - \frac{\pi}{4}\right)$$

$$\phi_2^{as} = \mp \frac{A_{\pm}|K|^{-1/2}}{2} e^{-t - j\pi\left(\frac{1}{2} \mp \frac{2}{3}\right)}$$

(4.145)

To derive the first connection formula we follow the procedure

$$\phi_{2+}^{as} + \phi_{2-}^{as} + \phi_{1+}^{as} + \phi_{1-}^{as}$$

$$A_+ = A_-$$

(4.146)

$$\text{thus} \to |K|^{-1/2} e^{-t} \to 2|K|^{-1/2} \cos\left(\xi - \frac{\pi}{4}\right)$$

The arrow indicates that the asymptotic solution on the left goes into the expression on the right as one crosses the turning point. These arrows are irreversible, or a small error in the phase of the cosine would be magnified by the positive exponential on the LHS of Equation (4.146).

Similarly we can write another set of connection formulae:

$$|K|^{-1/2} e^{-t} \leftarrow 2K^{1/2} \cos\left(\xi + \frac{\pi}{4}\right) \tag{4.147}$$

These formulae suffice for applications involving two turning points.

4.6.5 APPLICATION OF CORRECTION FORMULAE

There are three distinct categories of problems encountered in wave propagation in slab dielectric waveguides where the connection formulae obtained above can be applied to yield eigenvalue equations for bound modes. In the WKB context, a bound mode corresponds to a solution that is oscillatory between two turning points beyond which the solution is evanescent. These are treated separately.

4.6.5.1 Ordinary Turning Point Problem

The refractive index profile is illustrated in Figure 4.29(a) where the turning points are at x and x_2. For $x_1 < x \le x_2$, $K^2 > 0$, and the field is oscillatory. Elsewhere, the field is evanescent. The connection formulae connect solutions on both sides of a turning point, and these must be applied at both x_1 and x_2. Correct phase matching of the oscillatory fields yields the WKB eigenvalue equation. This equation contains the implicit prescription of the required propagation constant.

Thus in regions 2_a and 2_b as noted in Figure 4.29,

$$\varphi_{2a} = A_a |K|^{-1/2} e^{-t_1}$$

$$\varphi_{2b} = A_b |K|^{-1/2} e^{-t_2}$$

$$\text{with} \quad t_1 = \int_{x_1}^{x} |K| dx; \quad t_2 = \int_{x_2}^{x} |K| dx \tag{4.148}$$

By virtue of the first connection formulae given by (4.146), the solution in region 1 that connects with region 2a is

$$\varphi_{1a} = 2A_a K^{-1/2} \cos\left(\xi_1 - \frac{\pi}{4}\right)$$

$$\text{with} \quad \xi_1 = \int_{x_1}^{x} K \, dx \tag{4.149}$$

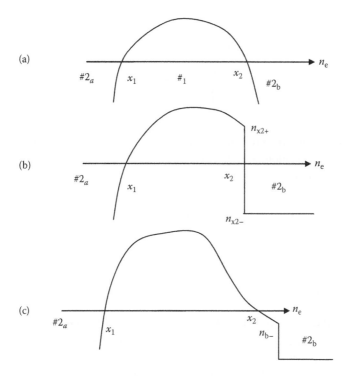

FIGURE 4.29 (a) Ordinary turning point with caustics at $x = x_1$ and x_2, where $n_2(x) = n_e$. (b) Step discontinuity at $x = x_2$. (c) Buried modes with mode index n_e close to step discontinuity at $x = b$.

And that which connects the solution in region 2b is

$$\varphi_{1b} = 2A_b K^{-1/2} \cos\left(\xi_2 - \frac{\pi}{4}\right)$$

(4.150)

$$\text{with} \quad \xi_2 = \int_x^{x_2} K \, dx$$

Now φ_{1a} and φ_{1h} represent one and only one solution in 1. Thus this consistency condition implies that

$$A_a = A_b$$

$$\xi_1 - \frac{\pi}{4} = \pm\left(\xi_2 - \frac{\pi}{4}\right)$$

(4.151)

The RHS of (4.151) can be written, taking the minus sign, as

$$-\left(\xi_2 - \frac{\pi}{4}\right) = \xi_1 - \frac{\pi}{4} - \eta$$

$$\eta = \int_{x_1}^{x_2} K\,dx - \frac{\pi}{2}$$

(4.152)

which must be a multiple of π to ensure correct phase matching of the WKB oscillatory solution. Equation (4.152) is expanded to give the familiar form of the WKB eigenvalue equation

$$\int_{x_1}^{x_2} [k_0^2 n^2(x) - \beta^2]^{1/2} = \left(m + \frac{1}{2}\right)\pi; \quad m = 0,1,2\ldots$$

(4.153)

This result fails when $x_1 = x_2$ or when there is an index discontinuity in the vicinity of the turning point.

4.6.5.2   Effect of an Index Discontinuity at a Turning Point

Figure 4.29(b) illustrates a typical example of a thin-film waveguide deposited on a substrate. The turning point at $x = x_2$ coincides with the film–air interface. At $x = x_2$, it is erroneous to apply the connection formulae, since they are derived on the assumption of a linear variation of $K^2(x_2)$. We resort, instead, to the boundary conditions imposed on the fields at $x = x_2$.

If we denote $n(x_{2-})$ and $n(x_{2+})$ as the values of the refractive index just before and after the step, then the standard phase shift at the discontinuity[60] is proportional to δ_{x_2} given by

$$\delta_{x_2} = \tan^{-1}\left(\frac{\beta^2 - k_0^2 n^2(x_{2+})}{-\beta^2 + k_0^2 n^2(x_{2-})}\right)^{1/2}$$

(4.154)

and the WKB eigenvalue equation transforms to

$$\int_{x_1}^{x_{2-}} [k_0^2 n^2(x) - \beta^2]^{1/2} = \left(m + \frac{1}{4}\right)\pi + \delta_{x_2}; \quad m = 0,1,2\ldots$$

(4.155)

Note that in the large step, $\delta_{x_2} \to \pi/2$, as in the case of a strongly asymmetrical waveguide.

4.6.5.3   Buried Modes Near an Index Discontinuity at a Turning Point

The analysis in the previous subsection treats only the turning point at $x = x_2$ fails directly on top of the index discontinuity. Thus strictly speaking, Equation (4.155) is only accurate for certain order modes (value of m) that satisfy this condition.

We denote the refractive index just before and after the step at $x = b$ by $n(b^-)$ and $n(b^+)$, respectively. In the region $x_2 < x < b$ the step at $x = b$ causes significant reflection of energy. It is no longer accurate to represent the fields in the region by a single decay exponential. Therefore, we include both decaying and growing exponentials and write

$$\varphi_1 = AK^{-1/2} \cos\left(\xi_1 - \frac{\pi}{4}\right); \quad x_1 < x < x_2$$

$$\varphi_{2a} = A|K|^{-1/2} \cos\left(\int_{x_1}^{x_2} K\, dx\right) e^{t_2} + \frac{1}{2} c \sin\left(\int_{x_1}^{x_2} K\, dx\right) e^{-t_2}; \quad x_2 < x < b \quad (4.156)$$

$$\varphi_{2b} = A|K|^{-1/2} e^{\left(\int_{xb}^{x} |K|\, dx\right)}; \quad x < b$$

where A is a constant. The coefficients of the growing and decaying fields in (4.156) have been chosen to satisfy the connection formulae at $x = x_2$. Thus, the equation of φ_1 in (4.156) is redundant as far as the eigenvalue equation is concerned. Continuity requirements on φ_{2a} and φ_{2b} at $x = b$ give the required eigenvalue equation:

$$\int_{x_1}^{x_{2-}} [k_0^2 n^2(x) - \beta^2]^{1/2} = \left(m + \frac{1}{4}\right)\pi + \delta_b; \quad m = 0,1,2\ldots \quad (4.157)$$

where

$$\delta_b = \tan^{-1}\left\{\left|\frac{|K(b^-)| + |K(b^+)|}{|K(b^-)| - |K(b^+)|}\right|\right\} e^{2\int_{x_2}^{b^-} |K|\, dx}$$

which is the corrected overall phase shift of the buried modes.

4.7 PROBLEMS

4.7.1 PROBLEM 1

A slab optical waveguide has a symmetrical refractive index based on pure silica as the substrate material. The core refractive index is 1.50 and its thickness is 4.00 μm. The cladding thickness is 20 μm. The refractive index difference is 0.09. The operating wavelength is 1300.0 nm.

   a. Is the operating wavelength in the UV, visible, near-infrared, infrared, or far-infrared region?

  b. Is the silica material less lossy at 1300 nm than at 1550 nm? Give reasons.

  c. Find the normalized frequency parameter V for the planar optical waveguide. Thence find the number of odd and even guided modes. If possible write a procedure in MATLAB to calculate the propagation constants of these guided modes.

  d. If the cladding refractive index is 1.515, would the optical waveguide support any guided mode at 1300 nm wavelength? If it does, find the number of guided modes for this structure. Sketch the field and intensity distribution of these modes across the waveguide cross section.

  e. Using the refractive index profile as in (d), design the geometrical structure of the waveguide so that it can support only one TE-even mode.

4.7.2   Problem 2

Assuming that the refractive indices at 1550 nm wavelength are the same, repeat problem 1 with an operating optical wavelength of 1550 nm.

4.7.3   Problem 3

The structure of an optical planar waveguide splitting junction Y is shown in Figure 4.26. The requirement is that the output of optical fields at the output ports of the Y-junction must be a single even TE mode. Using silica as the substrate material wih a refractive index of 1.500 at 1530 nm wavelength in vacuum, design the planar optical waveguide sections of the Y-junction. Assume that the splitting tilted junction area would support the same guided modes as that of the output straight branches.

4.7.4   Problem 4

  a. A slab optical waveguide has a symmetrical step refractive index profile based on pure silica as the cladding material with a refractive index of 1.480 at 1550 nm free-space wavelength.

    i. Write Maxwell's equations for lightwaves propagating in this medium. The wave equation for the electric field component of the electromagnetic lightwaves is given by

$$\nabla_t^2 E + (\beta^2 - n^2(\omega)k_0^2)E = 0$$

    where

$$\nabla_t^2 = \frac{\partial^2}{\partial x^2} + \frac{\partial^2}{\partial y^2}$$

    ii. For a planar optical waveguide with an infinite extension in the y-direction, show that the scalar wave equation for the TE modes is given by

$$\frac{d^2 E_y}{dx^2} + (k^2 n_j^2 - \beta^2) E_y = 0$$

where n_j ($j = 1$ or 2) represents the refractive index in either the core or the cladding regions, $k = 2\pi/\lambda$ is the free-space wavenumber, λ is the free-space wavelength, and β is the propagation constant of the light-waves along the z-direction.

b. Write the wave solutions for both the even and odd TE-guided modes in the core and cladding regions. The corresponding two eigenvalue equations of these guided modes are given by

$$v = u \tan u \qquad \text{for even TE-guided modes}$$

$$v = -\frac{u}{\tan u} \qquad \text{for odd TE-guided modes}$$

where u and v are defined by

$$u^2 = a^2(k^2 n_1^2 - \beta^2)$$

$$v^2 = a^2(-k^2 n_2^2 + \beta^2)$$

Prove just one of the above eigenvalue equations (either one of your choice). Give a physical interpretation of u and v.

 i. Obtain an expression for the normalized frequency V-parameter, and design a slab optical waveguide so that it can guide two TE optical-guided odd modes at an operating free-space wavelength of 1550 nm. It is recommended that the following parameters of the optical waveguide be specified: the slab core thickness, cladding thickness, relative refractive index difference between the core and cladding regions, and cutoff wavelength of TE modes of a order higher than the two TE odd modes.

 ii. Sketch the electric fields and intensity distributions of the guided modes in the transverse plane of the designed waveguide in part (i).

REFERENCES

1. S.E. Miller, "Integrated Optics: An Introduction," *Bell Syst. Technol. J.*, 48, 2059–2069, 1969.
2. C.R. Doerr and K. Okamoto, "Advances in Silica Planar Lightwave Circuits," *IEEE J. Lightw. Technol.*, 24(12), 4763–4770, 2006.
3. P.S. Chung, "Waveguide Modes, Coupling Techniques, Fabrication and Losses in Optical Integrated Optics," *J. Elect. Electron. Australia*, 5, 201–214, 1985.
4. http://en.wikipedia.org/wiki/Photonic_integrated_circuit (accessed December 2009); http://electron9.phys.utk.edu/optics421/modules/m10/integrated_optics.htm.
5. R.G. Hunsperger, "Integrated Optics Theory and Technology," in *Advanced Texts in Physics*, 6th ed., XXVIII, Springer, Berlin, 2009.

6. L.N. Binh and S.V. Chung, "Nonlinear Interactions in Thin Film Structures," at Proceedings of 8th Australian Workshop on Optical Communications, Adelaide, 1983, Session VII.

7. K. Uchiyama and T. Morioka, "All-Optical Time-Division Demultiplexing Experiment with Simultaneous Output of All Constituent Channels from 100Gbit/s OTDM Signal," *Electron. Lett.*, 37(10), 2001.

8. M. Nakazawa, T. Yamamoto, and K.R. Tamura, "1.28Tbit/s–70km OTDM Transmission Using Third- and Fourth-Order Simultaneous Dispersion Compensation with a Phase Modulator," *Electron. Lett.*, 36(24), 2000.

9. C. Schubert, R.H. Derksen, M. Möller, R. Ludwig, C.-J. Weiske, J. Lutz, S. Ferber, A. Kirstädter, G. Lehmann, and C. Schmidt-Langhorst. "Integrated 100-Gb/s ETDM Receiver," *IEEE J. Lightw. Technol.*, 25(1), 122–129, 2007.

10. H. Suzuki, M. Fujiwara, and K. Iwatsuki, "Application of Super-DWDM Technologies to Terrestrial Terabit Transmission Systems," *IEEE J. Lightw. Technol.*, 24(5), 1998–2005, 2006.

11. J.P. Turkiewicz, E. Tangdiongga, G. Lehmann, H. Rohde, W. Schairer, Y.R. Zhou, E.S.R. Sikora, A. Lord, D.B. Payne, G.-D. Khoe, and H. de Waardt, "160 Gb/s OTDM Networking Using Deployed Fiber," *IEEE J. Lightw. Technol.*, 23(1), 225–234, 2005.

12. R. Nagarajan, M. Kato, V.G. Dominic, C.H. Joyner, R.P. Schneider Jr., A.G. Dentai, T. Desikan, P.W. Evans, M. Kauffman, D.J.H. Lambert, S.K. Mathis, A. Mathur, M.L. Mitchell, M.J. Missey, S. Murthy, A.C. Nilsson, F.H. Peters, J.L. Pleumeekers, R.A. Salvatore, R.B. Taylor, M.F. Van Leeuwen, J. Webjorn, M. Ziari, S.G. Grubb, D. Perkins, M. Reffle, D.G. Mehuys, F.A. Kish, and D.F. Welch, "400 Gbit/s (10 Channel ´40 Gbit/s) DWDM Photonic Integrated Circuits," *Electronics Lett.*, 41(6), 2005.

13. T. Okoshi, "Recent Advances in Coherent Optical Fiber Communications," *IEEE J. Lightw. Technol.*, LT-5, 44–52, 1987.

14. L.G. Karzovky and O.K. Tonguz, "ASK and FSK Coherent Lightwave Systems: A Simplified Approximate Analysis," *IEEE J. Lightw. Technol.*, 8(3), 338–351, 1990.

15. S. Tsukamoto, D.-S. Ly-Gagnon, K. Katoh, and K. Kikuchi, "Coherent Demodulation of 40-Gbit/s Polarization-Multiplexed QPSK Signals with 16-GHz Spacing after 200-km Transmission," Paper PDP29 presented at Proceedings of CLEOS 2005, San Jose, 2005.

16. D.-S. Ly-Gagnon, S. Tsukamoto, K. Katoh, and K. Kikuchi, "Coherent Detection of Optical Quadrature Phase-Shift Keying Signals with Carrier Phase Estimation," *IEEE J. Lightw. Technol.*, 24(1), 12–21, 2006.

17. V. Ramaswamy and R.K. Lagu, "Numerical Field Solution for an Arbitrary Asymmetrical Gradient Index Planar Waveguide," *IEEE J. Lightw. Technol.*, LT-1, 408–417, 1983.

18. R.V. Schmidt and I.P. Kaminow, "Metal-Diffused Optical Waveguides in LiNbO3," *Appl. Phys. Lett.*, 25, 458–460, 1974.

19. W. Streifer, R.D. Burnham, and D.R. Scifres, "Modal Analysis of Separate-Confinement Heterojunction Lasers with Inhomegeneous Cladding Layers," *Opt. Lett.*, 8, 283–285, 1981.

20. D. Marcuse, *Light Transmission Optics*, Van Nostrand, New York, 1972, Chapter 8.

21. E.M. Conwell, "Modes in Optical Waveguides Formed by Diffusion," *Appl. Phys. Lett.*, 26, 328–329, 1973.

22. H. Kirchoff, "The Solution of Maxwell Equations for Inhomogeneous Dielectric Slabs," *AEU*, 26, 537–541, 1972.

23. D. Marcuse, "TE Modes of Graded Index Slab Waveguides," *IEEE J. Quant. Electron.*, QE-9, 1000–1006, 1973.

24. A.W. Snyder and J.D. Love, *Optical Waveguide Theory*, Chapman & Hall, London, 1983, Chapter 12.

25. T.R. Chen and Z.L. Yang, "Modes of a Planar Waveguide with Fermi Index Profile," *Appl. Opt.*, 24, 2809–2812, 1985.

26. H.F. Taylor, "Dispersion Characteristics of Diffused Channel Waveguides," *IEEE J. Quant Electron.*, QE-12, 748–752, 1976.

27. G.B. Hocker and W.K. Burns, "Mode Dispersion in Diffused Channel Waveguides by the Effective Index Method," *Appl. Opt.*, 16, 113–118, 1975.

28. L. Riviere, A. Yi-Yan, and H. Carru, "Properties of Single Mode Planar with Gaussian Index Profile," *IEEE J. Lightw. Technol.*, LT-3, 368–377, 1985.

29. S.K. Korotky et al., "Mode Size and Method for Estimating the Propagation Constant of Single Mode Ti:LiNbÒ Strip Waveguides," *IEEE Trans. Microw. Th. Technol.*, MTT-30, 1784–1789, 1982.

30. A.N. Kaul, S.L. Hossain, and K. Thyagarajan, "A Simple Numerical Method for Studying the Propagation Characteristics of Single-Mode Graded-Index Planar Optical Waveguides," *IEEE Trans. Microw. Th. Technol.*, MTT-34, 288–292, 1986.

31. L.N. Binh et al., "Design Considerations for Second Harmonic Generation in Thin Film Grown by Ion Beam Assisted Deposition Method," at Proceedings of 3rd Laser Conference, Melbourne, 1983, Session 10A.

32. J. Killingbeck, "A Pocket Calculator Determination of Energy Eigenvalues," *J. Phys. A*, 10, L09–L103, 1977.

33. R.A. Sammut and C. Pask, "Simplified Numerical Analysis of Optical Fibers and Planar Waveguides," *Electron. Lett.*, 17, 105–106, 1981.

34. J.D. Love and A.K. Ghatak, "Exact Solutions for TM Modes in Graded Index Slab Waveguides," *IEEE J. Quant. Electron.*, QE-15, 14–16, 1979.

35. D. Marcuse, "The Effects of the Ñ2n Term on the Modes of an Optical Square-Law Medium," *IEEE J. Quant. Electron.*, QE-9, 958–960, 1973.

36. H. Kolgenik and V. Ramaswamy, "Scaling Rule for Thin Film Optical Waveguides," *Appl. Opt.*, 13, 1857–1862, 1974.

37. P.K. Tien, "Integrated Optics and New Wave Phenomena," *Rev. Mod. Phys.*, 49, 361–420, 1977.

38. A. Watson, "Physics—Where the Action Is," *New Scientist*, January 30, 1986, pp. 42–44.

39. L. Mammel and L.G. Cohen, "Numerical Prediction of Fiber Transmission Characteristics from Arbitrary Refractive Index Profiles," *Appl. Opt.*, 21, 699–703, 1982.

40. K. Petermann, "Constraints for Fundamental Mode-Size for Broadband Dispersion-Compensated Single Mode Fibers," *Electron. Lett.*, 19, 712–714, 1983.

41. P. Sansonetti, "Modal Dispersion in Single Mode-Fibers: Simple Approximation Issued from Mode Spot Size Spectral Behavior," *Electron. Lett.*, 18, 647–648, 1982.

42. R.J. Black and C. Pask, "Slab Waveguides Characteristics by Moments of Refractive Index Profiles," *IEEE J. Quant. Electron.*, QE-20, 996–999, 1984.

43. S. Ruschin, "Approximate Formula for the Propagation Constant of the Basic Mode in Slab Waveguides of Arbitrary Index Profiles," *Appl. Opt.*, 24, 4189–4191, 1985.

44. R. Kiel and F. Auracher, "Coupling of Single-Mode Ti:Diffused LiNbO3 Waveguides to Single Mode Fibers," *Opt. Commun.*, 30, 23–28, 1979.

45. W.A. Gambling and H. Matsumara, "Propagation in Radially Inhomogeneous Single-Mode Fiber," *Opt. Quant. Electron.*, 10, 31–40, 1978.

46. E. Desurvire, *Erbium-Doped Fiber Amplifiers: Principles and Applications*, Wiley Series in Telecommunications and Signal Processing, J. Wiley, New York, September 2001.

47. E. Desurvire, J. Simpson, and P.C. Becker, "High-Gain Erbium-Doped Traveling-Wave Fiber Amplifier," *Opt. Lett.*, 12(11), 888–890, 1987.

48. J. Jeffery, "On Certain Approximate Solutions of Linear Differential Equations of Second Order," *Proc. London Math. Soc.*, 23, 428–436, 1923.

49. G. Wentzel, "Eine Verrallgemeirneung der Quanttenbedingungen fur die Zwecke der Wellemechanik," *Z. Phys.*, 38, 518–529, 1926.

50. H.A. Kramers, "Wellenmechanik und halbzahlig Quantisierung," *Z. Phys.*, 39, 828–840, 1926.
51. L. Brillouin, "Remarque's sur la mechanique ondulatoire," *J. Phys.*, 7, 353–368, 1926.
52. A. Ankiewicz, "Comparison of Wave and Ray Techniques for Solution of Graded Index Optical Waveguide Problems," *Opt. Acta*, 25, 361–373, 1978.
53. R.E. Langer, "On the Connection Formulae and the Solutions of the Wave Equation," *Phys. Rev.*, 51, 669–676, 1937.
54. J.M. Arnold, "Asymptotic Analysis of Planar and Cylindrical Inhomogeneous Waveguides," *Radio Sci.*, 16, 511–518, 1981.
55. J.P. Gordon, "Optics of General Guiding Media," *Bell Syst. Technol. J.*, 45, 321–332, 1966.
56. P.K. Tien and R. Ulrich, "Theory of Prism-Coupler and Thin Film Lightguides," *J. Opt. Soc. Am.*, 60, 1325–1337, 1966.
57. I.S. Gradshteyn and I.M. Ryzhik, *Tables of Integrals, Series, and Products,* Academic Press, New York, 1965.
58. P.K. Mitra and A. Sharma, "Analysis of Single Mode Inhomogeneous Planar Waveguides," *IEEE J. Lightw. Technol.*, LT-4, 204–212, 1986.
59. A.W. Snyder and R.A. Sammut, "Fundamental (HE11) modes of graded optical fibers," *J. Opt. Soc. Am.,* 69(12), 1663–1671, 1979.
60. M.J. Adams, *An Introduction to Optical Waveguides*, J. Wiley, New York, 1981.
61. M. Abramowitz and I.A. Stegun, *Handbook of Mathematical Functions*, Dover Publications, New York, 1972.
62. J.L. Dunham, *Phys. Rev.*, 41, 713, 1932.

5 Three-Dimensional Optical Waveguides

In modern optical communication systems, almost all optical devices, from passive to active components, such as lasers and waveguides, are in three-dimensional (3D) form. They ensure that the lightwaves are effectively trapped in planar structures, i.e., optical integrated circuit form, as shown in Figure 5.1, and thence manipulation of the lightwaves properties occurs by splitting, coupling, diffraction, and modulation by electro-optic, acousto-optic, or magneto-optic effects.

Therefore, following the fundamentals of planar optical waveguides, this chapter describes the 3D optical waveguides in which the waveguide is restricted in the transverse (x, y) plane, which is perpendicular to the propagation direction z-axis. A simplified analysis of these waveguides, the effective index method, and numerical techniques, the finite difference method (FDM), is described and examples are given. In this chapter we analyze the modes guided by 3D waveguides with rectangular geometries using mainly Marcatilli's method, and the effective index method as analytical techniques.

On the numerical method we select the FDM as the principal technique because it is simple and gives accurate results for optical waveguides operating in the linear region. We chose the FDM to study the quasi-TE- and quasi-TM-polarized waveguide modes due to its simplicity and plausible accuracy. We have employed the semivectorial analysis, which automatically takes full account of the discontinuities in the normal electric field components across any arbitrary distribution of internal dielectric interfaces. The eigenmodes of the Helmholz equation are solved by the application of the shifted inverse power iteration method. This method warrants the mode size and its relevant propagation constant, both of which are important parameters to the design of an optical waveguide. The grid size is nonuniform to maximize the accuracy of the optical guided modes and their propagation constants. Diffused waveguides and rib waveguides are designed with different parameters to demonstrate the effectiveness of the method, leading to an optimum design of waveguides of optical modulation and micro-ring resonators.

5.1 INTRODUCTION

To achieve efficient design of high-speed modulators and switches, especially micro-ring resonators, the fabrication of rib waveguides and $Ti:NbO_3$ waveguides with suitable mode sizes is essential to minimize waveguide insertion loss and also to maximize the overlap integral between the guided optical field and the applied modulating field. Furthermore, the bending or radius of curvature is so important for the ring resonator to keep the ring size as small as possible. Extensive studies have

FIGURE 5.1   Laser structure with optical waveguide as the gain and guided medium and electrodes placed at its two sides.

been devoted in recent decades to fabricating Ti:diffused LiNbO$_3$ waveguides that couple efficiently to single-mode fibers.[1,2–5] A major milestone was achieved when a total fiber–waveguide–fiber insertion loss of 1 dB was achieved for z-cut LiNbO$_3$ at 1.3 μm.[4] Such low loss was achieved by choosing fabrication parameters to yield a relatively deep, clean diffusion, which simultaneously minimized the fiber wave-guide mode mismatch loss and the propagation loss. Suchoski and Ramaswamy[6] have reported on the optimization of fabrication parameters to obtain Ti:LiNbO$_3$ single-mode waveguides that exhibit both minimum mode size and low propagation loss in the 1300 nm wavelength region. All these design requirements have led to the significance of the analysis of polarized modes in channel waveguides.

In general, the optical mode of the waveguide is acquired by solving the Helmholtz equation. However, only a few simple waveguide structures can be solved analytically. Extensive attempts have been made to obtain numerical solutions for a two-dimen-sional cross section of optical waveguides.[7–23] One method is the approximate mod-eling of a two-dimensional slab waveguide solution successively in both directions, following either the method of Marcatilli[5] or the effective index method (EIM).[24] However, these methods are not applicable to arbitrarily shaped optical waveguides, and neither do they handle waveguide mode near the cutoff region efficiently. A signif-icant number of numerical methods have been proposed to obtain rigorous solutions to the wave equation with pertinent boundary conditions. The popular techniques by far are the finite difference method (FDM),[10] finite element method (FEM),[22] and beam propagating method (BPM).[14] The application of different techniques based on the above methods, such as semivectorial E-field FDM,[12] semivectorial H-field FDM,[25] and Rayleigh quotient solution,[26] has been studied and reported. These meth-ods are applicable to arbitrarily shaped optical waveguides. In FEM and FDM, par-tial differential equations are discretized and then transformed to matrix equations. The calculations of mode indices and optical field distributions are then equivalent to obtaining eigenvalues and eigenfunctions of the coefficient matrices.

In this chapter, we first treat the 3D optical waveguide from an analytical point of view with the representation of a two distribution of the refractive index profile by two effective planar profiles. The propagation and mode guiding conditions obtained for these planar waveguides are then combined for the 3D waveguide.

Sections 5.2 and 5.3 describe the analytical estimation of a guided mode using Marcatilli's method and the effective index method. Section 5.4 then outlines the numerical formulation of the nonuniform finite difference scheme. Both quasi-TE- and quasi-TM-polarized modes are addressed. We also assess the accuracy of the numerical result of this scheme by computing the effective refractive index of rib and slab dielectric waveguides. The effect of grid spacing is also investigated. The effectiveness of the variable grid spacing in dealing with a waveguide mode near the cut-off region is also given. Sections 5.4 and 5.5 then give the treatment of the 3D optical waveguides by the FDM for uniform index regions, the rib waveguide, diffused index profiles, and diffused optical channel waveguides. Section 5.4 describes the modeling of a 3D optical waveguide with a graded index profile such as the Ti:LiNbO$_3$ channel waveguides. The effects of various waveguide fabrication parameters, such as the diffusion time, diffusion temperature, and thickness and width of the titanium strips, are studied. The accuracy of the numerical model is assessed by comparing our simulations with experimental and simulation results that are reported in literature. Section 5.5 describes the modeling of rib optical waveguides using the same finite difference method.

5.2 MARCATILLI'S METHOD

The cross section of typical 3D waveguides is shown in Figure 5.2, including raised strip or channel waveguide, strip loaded, rib or ridge, and embedded structures with a substrate and an overlay region.

Usually the raised channel waveguide is formed by depositing a thin-film layer, e.g., by molecular chemical vapor deposition (MOCVD) or by sputtering; then if we remove the film material in the outer regions by some means such as dry reactive etching, while keeping the film layer in the central portion intact, we have the raised strip or channel waveguides. The ridge or rib waveguides are similar to the raised strip waveguides except that the film layer on the two sides is partially removed, as shown in Figure 5.2(b). If we place a dielectric strip on the top of the film layer, as shown schematically in Figure 5.2(c), we have the strip-loaded waveguides. By embedding a high-index bar in the substrate region, we have the buried or embedded strip waveguides shown in Figure 5.2(d). Channel, ridge, strip-loaded, and buried strip waveguides are 3D waveguides with rectangular boundaries. Circular and elliptical fibers, 3D waveguides with curved boundaries, are discussed in Chapters 4 and 5. The refractive index of the 3D waveguide can vary with respect to the distance of depth. In this case we have a graded index channel waveguide, such as diffused channel optical waveguides formed by diffusion of impurity into LiNbO$_3$ substrate at a temperature around 1000°C.

In this section, we analyze the modes that are guided by 3D waveguides with rectangular geometries using mainly Marcatilli's method and the effective index method. The chapter consists of five sections. Since fields of 3D waveguides are complicated and difficult to analyze, we begin with a qualitative description.

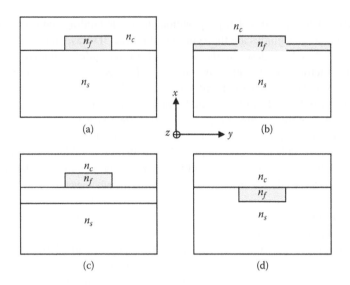

FIGURE 5.2 Channel 3D optical waveguides: (a) raised strip or channel, (b) ridge or rib, (c) strip loaded, and (d) embedded channel.

5.2.1 FIELD AND MODES GUIDED IN RECTANGULAR OPTICAL WAVEGUIDES

5.2.1.1 Mode Fields of H_x Modes

In 2D waveguides, one of the dimensions transverse to the direction of propagation is very large in comparison to the operating wavelength. This is the y-direction in Figure 5.1. The waveguide width in this direction is treated as infinitely large. As a result, fields guided by 2D dielectric waveguides can be classified as transverse electric (TE) or transverse magnetic (TM) modes, as discussed in Chapter 3. For TE modes, the longitudinal electric field component, E_z, is zero, and all other field components can be expressed in terms of H_z. For TM modes, H_z vanishes and all other field components can be expressed in terms of E_z. In 3D optical waveguides, the waveguide width and height are comparable to the operating wavelength. Neither the width nor height can be treated as infinitely large. Thus, neither E_z, H_z vanishes, except for some special cases. As a result, modes guided by 3D optical waveguides are neither TE nor TM modes, except for the special cases. In general, they are hybrid modes. A complicated scheme is needed to designate the hybrid modes. Since all field components are present, the analysis for hybrid modes is very complicated. Intensive numerical computations are often required.[27]

In many dielectric waveguide structures, the index difference is small. As a result, one of the transverse electric field components is much stronger than the other transverse electric field component. Goell has suggested a physically intuitive scheme to describe hybrid modes.[28] In Goell's scheme, a hybrid mode is labeled by the direction and distribution of the strong transverse electric field component. If the dominant electric field component is in the x- (or y-) direction, and if the electric field distribution has $p - 1$ nulls in the x-direction and $q - 1$ nulls in the y-direction, then the hybrid mode is identified as $E_{x,pq}$ (or $E_{y,pq}$) modes. The superscript denotes the direction of the *dominant transverse electric*.

Now consider a weakly guiding rectangular optical waveguide with a core of index n_1 and surrounded with lower indices n_j with $j = 2, 3, 4,$ and 5. The waveguide cross section is shown in Figure 5.2.

The rectangular waveguide can be considered to be equivalent to two slab waveguides, one extended in the x-direction and one in the y-direction. That means that the field is confined as a mode in the y-direction and the other in the x-direction. This is normally called the hybrid mode. Thus we can write the field component H_x in the five regions as portioned in Figure 5.2 as follows:

$$H_{x1} = C_1 \cos(\kappa_{x1}x + \phi_{x1})\cos(\kappa_{y1}y + \phi_{y1})e^{-j\beta z}; \quad \text{region 1}$$

$$H_{x2} = C_2 \cos(\kappa_{x2}x + \phi_{x2})e^{-j\kappa_{y2}y}e^{-j\beta z}; \qquad\qquad \text{region 2}$$

$$H_{x3} = C_3 e^{-j\kappa_{x3}x}\cos(\kappa_{y3}y + \phi_{y3})e^{-j\beta z}; \qquad\qquad \text{region 3} \qquad (5.1)$$

$$H_{x4} = C_4 e^{-j\kappa_{y4}y}\cos(\kappa_{x4}x + \phi_{x4})e^{-j\beta z}; \qquad\qquad \text{region 4}$$

$$H_{x5} = C_5 e^{-j\kappa_{x5}x}\cos(\kappa_{y5}y + \phi_{y5})e^{-j\beta z}; \qquad\qquad \text{region 5}$$

where C_j, ϕ_{xj}, ϕ_{yj} are the constants to be determined using the boundary conditions, κ_{xj}, κ_{yj} and are the propagation constants effective in the x and y transverse directions, respectively. For each region the propagation constants in the x-, y-, and z-directions, κ_{xj}, κ_{yj}, β, must satisfy

$$\kappa_{xj}^2 + \kappa_{yj}^2 + \beta^2 = k^2 n_j^2; \qquad j = 1,2,3,4,5 \qquad (5.2)$$

There are no additional constraints on the transverse propagation constant. In fact, the transverse propagation constants in regions 2–5 are imaginary; that is, the fields must decay to zero in these regions but in the rectangular core.

When expressed in terms of κ_{xj}, κ_{yj}, ϕ_x, ϕ_y, (5.1) can be simplified to

$$H_{x1} = C_1 \cos(\kappa_{x1}x + \phi_x)\cos(\kappa_{y1}y + \phi_y)e^{-j\beta z}; \quad \text{region 1}$$

$$H_{x2} = C_2 \cos(\kappa_{x2}x + \phi_x)e^{-j\kappa_{y2}y}e^{-j\beta z}; \qquad\qquad \text{region 2}$$

$$H_{x3} = C_3 e^{-j\kappa_x x}\cos(\kappa_y y + \phi_y)e^{-j\beta z}; \qquad\qquad \text{region 3} \qquad (5.3)$$

$$H_{x4} = C_4 e^{-j\kappa_y y}\cos(\kappa_x x + \phi_x)e^{-j\beta z}; \qquad\qquad \text{region 4}$$

$$H_{x5} = C_5 e^{-j\kappa_x x}\cos(\kappa_y y + \phi_y)e^{-j\beta z}; \qquad\qquad \text{region 5}$$

5.2.1.2   Boundary Conditions at the Interfaces

5.2.1.2.1   Horizontal Boundary $y = \pm h/2$; $|x| < w/2$

Along the horizontal boundaries, the tangential components are E_x, E_z, H_x, H_z; the x-components are ignored, as their amplitudes are extremely small compared with other components. Using Maxwell's equation we can observe that

- E_z is continuous at the boundary and tangential, implying that

$$\frac{1}{n_j^2} \frac{\partial H_x}{\partial y}$$

- Being tangential to the horizontal lines, H_x must be continuous everywhere along the horizontal lines. The tangential derivative $\partial H_x / \partial c$, and therefore H_z, must also be continuous on the horizontal lines. In other words, if H_x is continuous at the horizontal lines, so is H_z.

Thus all the boundary conditions are met if we have continuity of the term

$$\frac{1}{n_j^2} \frac{\partial H_x}{\partial y}$$

5.2.1.2.2   Vertical Boundary $x = \pm w/2$; $|y| < h/2$

Along this boundary we have the tangential components in the y- and z-directions, and the normal direction is x. Only the components E_y, H_x are significant, and E_y is continuous if H_x is continuous. Applying these conditions for the field components at $x = \pm w/2$ we obtain

$$E_{z1} - E_{z3} = \frac{j\eta_0}{k} \left(\frac{1}{n_1^2} \frac{\partial H_{x1}}{\partial y} - \frac{1}{n_3^2} \frac{\partial H_{x3}}{\partial y} \right) + O(\partial^2)$$

$$= \frac{j\eta_0}{k} \frac{1}{n_1^2} \left(\frac{\partial(H_{x1} - H_{x3})}{\partial y} \right) - \frac{j\eta_0}{n_3} \frac{n_1^2 - n_3^2}{n_1^2} \frac{1}{kn_3} \frac{\partial(H_{x3})}{\partial y}$$

(5.4)

The second term of (5.4) can be ignored due to the very small difference in the refractive index terms. Thus it can be written as

$$E_{z1} - E_{z3} = \frac{j\eta_0}{k} \frac{1}{n_1^2} \left(\frac{\partial(H_{x1} - H_{x3})}{\partial y} \right) + O(\partial^2)$$

(5.5)

In other words the component E_z is continuous if H_x is continuous there.

5.2.1.2.3 *Transverse Vector* κ_x, κ_y

The transverse momentum vector κ_x can now be determined from the boundary conditions discussed above. One would seek an oscillating behavior of the waves in the waveguide region and exponentially decay to zero in the cladding regions. At $y = \pm h/2$ the continuity of H_x and

$$\frac{1}{n_j^2}\frac{\partial H_x}{\partial y}$$

leads to

$$C_1 \cos\left(\frac{1}{2}\kappa_y h + \phi_y\right) = C_2 e^{-j\kappa_{y2}h/2}$$

$$-\frac{\kappa_y}{n_1^2} C_1 \sin\left(\frac{1}{2}\kappa_y h + \phi_y\right) = -\frac{j\kappa_{y2}}{n_2^2} C_2 e^{-j\kappa_{y2}h/2}$$

(5.6)

Combining these equations we obtain the relation

$$\tan\left(\frac{1}{2}\kappa_y h + \phi_y\right) = -\frac{j\kappa_{y2}n_1^2}{\kappa_y n_2^2}$$

(5.7)

From (5.2) we can deduce that

$$j\kappa_{y2} = \sqrt{k^2(n_1^2 - n_2^2) - \kappa_y^2}$$

(5.8)

Thus (5.7) becomes

$$\tan\left(\frac{1}{2}\kappa_y h + \phi_y\right) = \frac{\sqrt{k(n_1^2 - n_2^2) - \kappa_y^2}}{\kappa_y n_2^2}$$

(5.9)

Or alternatively we have

$$\frac{1}{2}\kappa_y h + \phi_y = q'\pi + \tan^{-1}\left(\frac{\sqrt{k^2(n_1^2 - n_2^2) - \kappa_y^2}}{\kappa_y n_2^2}\right)$$

$$\frac{1}{2}\kappa_y h + \phi_y = q''\pi + \tan^{-1}\left(\frac{\sqrt{k^2(n_1^2 - n_4^2) - \kappa_y^2}}{\kappa_y n_4^2}\right); \quad \text{at } y = -h/2$$

(5.10)

where q', q'', and q are integers. Then eliminating ϕ_y we can rewrite as

$$\kappa_y h_y = q\pi + \tan^{-1}\left(\frac{\sqrt{k^2(n_1^2 - n_2^2) - \kappa_y^2}}{\kappa_y n_2^2}\right) + \tan^{-1}\left(\frac{\sqrt{k^2(n_1^2 - n_4^2) - \kappa_y^2}}{\kappa_y n_4^2}\right) \quad (5.11)$$

This is the dispersion relation for the TM modes guided in the channel waveguide, and it is also similar to that for a planar waveguide. The two terms on the right-hand side (RHS) of (5.11) represent the phase shift, normally called the Goos–Hanchen shift for the "rays" penetrating into the cladding of the guided fields. Similar to this boundary condition and the dispersion relationship, the dispersion characteristics for the transverse vector can be written as

$$\kappa_x w = p\pi + \tan^{-1}\left(\frac{\sqrt{k^2(n_1^2 - n_3^2) - \kappa_x^2}}{\kappa_x}\right) + \tan^{-1}\left(\frac{\sqrt{k^2(n_1^2 - n_5^2) - \kappa_x^2}}{\kappa_x}\right) \quad (5.12)$$

with p as an integer.

5.2.1.3 Mode Fields of E_y Modes

Similar to the analysis given for the H_x modes, the E_x modes can be found with the dispersion relation by using the continuity properties of the field components H_x; $\partial H_y/\partial y$. We then obtain

$$\kappa_y h = q\pi + \tan^{-1}\left(\frac{\sqrt{k^2(n_1^2 - n_2^2) - \kappa_y^2}}{\kappa_y}\right) + \tan^{-1}\left(\frac{\sqrt{k^2(n_1^2 - n_4^2) - \kappa_y^2}}{\kappa_y}\right) \quad (5.13)$$

$$\kappa_x w = p\pi + \tan^{-1}\left(\frac{n_1^2\sqrt{k^2(n_1^2 - n_3^2) - \kappa_x^2}}{\kappa_x n_3^2}\right) + \tan^{-1}\left(\frac{n_1^2\sqrt{k^2(n_1^2 - n_5^2) - \kappa_x^2}}{\kappa_x n_5^2}\right) \quad (5.14)$$

Equations (5.13) and (5.14) specify the dispersion relationship for the TM modes with a planar waveguide thickness of W.

Thus, Marcatilli's method is modeled for two equivalent planar waveguides in the horizontal and vertical directions. It corresponds to the dispersion relations (5.11) and (5.12) for TM modes guided by a planar waveguide of thickness W. The dominant electric field of E_x modes is in parallel with the horizontal boundaries. Thus we use the dispersion equation of TE modes guided by waveguide H to determine κ_y. The dominant electric field component of E_x modes is perpendicular to the vertical boundaries of waveguide W. Therefore we use the dispersion for TM modes guided by the 2D waveguide to evaluate κ_x. With κ_x, κ_y known, the propagation constant can be determined from (5.2).

5.2.2 Dispersion Characteristics

As an example, we consider a dielectric bar of index n_1 immersed in a medium with index n_2, as shown in Figure 5.3, with uniform refractive indices in the regions

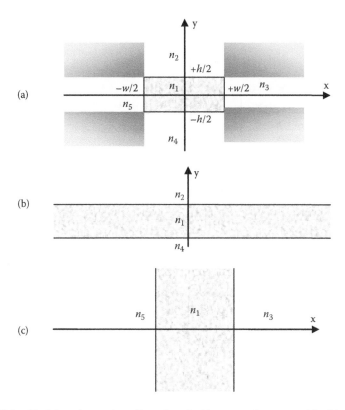

FIGURE 5.3 Model used to analyze E_y modes of a (a) rectangular waveguide, (b) waveguide H, and (c) waveguide W.

surrounding the channel waveguiding region. To facilitate comparison, we define the normalized frequency parameter V and the normalized guide index b, or normalized propagation constant in terms of n_1, n_2, h:

$$V = kh\sqrt{n_1^2 - n_2^2} \simeq \frac{2\pi}{\lambda} hn\sqrt{2\Delta}$$

$$b = \frac{\beta^2 - k^2 n_2^2}{k^2(n_1^2 - n_2^2)}$$

(5.15)

The normalized effective refractive index can be evaluated as a function of the normalized frequency parameter V to give the dispersion curves as shown in Figure 5.4, in which the curves obtained from the finite element method and the Marcatilli methods are also contrasted with agreement.

A numerical evaluation for silica doped with a GeO_2 waveguide and cladding region is pure silica. The relative refractive index of the core and the pure silica cladding is 0.3 or 0.5%; then using the single-mode operation given in Figure 5.4, we can select $V = 1$, and using (5.15), the cross section of the rectangular waveguide is 3 ×

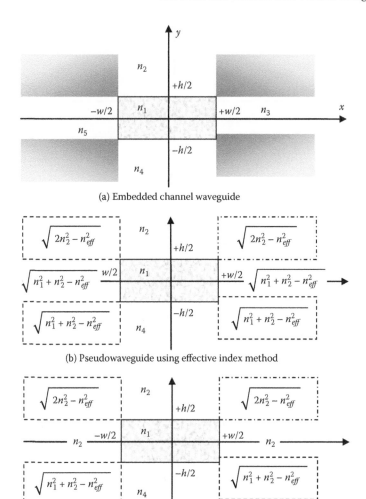

(a) Embedded channel waveguide

(b) Pseudowaveguide using effective index method

(c) Equivalent waveguide using Marcatilli method

FIGURE 5.4   An embedded channel optical waveguide: (a) waveguide structure and its representation using (b) the effective index method and (c) the Marcatilli method.

3 μm² for 0.5% relative refractive index, and for 0.3% the dimension is 6 × 6 micron². The refractive index of pure silica is 1.448 for an operating wavelength of 1550 nm.

5.3   EFFECTIVE INDEX METHOD

5.3.1   General Considerations

Similar to the Marcatilli method discussed in the previous section, the effective index method is also an approximate method for analyzing rectangular waveguides. In the Marcatilli method, a 3D waveguide (see Figure 5.1) is replaced by two 2D waveguides: waveguides H and W depicted in Figure 5.3. The two 2D waveguides

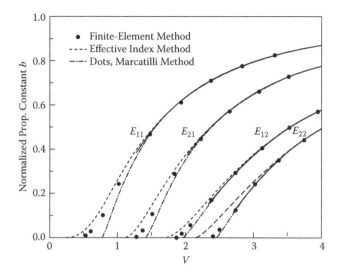

FIGURE 5.5 Dispersion characteristics, dependence of the normalized propagation constant of the guided modes as a function of the parameter *V*, and the normalized frequency: comparison of three numerical, analytical methods for rectangular optical waveguides consisting of uniform core and cladding.

are mutually independent in that the waveguide parameters of the two 2D waveguides come directly from the original 3D waveguide.

To provide a theoretical basis for the effective index method, in lieu of the original 3D waveguide, we consider a pseudowaveguide that can be resolved into aforementioned waveguides I and II, or I′ and II′. The pseudowaveguide is chosen such that waveguides I and II, or I′ and II′, can be easily identified and analyzed. The dispersion of waveguide II, or II′, is used as an approximation for β of the original 3D waveguide. The structures of these waveguides are shown in Figure 5.5.

Consider E_y modes guided by a 3D waveguide shown in Figure 5.5(a). As discussed above, all field components of E_y modes can be expressed in terms of H_x, which can be written as $h_x(x, y)e^{-j\beta z}$, with $h_x(x, y)$ the field distribution in the transverse plane and $n(x, y) = n_j$; $j = 1 - 5$. The wave equation in the transverse plane can be obtained as

$$\left[\frac{\partial^2}{\partial x^2} + \frac{\partial^2}{\partial y^2} + k^2 n^2(x,y) - \beta^2\right] h_x(x,y) = 0 \qquad (5.16)$$

Instead of considering the 3D waveguide problem, we can modify the refractive index distribution so that a pseudowaveguide structure can be obtained as shown in Figure 5.5. The refractive index of the pseudowaveguide can be written as

$$n_{ps}^2 = n_x^2(x) + n_y^2(y) \qquad (5.17)$$

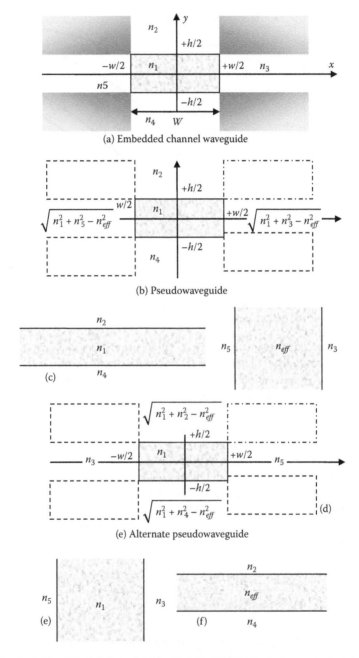

FIGURE 5.6 Model of pseudowaveguides used in the effective index method: (a) rectangular channel waveguide, (b) pseudowaveguide, (c) waveguide I, (d) waveguide II, (e) alternate pseudowaveguide, (f) waveguide I′, and (g) waveguide II″.

Then one can determine $n(x, y)$ by the common method of separation of variables; that is, the distribution $h_x(x, y)$ can be represented as the product of two distribution functions $X(x)$ and $Y(y)$, and the wave equation in the transverse plane can be written as

$$\frac{1}{X(x)}\frac{\partial^2 X(x)}{\partial x^2} + \frac{1}{Y(y)}\frac{\partial^2 Y(y)}{\partial y^2} + [k^2 n_x^2(x) + k^2 n_y^2(y) - \beta^2]h_x(x,y) = 0$$

or   (5.18)

$$\frac{1}{Y(y)}\frac{\partial^2 Y(y)}{\partial y^2} + k^2 n_y^2(y) = -\frac{1}{X(x)}\frac{\partial^2 X(x)}{\partial x^2} - k^2 n_x^2(x) - \beta^2$$

This equation is physically possible when the two sides equate to nil. So we have

$$\frac{1}{Y(y)}\frac{\partial^2 Y(y)}{\partial y^2} + k^2[n_y^2(y) - n_{eff}^2] = 0$$

(5.19)

$$-\frac{1}{X(x)}\frac{\partial^2 X(x)}{\partial x^2} - k^2[n_x^2(x) + n_{eff}^2] - \beta^2 = 0$$

These two equations can be solved subject to the boundary conditions to derive the propagation constant along the z-direction of the waveguide. Thus, the complete solution is the product of the two functions $X(x)$ and $Y(y)$ and the phase term representing the propagation of the field along the z-direction.

5.3.2 PSEUDOWAVEGUIDE

Consider the waveguide structure shown in Figure 5.5. The refractive index distribution of the channel waveguide core and cladding is shown and given as

$$n_{ps}^2(x,y) = \begin{cases} n_1^2 & \text{region 1} \\ n_2^2 & \text{region 2} \\ n_1^2 + n_3^2 - n_{eff}^2 & \text{region 3} \\ n_4^2 & \text{region 4} \\ n_1^2 + n_5^2 - n_{eff}^2 & \text{region 5} \end{cases}$$ (5.20)

This distribution can be considered the superposition of two distributed functions:

$$
n_y^2(y) = \begin{cases} n_2^2 & \text{region } y > h/2 \\ n_1^2 & \text{region } -h/2 \le y \le h/2 \\ n_4^2 & \text{region } y < -h/2 \end{cases}
$$

$$(5.21)$$

$$
n_x^2(x) = \begin{cases} n_3^2 - n_{\text{eff}}^2 & \text{region } x > w/2 \\ 0 & \text{region } -w/2 \le x \le w/2 \\ n_5^2 - n_{\text{eff}}^2 & \text{region } x < -w/2 \end{cases}
$$

Then, similar to the method obtained in Marcatilli's method, the dispersion relation can be obtained as

$$
kh\sqrt{n_1^2 - n_{\text{eff}}^2} = q\pi + \tan^{-1}\left(\frac{n_1^2}{n_2^2} \frac{\sqrt{n_{\text{eff}}^2 - n_2^2}}{\sqrt{-n_{\text{eff}}^2 + n_1^2}} \right) + \tan^{-1}\left(\frac{n_1^2}{n_4^2} \frac{\sqrt{n_{\text{eff}}^2 - n_4^2}}{\sqrt{-n_{\text{eff}}^2 + n_1^2}} \right) \quad (5.22)
$$

$$
kw\sqrt{n_{\text{eff}}^2 - N} = p\pi + \tan^{-1}\left(\frac{\sqrt{N^2 - n_3^2}}{\sqrt{-n_{\text{eff}}^2 + N^2}} \right) + \tan^{-1}\left(\frac{\sqrt{N^2 - n_5^2}}{\sqrt{n_{\text{eff}}^2 - N^2}} \right) \quad (5.23)
$$

Using these dispersion relations the dispersion characteristics of an embedded channel waveguide with cladding as shown in Figure 5.3 can be obtained very close to that given in Figure 5.6 as the dashed curves.

5.4   FINITE DIFFERENCE NUMERICAL TECHNIQUES FOR 3D WAVEGUIDES

The main purpose of selecting the FDM to study the quasi-TE- and quasi-TM-polarized waveguide modes is its simplicity and plausible accuracy. We have employed semivectorial analysis,[12,23,25] which automatically takes full account of the discontinuities in the normal electric field components across any arbitrary distribution of internal dielectric interfaces. The semivectorial FDM, despite its simplicity and being free from troublesome spurious solutions, has two major disadvantages of being computationally intensive and requiring a large amount of memory. Hence, it is necessary to introduce the discretization scheme on the nonuniform mesh, in which mesh intervals can be changed arbitrarily depending on waveguide structures. For this reason, we have modeled the waveguide mode with the finite difference method, which employs a nonuniform discretization scheme.[9,23] Such a discretization scheme enables us to increase the size of the problem space so that the field component at the boundary can be assumed to have vanished. The grid spacing increases monotonically with increasing distance from the guiding region. The grid lines can also be aligned with the boundaries of the step index changes in conventional structures, such as rib, ridge, and strip-loaded waveguides, as well as quantum well structures. Furthermore, by judiciously placing the grid lines and corresponding cell structure efficiently, we can reduce the required matrix size and hence redundant computer calculations, while preserving the accuracy of the calculations. The non-uniform discretization scheme also enables us to handle the waveguide mode near the cutoff region with a relative simple boundary condition. The eigenmodes of the Helmholz equation are solved by the application of the shifted inverse power iteration method. This method warrants both the mode size and its relevant propagation constant, which are important parameters to the design of an optical waveguide.

Apart from being able to access the accuracy of the final product of our work, which is the Semivectorial Mode Modeling (SVMM) computer program, we also present an overview of its application in modeling a Ti:LiNbO$_3$ channel waveguide for optical devices such as modulators and switches.

5.4.1 Nonuniform Grid Semivectorial Polarized Finite Difference Method for Optical Waveguides with Arbitrary Index Profile

5.4.1.1 Propagation Equation

For harmonic wave propagation in the z-direction along a rib or channel waveguide, we consider the following fields:

$$E(x,y,z) = (E_x, E_y, E_z) \exp j(\omega t - \beta z) \tag{5.24}$$

$$H(x,y,z) = (H_x, H_y, H_z) \exp j(\omega t - \beta z) \tag{5.25}$$

$$D = \varepsilon(x,y)E, \, B = \mu H \tag{5.26}$$

where the dielectric permittivity $\varepsilon(x,y)$ is piecewise constant and the magnetic permeability μ is completely constant throughout the solution domain. The components of the electric and magnetic fields in Equation (5.1) are functions of x and y only. Then, applying the Maxwell equations in the magnetic and charge-free media and taking appropriate algebra, we obtain the wave equation

$$\nabla \times (\nabla \times E) = \nabla(\nabla \cdot E) - \nabla^2 E = \omega^2 \varepsilon \mu E = k^2 n^2 E \tag{5.27}$$

in which $k = \omega(\varepsilon_0 \mu_0)^{1/2} = 2\pi/\lambda$ and $\varepsilon = \varepsilon_0 n^2(x,y)$ with λ being the free-space wavelength. With the divergence of $\nabla \cdot D = 0$ and $\nabla \log_e \varepsilon = \nabla \varepsilon/\varepsilon$ we get

$$\nabla \cdot E = -E \cdot \nabla \log_e \varepsilon = -E \cdot \nabla n^2 / n \tag{5.28}$$

This may be substituted into (5.4) to yield the wave equation

$$\nabla^2 E + k^2 E + \nabla(E \cdot \nabla n^2/n) = 0 \tag{5.29}$$

As $n(x, y)$ is a piecewise constant, $\nabla n^2/n = 0$, and it should be noted that $\nabla n^2/n$ is undefined at internal dielectric interfaces where $n(x, y)$ is discontinuous. With the assumption that the fields are polarized either perpendicular (quasi-TM) to or parallel (quasi-TE) to the crystal surface, and that the major field components of the modes are perpendicular to the direction of the propagation, Equation (5.7) can be reduced to

$$(\nabla_T^2 + k^2 n^2)E = \beta^2 E \tag{5.30}$$

in which $\nabla_T^2 = \partial^2/\partial x^2 + \partial^2/\partial y^2$, the transverse Laplacian, and β is the propagation constant. This is essentially the Helmholz wave equation.

5.4.1.2 Formulation of Nonuniform Grid Difference Equation

Figure 5.7(a) shows the grid lines used in the finite difference method formulation. The grid lines are chosen in such a way that denser grids are allocated around the guiding region, while coarser grids are assigned to regions farther away from the waveguide. Boundaries of abrupt index changes are straddled by the grid lines wherever necessary. Figure 5.7(b) shows the magnified view of a portion of the grid for a more detailed illustration. Each cell point is located in the center of each rectangular cell; h_i and h_j are the horizontal and vertical grid sizes. The refractive index within each cell is assumed to be uniform. $n_{i,j}$ and $n_{i+1, j}$ represent the values of the refractive index of each small cell as an approximation and are taken from the continuous refractive index profile $n(x, y)$. Nonuniform spacing of the grid lines provides some flexibility in setting up the nonuniform grid FDM. The nonuniform discretization with increasing spacing away from the guiding region permits sufficient extension of the boundary. This enables us to assume a Dirichlet boundary condition (metal box) where all fields have vanished.

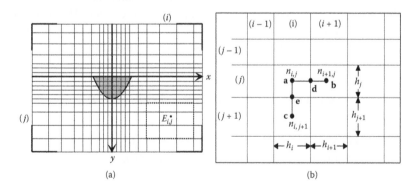

FIGURE 5.7 (a) Nonuniform discretized grid for FDM scheme. (b) A magnified portion of the grid lattice and cell structure of point i, j.

5.4.1.2.1   *Quasi-TE Mode*

For quasi-TE-polarized mode, E_y is assumed to be zero. E_x is continuous across the horizontal interfaces but discontinuous across vertical interfaces. Therefore, the quasi-TE modes are the eigensolutions of the equation

$$\nabla_t^2 E_x + k^2 n^2 E_x = \beta^2 E_x \qquad (5.31)$$

The discontinuity across the vertical interface will need to be taken into account when formulating the difference equation.

Figure 5.8 illustrates the quasi-TE field discontinuity at the boundary between cells (i, j) and $(i + 1, j)$. Consider the points a, d, and b, with d being at the boundary of the dielectric interface. The horizontal axis is the x-axis, while the vertical axis is the electric field amplitude of the respective position of the cell. Assume that the x-axis is pointing toward the east. So, E_E and E_W are the field amplitudes just to the east and the west of the boundary between the cells (i, j) and $(i +1, j)$. $E_{i,j}^v$ is the virtual field in cell (i, j), which is the extension of the actual field $E_{i+1,j}$. In other words, $E_{i,j}^v$ is the field seen by the cell $(i + 1, j)$. Similarly, $E_{i+1,j}^v$ is the extension of $E_{i,j}$. n_E and n_W are the refractive indices just to the east and west of the boundary. Since we consider a slowly varying index distribution, we assume that n_E and n_W are approximately equal to $n_{i,j}$ and $n_{i+1,j}$, respectively. The boundary conditions between the cells (i, j) and $(i + 1, j)$ are given as follows:

$$n_E^2 E_E = n_W^2 E_W \Rightarrow n_{i,j}^2 E_E = n_{i+1,j}^2 E_W \qquad (5.32)$$

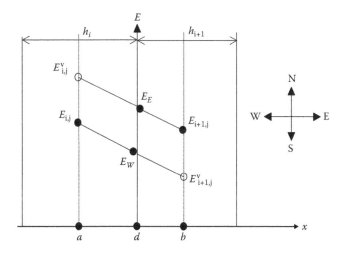

FIGURE 5.8 Quasi-TE electric field discontinuity at the boundary between cells (i, j) and $(i + 1, j)$. Solid lines are the actual field profiles along the x-axis, while $E_{i,j}^v$ and $E_{i+1,j}^v$ are virtual fields.

$$\frac{\partial}{\partial x} E_E = \frac{\partial}{\partial x} E_W = p^+ \tag{5.33}$$

where p^+ represents the field gradient at the boundary between the cells. We can then use the approximate relationship between $E_{i,j}$, $E_{i+1,j}$, $E_{i,j}^v$, $E_{i+1,j}^v$, and obtain the following equations for E_E and E_W:

$$E_{i+1,j} \approx E_E + (h_{i+1}/2) \cdot p^+; \quad E_{i,j}^v \approx E_E - (h_i/2) \cdot p^+; \quad E_{i+1,j}^v \approx E_W + (h_{i+1}/2) \cdot p^+ \tag{5.34}$$

and

$$E_{i,j}^v \approx E_W - (h_i/2) \cdot p^+ \tag{5.35}$$

where h_i and h_{i+1} are the horizontal lengths of cells (i, j) and $(i + 1, j)$. The four equations above are in fact redundant. Therefore we need to consider only either $E_{i,j}^v$ or $E_{i+1,j}^v$, and we choose $E_{i+1,j}^v$. The following shows the algebraic manipulation (5.11):

$$p^+ = 2(E_{i+1,j}^v - E_{i,j})/(h_i + h_{i+1}); \quad E_{i+1,j}^v = E_{i+1,j} + (E_W - E_E);$$
$$h_{i+1}(E_W - E_{i,j}) = h_i(E_{i+1,j} - E_E) \tag{5.36}$$

and then

$$E_{i+1,j}^v = \frac{n_{i+1,j}^2(h_i + h_{i+1})E_{i+1} + h_{i+1}(n_{i+1,j}^2 - n_{i,j}^2)E_{i,j}}{(n_{i,j}^2 h_i + n_{i+1,j}^2 h_{i+1})} \tag{5.37}$$

With a similar procedure between cells (i, j) and $(i + 1, j)$, we can obtain

$$p^- = 2\frac{E_{i,j} - E_{i-1,j}^v}{h_i + h_{i-1}} \quad \text{and} \quad E_{i-1,j}^v = \frac{n_{i-1,j}^2(h_i + h_{i-1})E_{i-1,j} + h_{i-1}(n_{i-1,j}^2 - n_{i,j}^2)E_{i,j}}{(n_{i,j}^2 h_i + n_{i-1,j}^2 h_{i-1})} \tag{5.38}$$

where p^- is now the field gradient at the boundary between the cells $(i-1, j)$ and (i, j). Note that the quasi-TE electric field is continuous in terms of the y-direction even if there are discontinuities in the refractive index. Therefore $E_{i,j+1}^v = E_{i,j+1}$, $E_{i,j-1}^v = E_{i,j-1}$.

The second derivative can be derived as

$$\frac{\partial^2}{\partial x^2} E_{i,j} = \frac{1}{h_i}[p^+ - p^-] = \frac{1}{h_i}\left[\frac{2(E_{i+1,j}^v - E_{i,j})}{h_{i+1} + h_i} - \frac{2(E_{i,j} - E_{i-1,j}^v)}{h_i + h_{i-1}}\right] \tag{5.39a}$$

$$\frac{\partial^2}{\partial y^2} E_{i,j} = \frac{1}{h_j}[p^+ - p^-] = \frac{1}{h_j}\left[\frac{2(E_{i,j+1} - E_{i,j})}{h_{j+1} + h_j} - \frac{2(E_{i,j} - E_{i,j-1})}{h_j + h_{j-1}}\right] \qquad (5.39b)$$

Thence we get the discrete wave equation as

$$\frac{\partial^2}{\partial x^2} E_{i,j} = \frac{2n_{i-1,j}^2}{h_i(n_{i,j}^2 h_i + n_{i-1,j}^2 h_{i-1})} E_{i-1,j} + \frac{2n_{i+1,j}^2}{h_i(n_{i,j}^2 h_i + n_{i+1,j}^2 h_{i+1,j})} E_{i+1,j}$$

$$- \left[\frac{2n_{i,j}^2}{h_i(n_{i,j}^2 h_i + n_{i-1,j}^2 h_{i-1})} + \frac{2n_{i,j}^2}{h_i(n_{i,j}^2 h_i + n_{i+1,j}^2 h_{i+1})}\right] E_{i,j} \qquad (5.40a)$$

$$\frac{\partial^2}{\partial y^2} E_{i,j} = \frac{1}{h_j}\left[\frac{2(E_{i,j+1} - E_{i,j})}{h_{j+1} + h_j} - \frac{2(E_{i,j} - E_{i,j-1})}{h_j + h_{j-1}}\right] \qquad (5.40b)$$

Substituting these into the Helmholtz equation,

$$C_{i-1,j}E_{i-1,j} + C_{i+1,j}E_{i+1,j} - C_{i,j}E_{i,j} + C_{i,j-1}E_{i,j-1} + C_{i,j+1}E_{i,j+1} = \beta^2 E_{i,j} \qquad (5.41)$$

where

$$C_{i-1,j} = \frac{2n_{i-1}^2}{h_i(n_{i,j}^2 h_i + n_{i-1,j}^2 h_{i-1})}; \quad C_{i+1,j} = \frac{2n_{i+1}^2}{h_i(n_{i,j}^2 h_i + n_{i+1,j}^2 h_{i+1})};$$

$$C_{i,j-1} = \frac{2}{h_j(h_j + h_{j-1})}; \quad C_{i,j+1} = \frac{2}{h_j(h_j + h_{j+1})}; \qquad (5.42)$$

$$\text{and} \quad C_{i,j} = C_{i-1,j} + C_{i+1,j} + C_{i,j-1} + C_{i,j+1} - k^2 n_{i,j}^2$$

The above equations are essentially eigenvalue equations of

$$\mathbf{C}_{TE}\mathbf{E}_{TE} = \beta_{TE}^2 \mathbf{E}_{TE} \qquad (5.43)$$

in which $\mathbf{C}_{TE}$ is a nonsymmetric band matrix that contains the coefficient of the above equations, β_{TE}^2 is the TE propagation eigenvalue, and $\mathbf{E}_{TE}$ is the corresponding normalized eigenvector representing the field profile $E_x(x, y)$.

5.4.1.2.2 Quasi-TM Mode

The quasi-TM mode can be formulated in a similar fashion. The only difference is that for the quasi-TM-polarized mode, E_x is assumed to be zero and E_y is continuous across the vertical interfaces but discontinuous across horizontal interfaces. Essentially, the quasi-TM modes are the eigensolutions of the equation

$$\nabla_t^2 E_y + k^2 n^2 E_y = \beta^2 E_y \tag{5.44}$$

The detailed derivation of the equation can be found in the literature.[23] The following are the derivatives and their relevant difference with reference to Figure 5.9 for the discretized fields:

$$\frac{\partial^2}{\partial y^2} E_{i,j} = \frac{2n_{i,j-1}^2}{h_i(n_{i,j}^2 h_i + n_{i,j-1}^2 h_{j-1})} E_{i,j-1} + \frac{2n_{i,j+1}^2}{h_i(n_{i,j}^2 h_i + n_{i,j+1}^2 h_{j+1})} E_{i,j+1}$$
$$- \left[\frac{2n_{i,j}^2}{h_i(n_{i,j}^2 h_i + n_{i,j-1}^2 h_{j-1})} + \frac{2n_{i,j}^2}{h_i(n_{i,j}^2 h_i + n_{i,j+1}^2 h_{j+1})} \right] E_{i,j} \tag{5.45a}$$

$$\frac{\partial^2}{\partial x^2} E_{i,j} = \frac{1}{h_i} \left[\frac{2(E_{i+1,j} - E_{i,j})}{h_{i+1} + h_i} - \frac{2(E_{i,j} - E_{i-1,j})}{h_i + h_{i-1}} \right] \tag{5.45b}$$

Substituting these into the Helmholtz equation we get

$$C_{i-1,j} E_{i-1,j} + C_{i+1,j} E_{i+1,j} - C_{i,j} E_{i,j} + C_{i,j-1} E_{i,j-1} + C_{i,j+1} E_{i,j+1} = \beta^2 E_{i,j} \tag{5.46}$$

where

$$C_{i-1,j} = \frac{2}{h_i(h_i + h_{i-1})}; \quad C_{i+1,j} = \frac{2}{h_i(h_i + h_{i+1})}; \quad C_{i,j-1} = \frac{2n_{j-1}^2}{h_j(n_{i,j}^2 h_j + n_{i,j-1}^2 h_{j-1})};$$

$$C_{i,j+1} = \frac{2n_{j+1}^2}{h_j(n_{i,j}^2 h_j + n_{i,j+1}^2 h_{j+1})}; \quad \text{and} \tag{5.47}$$

$$C_{i,j} = C_{i-1,j} + C_{i+1,j} + C_{i,j-1} + C_{i,j+1} - k^2 n_{i,j}^2$$

5.4.1.2.3 Eigenvalue Matrix

To solve the difference equation, we need first to discretize the problem space. We assume that the space is sliced into NX pieces along the x-direction and NY pieces along the y-direction. This will give us a total of $N (= NX \times NY)$ grid points. The refractive index of each cell is then allocated according to the relevant index distribution.

When the finite difference wave equation is evaluated at a grid point, say $E_{i,j}$, it will yield a 5-point linear equation in terms of the E-field of the immediate neighbors, namely, $E_{i-1,j}$, $E_{i+1,j}$, $E_{i,j-1}$, $E_{i,j+1}$, each with its relevant coefficient as shown in Equations (5.45) and (5.46). For a cross-sectional area of a waveguide with N such grid points, we would end up with N linearly dependent algebraic equations.

We will now scan through the grid points row after row, at the same time relabeling the subscripts of E from 1 to N. Consider the original grid point (i, j). Assuming that the new sequence number is k, then (5.46) can be rewritten as

$$p_k E_k + l_k E_{k-1} + r_k E_{k+1} + t_k E_{k-Nx} + b_k E_{k+Nx} = \beta^2 E_k \tag{5.48}$$

where p_k, l_k, r_k, t_k, b_k, are the coefficients $C_{i,j}, C_{i-1,j}, C_{i+1,j}, C_{i,j-1}, C_{i,j+1}$, respectively. We can then collect terms and write the equations in a matrix form.

For a 3×3 grid of the refractive index profile, we can write the matrix equations as an eigenvalue equation of the form $[C] \cdot [E] = \beta^2[E]$, in which $[C]$ is a nonsymmetric band matrix that contains the coefficient of the above equations, β^2 is the propagation eigenvalue, and $[E]$ is the corresponding normalized eigenvector representing the field profile $E(i, j)$. In the next section we will discuss the approach that we adopt in solving the eigenvalue problem given as

$$
\begin{bmatrix}
p_1 & r_1 & 0 & b_1 & 0 & 0 & 0 & 0 & 0 \\
l_2 & p_2 & r_2 & 0 & b_2 & 0 & 0 & 0 & 0 \\
0 & l_3 & p_3 & r_3 & 0 & b_3 & 0 & 0 & 0 \\
t_4 & 0 & l_4 & p_4 & r_4 & 0 & b_4 & 0 & 0 \\
0 & t_5 & 0 & l_5 & p_5 & r_5 & 0 & b_5 & 0 \\
0 & 0 & t_6 & 0 & l_6 & p_6 & r_6 & 0 & b_6 \\
0 & 0 & 0 & t_7 & 0 & l_7 & p_7 & r_7 & 0 \\
0 & 0 & 0 & 0 & t_8 & 0 & l_8 & p_8 & r_8 \\
0 & 0 & 0 & 0 & 0 & t_9 & 0 & l_9 & p_9
\end{bmatrix}
\begin{bmatrix}
E_1 \\ E_2 \\ E_3 \\ E_4 \\ E_5 \\ E_6 \\ E_7 \\ E_8 \\ E_9
\end{bmatrix}
= \beta^2
\begin{bmatrix}
E_1 \\ E_2 \\ E_3 \\ E_4 \\ E_5 \\ E_6 \\ E_7 \\ E_8 \\ E_9
\end{bmatrix}
$$

$$\tag{5.49}$$

There are a few major features of the matrix equation above: (1) This type of matrix is often referred to as tridiagonal matrix with fringes. The order of the matrix is $N \times N$, the square of the total number of grid points. Most of the terms in the matrix are zeros. (2) The matrix is nonsymmetrical relative to the diagonal term. (3) The central three diagonal terms always exist and are always nonzero. (4) The coefficients p, l, r, t, b make up the five bands of the matrix, with p being the main diagonal, l and r the subdiagonals, and t and b the superdiagonal. (5) The subdiagonal terms are just one term away from the main diagonal, while the superdiagonal terms are NX terms away from the main diagonal. The distance between the main diagonal and the last nonzero superdiagonal band is commonly referred to as the half-bandwidth of a band matrix. (6) Terms such as $l_1, r_N, t_1 - t_{NX}$, and $b_{N-Nx} - b_N$ are missing. This is so because the evaluations of these terms require the E-values outside the boundary area, and these values have been assumed to be zero. Therefore, they need not be represented.

5.4.1.3 Inverse Power Method

The properties and characteristics of the eigenvalue problem are well known and have been addressed rather extensively in many textbooks.[29,30] This section provides only a brief overview to highlight the more specific points related to our particular approach.

An $N \times N$ matrix A is said to have an eigenvector x and a corresponding eigenvalue λ if the following condition is satisfied:

$$A \cdot x = \lambda x \tag{5.50}$$

There can be more than one distinct eigenvalue and eigenvector corresponding to a given matrix. The zero vector is not considered to be an eigenvector at all. The above equation holds only if

$$\det|A - \lambda I| = 0 \tag{5.51}$$

which is known as the characteristic equation of the matrix. If this is expanded, it becomes an Nth degree polynomial in λ whose roots are the eigenvalues. This is an indication that there are always N, though not necessarily distinct, eigenvalues. Equal eigenvalues coming from multiple roots are called degenerate. Root searching in the characteristic equation, however, is usually a very poor computational method for finding eigenvalues. There are many more efficient algorithms available in locating the eigenvalues and their corresponding vectors.

Unfortunately, there is no universal method for solving all matrix types. For certain problems, either the eigenvalues or eigenvectors are needed, while others require both. Furthermore, some problems may only need a small number of solutions out of the total N solutions available, while others need all. To complicate the matter even further, the eigensolutions could be complex, and some matrices can be so ill-behaved that round-off errors in computing can lead to a nonconvergence of the solution. Therefore, it is of vital importance to be able to choose the right approach in solving an eigenproblem. Choosing an algorithm often involves the classification of matrices into types such as symmetry, nonsymmetry, tridiagonal, banded, positive definite, definite, Heisenberg, sparse, and random. The matrix in our problem is a nonsymmetric banded matrix with bandwidth equal to twice the number of columns in the grid profile. It has great sparsity, for most of the elements are zeros. Also, we need only a few eigenvalues that correspond to the guided modes of the waveguide. In other words, there are only a limited number of guided modes—hence the number of eigenvalue λ. The number of eigensolutions required is small compared with the size of the matrix (often in the order of tens of thousands). All these different factors have led to the choice of the approach called the inverse iteration method.[29,30]

The basic idea behind the inverse iteration method is quite simple. Let y be the solution of the linear system

$$(A - \tau I) \cdot y = b \tag{5.52}$$

where b is a random vector and τ is close to some eigenvalue λ of A. Then the solution y will be close to the eigenvector corresponding to λ. The procedure can be iterated: replace b by y and solve for a new y, which will be even closer to the true eigenvector. We can see why this works by expanding both y and b as linear combinations of the eigenvectors x_j of A:

$$y = \sum_j \alpha_j x_j \quad \text{and} \quad b = \sum_j \beta_j x_j \tag{5.53}$$

Then we have

$$\sum_j \alpha_j (\lambda_j - \tau) x_j = \sum_j \beta_j x_j \tag{5.54}$$

so that

$$\alpha_j = \frac{\beta_j}{\lambda_j - \tau} \quad \text{and} \quad y = \sum_j \frac{\beta_j x_j}{\lambda_j - \tau} \tag{5.55}$$

If τ is close to λ_n, say, then provided β_n is not accidentally too small, y will be approximately x_n, up to a normalization. Moreover, the iteration of this procedure gives another power of $\lambda_j - \tau$ in the denominator of (5.55). Thus the convergence is rapid for well-separated eigenvalues.

Suppose at the ith stage of iteration we are solving the equation

$$(A - \lambda_i I) \cdot y = x_i \tag{5.56}$$

where x_i and λ_i are our current guesses for some eigenvector and eigenvalue of interest (we shall see below how to update λ_i). The exact eigenvector and eigenvalue satisfy

$$A \cdot x = \lambda x \rightarrow (A - \lambda_i I) \cdot x = (\lambda - \lambda_i) x \tag{5.57}$$

Since y of (5.57) is an improved approximation to x, we normalize it and set

$$x_{i+1} = \frac{y}{|y|} \tag{5.58}$$

We get an improved estimate of the eigenvalue by substituting our improved guess y in (5.57). By (5.58), the left-hand side is x_i, and so calling λ our new value λ_{i+1}, we find

$$\lambda_{i+1} = \lambda_i + \frac{|x|^2}{|x_i \cdot y|}$$

Although the formulae of the inverse iteration method seem to be rather straightforward, the actual implementation can be quite tricky. Most of the computational load occurs in solving the linear system of equations. It would be advantageous if we could solve (5.32) quickly. Remember that the size of the matrix in our case is dependent upon the total grid size of the problem space. For a typical grid size of 100 by 100, for example, the coefficient matrix would be of size 10,000 by 10,000. The core memory required in a digital computer to store the entire matrix would be phenomenal. A linear system solver such as routines that are available in LINPACK employs a common LU factorization (Gaussian elimination) plus a backward substitution combination algorithm, much like the manual way of solving linear equations. There is extensive coverage on this topic in most numerical textbooks.[30] We will therefore not discuss it further except to mention that the LU factorization needs only to be done before the first iteration. When the iteration starts, we already have the steps involved in elimination stored away in an array, and only backward substitution is necessary. This approach, even with a storage-optimized mode in the LINPACK routine, still has a storage requirement of about $3 \times$ (bandwidth of matrix $\times$ matrix size). Even though this would mean a considerable reduction in memory storage, it still amounts to a rather substantial memory size.

Also, the preconditioner that employs the incomplete Cholesky conjugate gradient method and the Orthomin accelerator[31] has been found to be most stable and converges most quickly for our matrix. On average, the combination of the preconditioner and accelerator enables us to complete a simulation of a typical waveguide in 3 to 5 min on a PC Pentium 4. The same simulation that incorporates the LINPACK LU decomposition routine would take 25 min on the same computer with a substantially greater amount of memory. Since the zero elements are no longer involved in the calculations, it is understandable that the Non-Symmetric Preconditioned Conjugate Gradient (NSPCG) iterative method will perform more efficiently.

By incorporating the NSPCG numerical solver and the inverse iterative method, we have successfully implemented a mode modeling program, SVMM, capable of modeling a channel waveguide of an arbitrary index profile. The inverse iterative method also enables us to model the higher-order modes that are supported by the waveguide structure.

5.4.2   Ti:LiNbO₃-Diffused Channel Waveguide

The modeling of a Ti:LiNbO$_3$ channel waveguide, a graded index waveguide, plays a significant role in the design of the optical modulators and switches. Efficient design of such optical devices requires good knowledge of the modal characteristics of the relevant channel waveguide. In Chiang,[7] we outlined the general overview of the waveguide fabrication process. In this section, we will attempt to employ our SVMM program to simulate the waveguide mode of the Ti:LiNbO$_3$ waveguide and compare the results with published experimental results. Our objective, apart from assessing the usefulness of SVMM, is to understand the key features in the fabrication of a Ti:LiNbO$_3$ waveguide for a Mach–Zehnder optical modulator.

To achieve our purpose, a good knowledge of the refractive index profile of the diffused waveguide is required. Over the past decades, much work has been done in fabricating a low-loss, minimum mode size, Ti-diffused channel waveguide.[6,32] From these references, we can gather our knowledge of the diffusion process involved in the fabrication of a LiNbO$_3$ waveguide and its relevant diffusion profile. Based on this knowledge, we can then profess to model the modal characteristics of the waveguide by SVMM. The following section shows how SVMM can be used for the design of practical waveguides.

5.4.2.1   Refractive Index Profile of Ti:LiNbO$_3$ Waveguide

When Ti metal is diffused, the Ti ion distribution spreads more widely than the initial strip width. The profiles can be described by the sum of an error function, while the Ti ion distributions perpendicular to the substrate surface can be approximated by a Gaussian function.[1,6,33] This of course is true only if the diffusion time is long enough to diffuse all the Ti metal into the substrate. We consider this case as having the finite diffusant source. However, if the total diffusion time is shorter than needed to exhaust the Ti source, the lateral diffusion profile would take up the sum of the complementary error function, while the depth index profile is given by the complementary function.[34] This case is considered to have an infinite diffusant source.[23] In our study, we would assume that there is sufficient time for the source to be fully diffused because in most practical waveguides, it is undesirable to have Ti residue deposited on the surface of the waveguide as this will increase the propagation loss.[6] This increase in propagation loss is a result of stronger interaction with the LiNbO$_3$ surface (and thus an increased scattering loss) as the modes become more weakly guided.

In general, the refractive index distribution of a weakly guiding channel waveguide is

$$n(x,y) = n_b + \Delta n(x,y) = n_b + \Delta n_0 \cdot f(x) \cdot g(y) \qquad (5.59)$$

where n_b is the refractive index of the bulk (substrate) and $\Delta n(x, y)$ is the variation of the refractive index in the guiding region. $\Delta n(x, y)$ in our diffusion model is essentially a separable function where $f(x)$ and $g(y)$ are the functions that describe the lateral and perpendicular diffusion profiles, while Δn_0 is known as the surface index change after diffusion. The surface index change is defined as the change of refractive index on the substrate just below the center of the Ti strip. In other words, it is the refractive index when both $f(x)$ and $g(y)$ assume the value of unity.

The variation of the refractive index can be modeled as below[23]

$$\Delta n(x,y) = \frac{dn}{dc} \tau \int_{-w/2}^{w/2} \frac{2}{d_y \sqrt{\pi}} \exp\left[-\left(\frac{y}{d_y}\right)^2\right] \frac{1}{d_x \sqrt{\pi}} \cdot \exp\left[-\left(\frac{x-u}{d_x}\right)^2\right] du \qquad (5.60)$$

$$= \Delta n_0 \cdot f(x) \cdot g(y)$$

where

$$f(x) = \frac{1}{2}\left[erf\left(\frac{x+\frac{w}{2}}{d_x}\right) - erf\left(\frac{x-\frac{w}{2}}{d_x}\right)\right] \Bigg/ erf\left(\frac{w}{2d_x}\right) \qquad (5.61)$$

$$g(x) = \exp\left[-\left(\frac{y}{d_y}\right)^2\right] \qquad (5.62)$$

and

$$\Delta n_0 = \frac{dn}{dc} \frac{2}{\sqrt{\pi}} \frac{\tau}{d_y} erf\left(\frac{w}{2d_x}\right)$$
(5.63)

with

$$d_x = 2\sqrt{D_x t}, d_y = 2\sqrt{D_y t}$$
(5.64)

In the above expressions, t is the total diffusion time, c is the Ti concentration, d_x and d_y are the diffusion lengths, and D_x and D_y are the diffusion constants in each direction. τ and w are the initial Ti strip thickness. dn/dc is the change of index per unit change in Ti metal concentration. The change of surface index would approach the value where

$$\Delta n_0 = \frac{dn}{dc} \frac{2}{\sqrt{\pi}} \frac{\tau}{d_y}$$

Any increase in the surface index will have to come from a thicker Ti strip, or a decrease in diffusion depth, d_y, which involves an increase or decrease in diffusion temperature. According to the work of Fukuma and Noda[1] the diffusion lengths are very close to one another in both lateral and depth directions (isotropic diffusion) at 1025°C for z-cut crystal. An increase in temperature greater than that would result in a higher diffusion constant in the depth direction and a lower value for lateral diffusion and vice versa for diffusion temperature lower than 1025°C. The diffusion length can also be changed by monitoring the diffusion time. Essentially, a longer diffusion time means a lower surface index change, as most of the Ti source would be diffused deeper into the substrate. It is expected that a higher change of surface index since not all the Ti metal is completely diffused. The following graphs (see Figures 5.9, 5.10, and 5.11) show the variation of the diffusion profile as we vary both the initial titanium width and the diffusion time. The fabrication condition and parameters are assumed to have $T = 1025°C$, $\tau = 1100$ Å, $dn/dc = 0.625$, and $d_x = d_y = 2$ μm.

We can see in these figures that by controlling the width of the initial Ti strip width, we can vary the change of the refractive index and the relative size of the channel waveguide, thus enabling us to control the number of modes that can be supported by the waveguide.

In general, a narrow initial Ti width would give a near cutoff mode, for the refractive index change would be too small. The optical mode would be weakly confined, thus giving a larger mode size. As we increase the Ti width, the refractive index change would be higher and the waveguide mode would be better confined and have a smaller mode size. However, the mode size would increase with a further increase in Ti width due to a larger physical size of the waveguide. The change of the surface index can also be controlled by varying the thickness of the Ti strip. As (5.53) implies, the surface index change is proportional to the strip thickness, τ. The Ti

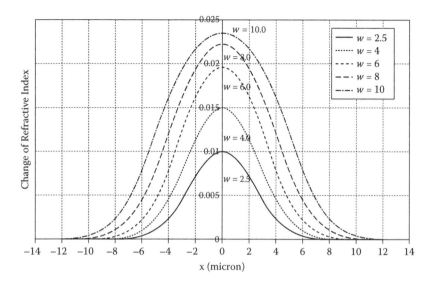

FIGURE 5.9 Lateral diffusion variation with increasing Ti strip width.

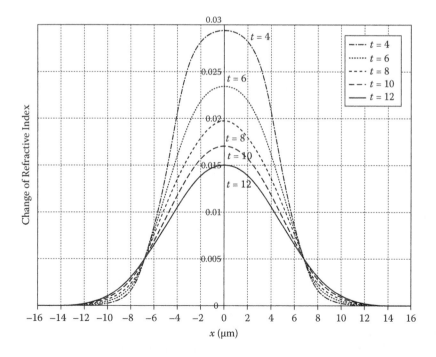

FIGURE 5.10 Lateral diffusion variation with increasing diffusion time.

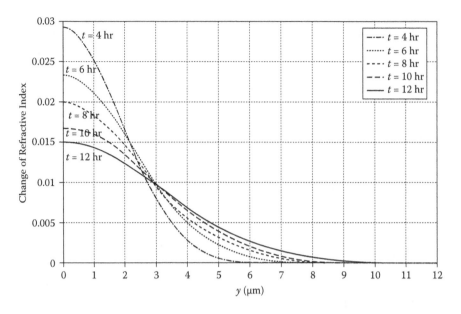

FIGURE 5.11 Depth index variation with increasing diffusion time.

thick film can be diffused at 1000–1050°C for 6 h at around 500–800 Å.[6] If the Ti strip is too thin, the refractive index change approaches cutoff conditions. All these characteristics are illustrated in the next section when we model the waveguide mode with SVMM.

5.4.2.2 Numerical Simulation and Discussion

With the above knowledge of the diffusion profile, we are now in a good position to feed these models into the SVMM program to investigate the modal characteristics of the Ti:LiNbO$_3$ waveguide. In this section, we attempt to simulate the experimental work reported in Suchoski and Ramaswamy[6] in fabricating a minimum mode size, low-loss Ti:LiNbO$_3$ channel waveguide. We will restrict our analysis to the z-cut by propagating material, since this would be the substrate cut for the optical modulator. For this particular substrate cut, the relevant optical field would be TM polarized, which correspond to the polarization along the extraordinary index axis of the crystal. Hence the change of refractive index concern would be the extraordinary index, n_e.

In Suchoski and Ramaswamy's work, the TM-polarized mode width and depth, which are defined as 1/e intensity full width and full depth, are measured for Ti:LiNbO$_3$ waveguides fabricated under the condition where $T = 1025°C$ for 6 h. The sample waveguides have Ti thicknesses ranging from 500 to 1100 Å and Ti strip widths ranging from 2.5 to 10 μm.

The laser source wavelength is assumed to be at 1.3μm. The following graph, Figure 5.12, is extracted from Suchoski and Ramaswamy[6] and Fukuma and Noda.[29] In view of these experimental results, we can see that the mode size increases as the Ti strip width is decreased from 4 to 2.5 μm. This increase is more pronounced,

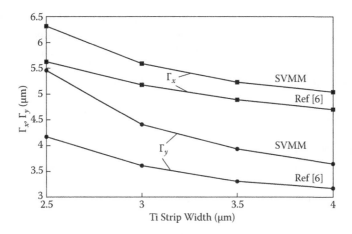

FIGURE 5.12 Simulation of mode sizes with nominal diffusion parameters.

especially with the thinner Ti films, because the waveguides become closer to cut-off, as thinner Ti films result in a lower value of Δn. The TM mode depth and width decrease as the Ti thickness is increased from 500 to 800 å. However, for 4 μm strip widths, the mode size does not decrease further for Ti films thicker than 800 Å. This is an indication that it is not possible to diffuse any more Ti into the substrate for Ti thicknesses of more than 800 Å for a 6 h diffusion time. We now proceed to simulate the above experiment with our program. We will focus on Ti thicknesses that range between 700 and 800 Å because it is the thickness that gives the minimum mode sizes, which is ideal for the design of an optical modulator for maximizing the overlap integral between guided optical modes and an applied modulating field. To achieve that, we must first work out the suitable diffusion parameter to be used in our program. Various values of *dn/dc* have been reported.[1] Measurements reported by Minakata et al.[35] show the change of extraordinary index n_e per Ti concentration as

$$\frac{dn_e}{dc} = 0.625$$

The nominal values for diffusion constants D_x and D_y are from the work of Fukuma and Noda,[1] and were both measured to be 1.2×10^{-4} μm²/s at the nominated temperature, which is 1025°C. This makes both diffusion lengths of d_x and d_y the value of 2 μm.

With these nominal parameters, we simulate the waveguide with a Ti thickness, τ, of 700 Å. Figure 5.12 shows the result of our simulation compared to the experimental one, and some illustrations of the TM mode profile.

As it turns out, the simulated results appear to have overestimated both Γ_x and Γ_y. Such a discrepancy in anticipated fabrication of the diffused waveguide is subjected to many changes. Various reports[1,6,32,34–37] have shown that even though the nominal diffusion condition can be very much the same, the measured diffusion parameters can still differ greatly from one another due to possible differences in stoichiometry between different crystals and measurements techniques. Therefore, there would certainly be some uncertainties that lie in fabrication parameters, and also in the application of the refractive index model described in Ramaswamy et al.[3] Such uncertainty can be compensated by adjusting the values of *dn/dc* and also D_x and D_y. We find that by adjusting the following diffusion parameters, where

$$dn/dc = 0.8, \quad D_x = 1.4 \times 10^{-4}\,\mu\text{m/s}^2, \quad \text{and} \quad D_y = 1.1 \times 10^{-4}\,\mu\text{m/s}^2$$

our simulation results correspond well within design limit with the experimental work done by Suchoski for the case where the waveguide is well guided. The result is shown in Figure 5.12.

Having found the suitable diffusion parameter, the simulation of another experimental result from Suchoski and Ramaswamy[6] is conducted for $\tau = 750$ Å, and the Ti strip width, *w*, ranges from 2.5 to 10 µm. Figure 5.18 depicts the comparison of both simulated and experiment results of the mode size variation with respect to the width of the Ti strip.

Figures 5.13, 5.14, and 5.15 show the simulated field distributions of the intensity and contour profiles, along the *x*-direction respectively, of the fundamental guided mode.

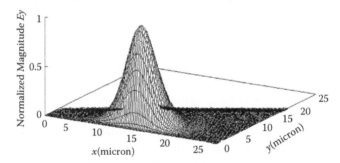

FIGURE 5.13   Typical modal field distribution of a diffused channel waveguide.

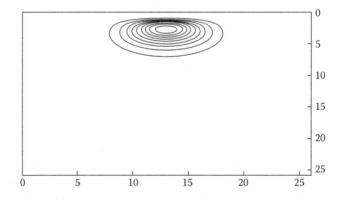

FIGURE 5.14   Contour plot of the modal field of a diffused channel waveguide—vertical and horizontal dimensions in micrometers.

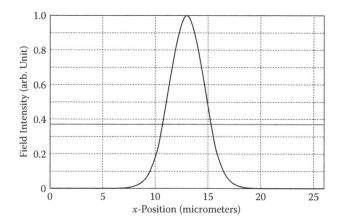

FIGURE 5.15 Horizontal mode profile of a diffused channel waveguide.

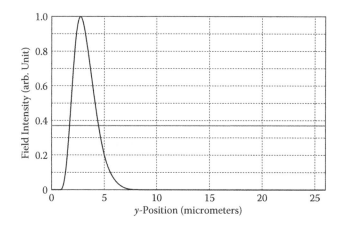

FIGURE 5.16 Vertical mode profile of a diffused channel waveguide.

The results in Figure 5.18 show that the mode width, Γ_x, corresponds well with the experimental result, with differences of less than 3%. The mode depth, Γ_y, however, matches only to within 8%. Despite the slight discrepancy, the SVMM's result still shows the qualitative characteristic of the diffused waveguide. We can also observe that the modal width, Γ_x, started at a large value and then decreased with wider Ti strip width before reaching higher values. Effectively, the larger initial mode size is due to the lower refractive index change resulting from a much narrower Ti width, thus causing the optical mode to be less confined. As the Ti strip becomes wider, it gives a higher change of refractive index, and hence a better confined optical mode. The mode width, however, would increase further as we increase the Ti width simply because of the increase in the physical width of the waveguide. At the same time, the larger physical width would enable the waveguide to support a higher-order mode.

Figure 5.19 depicts the variation of the normalized mode index b defined as[38] $b = (n_{eff}^2 - n_s^2)/(2\Delta n \cdot n_s)$ with respect to the variation of the width of the Ti strip. The

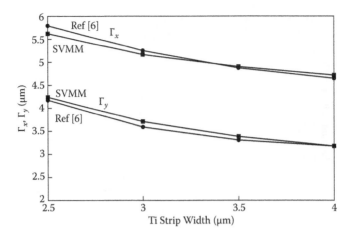

FIGURE 5.17   Comparison of simulated and experimental mode sizes for $\tau = 700$ Å.

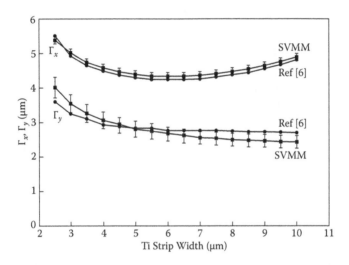

FIGURE 5.18   Comparison of simulated and experimental mode sizes for $\tau = 700$ Å.

waveguide becomes more strongly guided, i.e., has a higher effective index, as the Ti width is increased. At the same time, higher-order modes begin to appear as the strip width gets significantly larger than 6 µm. Figure 5.19 shows the distribution of higher-order modes of a waveguide diffused with a 10 µm Ti strip width.

The modal depth, however, decreases with wider Ti strip width because any wider Ti strip width does not affect the diffusion depth, but lateral mode distribution would support higher-order modes. The surface index would increase with thicker Ti film, thus leading to smaller modal depth. The surface index, however, reaches a maximum value only as we increase w. Therefore by increasing the Ti strip width to a certain point, the modal depth would cease to decrease further, as observed by both the experimental and simulated results. At this point, lateral diffusion would dominate.

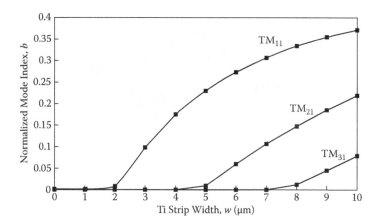

FIGURE 5.19   Normalized mode index, b, as a function of Ti strip width, w.

It is worth being reminded that the limiting case of increasing width in the Ti width is a planar waveguide.

Figure 5.21 shows simulation results with waveguides of the same diffusion parameter but with the Ti thickness for the diffusion of 800 Å. It shows that SVMM overestimates the mode size of the diffused waveguide. In fact, the thickness of 800 Å, as mentioned before, corresponds to the case where the Ti thickness has just depleted. In other words, the change of surface index is at its highest point for that diffusion time. The mode size would therefore appear to be much smaller than those waveguides in which the Ti has been diffused sufficiently longer than the time needed to just deplete all the Ti. This explains why the simulated modal width and depth are larger than the practical one. Figures 5.22 and 5.23 summarize the simulated results for a waveguide for a range of Ti thicknesses.

From the curves of Figures 5.22 and 5.23, we can see that the modal width and depth increase monotonically with the Ti thickness. It is not difficult to see that from the experimental results the diffusion model and diffusion parameters are no longer valid when the Ti film exceeds the thickness that is fully diffused of around 800 Å. For any thickness beyond that, we will need to resort to another diffusion model. In our case, this isn't necessary because having a Ti film thicker than the diffusable amount would lead to scattering loss, thus increasing the total insertion loss of the device.

In this section thus far, we have demonstrated how SVMM can be used apart from a simulating rib waveguide, to simulate a diffused channel waveguide. In fact, the finite difference method can be employed to obtain reasonable accuracy of the mode index and its distribution, as well as the evolution with different diffusion parameters for optical waveguides having an arbitrary index profile. Simulation of the Ti:LiNbO$_3$ waveguide, however, is not a straightforward matter because fabrication of such a waveguide is subjected to many changes, such as differences in crystal quality, diffusion process, density variations of the deposited titanium films, and measurement techniques. As a result, we can see inconsistencies in the published literature. Fouchet et al.[34] have shown the relation between the refractive index change $\Delta n_{e,o}(Z)$ and the Ti concentration $C(Z)$ in the mathematical form of

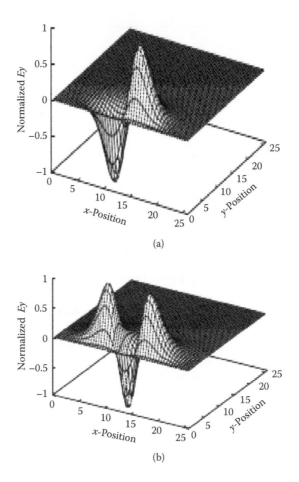

(a)

(b)

FIGURE 5.20   (a) TM$_{21}$ mode: $\tau = 750$ Å, $w = 10\,\mu$m; (b) TM$_{31}$ mode: $\tau = 750$ Å, $w = 10\,\mu$m.

$$\Delta n_{e,o}(Z) = A_{e,o}(C_o,\lambda) \cdot (C(Z))^{\alpha_{e,o}} \qquad (5.65)$$

The expression shows that the proportionality coefficient $A_{e,o}$ depends not only on the wavelength λ, but also on the diffusion parameters that are characterized by C_o, the Ti surface concentration.

In other words, the diffusion model that we used in our simulation is only a crude representation of the diffused waveguide. To enhance the accuracy of the simulation, we will need to provide a more accurate diffusion model that takes into account the dispersion relationship of the change in refractive index profile in Ti:LiNbO$_3$; despite being a crude representation of the diffusion process, it is still sufficient to demonstrate the credibility of SVMM in modeling a diffused waveguide. The simulations that we did in this particular section not only have shown the usefulness of SVMM, but also have provided a qualitative overview of the design of the Ti:LiNbO$_3$ waveguide. At this point, this program can surely calibrate against diffusion data that are

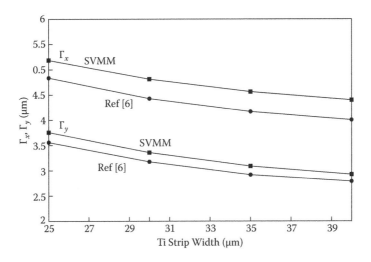

FIGURE 5.21   Comparison of simulated and experimental mode sizes for $\tau = 800$ Å.

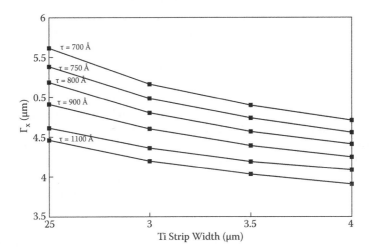

FIGURE 5.22   Simulated modal width Γ_x for a range of Ti thickness, τ.

measured in house and be used as a tool in the design of a Ti:LiNbO$_3$ waveguide for optical modulators.

5.5   MODE MODELING OF RIB WAVEGUIDES

In every finite difference approach, a few approximations are made and will therefore introduce some error into the final result. The following are a few approximations that are likely to introduce some error in our calculation: (1) the approximation of the full vectorial wave equation by the semivectorial one, (2) the replacement of the differential equation with the difference equations, (3) discretization error,

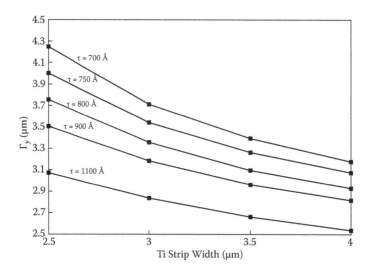

FIGURE 5.23 Simulated modal width Γ_x for a range of Ti thickness, τ.

(4) round-off error, and (5) the errors that are introduced by the NSPCG numerical solver itself.

To assess the accuracy, capability, and limitations of our program, we have calculated fundamental mode indices of three well-known rib waveguides that are often used as waveguide modeling benchmarks. Results of polarized modes have been published.[10–22] The geometry of the rib waveguide is shown in Figure 5.24. Parameters include the width of the rib w, height of the rib h, thickness of the guiding layer underneath the rib d, index of the substrate n_s, and index of the guiding layer n_g, listed in Table 5.1. The refractive index of the air cladding region, n_c, is unity.

The three waveguides each have a different characteristic. Structure 1 has relatively large vertical refractive index steps ($\Delta n = 2.44$ and 0.1), which could, for example, correspond to a GaAs guiding layer bound by air and a $Ga_{0.75}Al_{0.25}As$ confining layer. In the lateral direction, the rib height is large and the width narrow. This structure, with strong light confinement in both the lateral and vertical directions, is useful for curved guides, as radiation loss is minimized. This structure does not allow application of the effective index method because the slab outside the rib is cut off.

TABLE 5.1

Parameters of Rib Waveguide for Calculation Benchmark

Guide	n_g	n_s	d (μm)	h (μm)	w (μm)
1	44	34	0.2	1.1	2
2	44	36	0.9	0.1	3
3	44	435	5	2.5	4

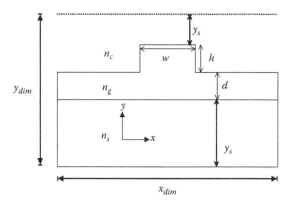

FIGURE 5.24   Typical structure of rib waveguide.

Structure 2 shows a weakly guiding feature. In this case the rib height is much less, allowing the mode to extend laterally. This is particularly useful for directional coupler structures, as strong coupling between adjacent guides will result in short coupling lengths. The guiding layer thickness is made small to give a thin mode shape in the vertical direction, and thus low-voltage operation. Essentially, this structure is tightly confined vertically and weakly confined horizontally. Such features enable the application of the effective index method[1,16,24] because the small etch step and large width-to-height ratio are the conditions of validity of this approximate method.

Structure 3 gives a good coupling to an optical fiber. Insertion loss is a crucial parameter for most waveguide devices and is determined by propagation loss and losses due to mode mismatch. Fresnel reflection loss is also important, but can be reduced to insignificant levels by using $\lambda/4$ antireflection coatings. Mode profiles of a circularly symmetric optical fiber and a waveguide will, in general, be different, due to the differing refractive indices of the semiconductor and the fiber, and also the differing shapes of the modes. The effects of both factors may be alleviated by the use of appropriate waveguide designs. In structure 3 the guiding layer is relatively thick, and the strip width and height are adjusted to give a more symmetric mode shape. In this structure the slab mode is near cutoff. Again, it should be pointed out that because the rib height is nearly twice the slab thickness and the rib width is less than the rib height, the accuracy of the effective index method is expected to be poor. Figures 5.25 to 5.28 are the contour plot and 3D plot of the TE-polarized mode of the three-waveguide structure calculated by the SVMM program. Similarly, as an example, Figure 5.28(a) shows the three-dimensional plot of TE-polarized mode profile for waveguide of structure 3 listed in Table 5.1; Figure 5.28(b) plots the field contour of TE-polarized mode profile for such waveguide.

The grid sizes h_x and h_y are 0.1. Since we assume that the field value around the computational boundary is zero, this would mean that we require a much larger computational window for both structures 2 and 3, so that the assumption would be valid. This, however, would mean that we can either use a coarser grid, leading to reduction in computing accuracy, or maintain the grid size but face up with a huge eigenmatrix to solve. For that reason, the variable grid size comes in handy. We can avoid a severe

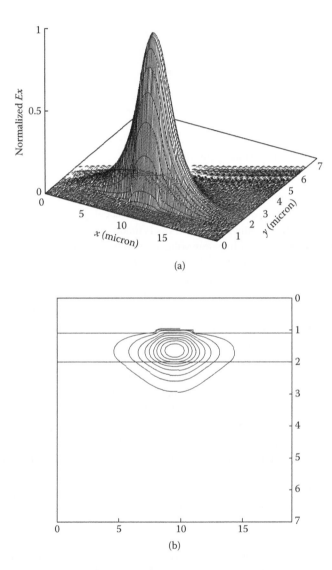

FIGURE 5.25 (a) Three-dimensional plot of TE-polarized mode profile for waveguide structure with low rib; (b) contour plot of TE-polarized mode profile.

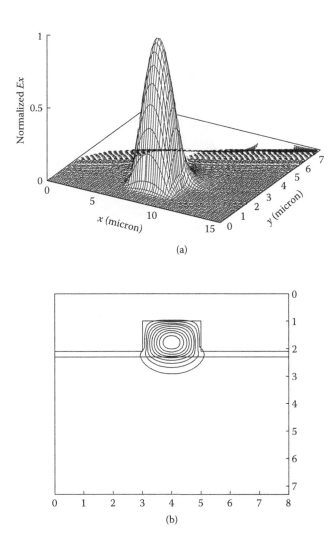

(a)

(b)

FIGURE 5.26 (a) Three-dimensional plot of TE-polarized mode profile for waveguide structure 1; (b) contour plot of TE-polarized mode profile for waveguide structure 1.

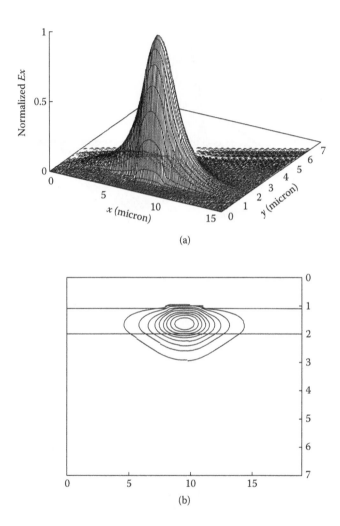

FIGURE 5.27 (a) Three-dimensional plot of TE-polarized mode profile for waveguide structure 2; (b) contour plot of TE-polarized mode profile for waveguide structure 2.

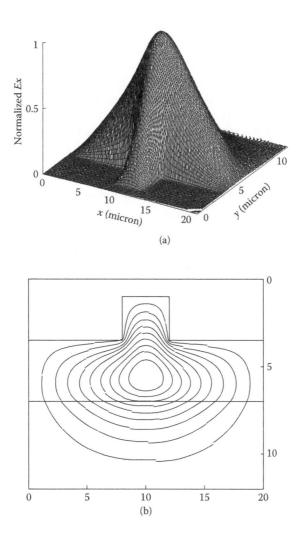

(a)

(b)

FIGURE 5.28   (a) Three-dimensional plot of TE-polarized mode profile for waveguide structure 3; (b) contour plot of TE-polarized mode profile for waveguide structure 3.

storage penalty by judiciously placing the denser mesh around the area the higher field value is assuming and coarser mesh at the region of the much lower field value. This would thus allow us to extend the boundary of the computation without incurring a severe storage problem while preserving the accuracy of the computation. The choice of grid size and its influence on the accuracy of the final results are discussed and illustrated in the next section.

5.5.1 CHOICE OF GRID SIZE

A judicious choice of grid size is likely to produce a plausible simulation result. To assess the effect of grid size on the accuracy of our simulation program, we compute the effective index for the TE-polarized mode of structure 1 by varying the grid size in both the x- and y-directions, namely, h_x and h_y. We compare our result with the one simulated by Lusse et al.,[11] which uses a dense mesh of 508×394 mesh points, the full vectorial finite difference method.

In simulations 1 to 6 (see Table 5.2), the value of $h_y = 0.1$ is kept constant while reducing h_x from 0.5 down to 0.025. As we can see, as h_x reaches 0.025, we can no

TABLE 5.2

Calculation of Effective Index with Different Choices of Grid Size

Sim. No.	h_x (μm)	h_y (μm)	xdim (μm)	ydim (μm)	Total Grid	Effective Index
1	0.5	0.1	8.0	7.3	16×73	3,9134,74
2	0.25	0.1	8.0	7.3	32×73	3,899,896
3	0.125	0.1	8.0	7.3	64×73	3,895,512
4	0.1	0.1	8.0	7.3	80×73	3,894,906
5	0.05	0.1	8.0	7.3	160×73	3,894,048
6	0.025	0.1	8.0	7.3	320×73	3,893,836
7	0.025	0.05	8.0	7.3	320×146	3,888,583
8	0.025	0.025	8.0	7.3	320×292	3,887,148
9	0.0–2.0:0.1 2.0–2.5:0.05 2.5–0:0.025 0.0–4.0:0.05 4.0–5.5:0.025 5.5–6.0:0.05 6.0–8.0:0.1	0.0025	8.0	7.3	240×292	3,887,162
10	0.0–2.0:0.1 2.0–2.5:0.05 2.5–0:0.025 0.0–4.0:0.05 4.0–5.5:0.025 5.5–6.0:0.05 6.0–8.0:0.1	0.0–4.0:0.025 4.0–7.3:0.05	8.0	7.3	240×226	3,887,165
P. Lusse[11]	—	—	—	—	508×394	88,687

longer get a significant improvement on the accuracy. Further reduction of grid size down to 0.01 would be highly impractical because we would end up with 800 grid points along the x-direction, thus paying a high penalty in terms of computer memory. In simulations 7 and 8, we keep h_x at 0.025 while reducing h_y from 0.1 down to 0.025; another significant improvement in accuracy is shown and the results get very close to those simulated by Lusse et al.[11] with both h_x and h_y equal to 0.025, a grid size of 320 × 292; the difference of our calculated effective index from that of Lusse et al.[11] is 2.78×10^{-5}. Simulations 9 and 10 show how the nonuniform scheme could economize storage usage while preserving the desired accuracy. By placing a denser grid mesh around the region where higher field values are assumed and a coarser mesh for regions farther away, we manage to reduce our mesh size from 320 × 292 to 240 × 226 (a total reduction of 39,200 points) without significant loss in accuracy, as can be seen from the graph. The nonuniform grid allocation scheme has in this particular case shown its usefulness. (Note that each reduction of grid size needs to be multiplied by 26, for that is the amount of work space required by the coefficient matrix, eigenvector, and NSPCG numerical solver.)

5.5.2 NUMERICAL RESULTS

Table 5.3 shows the values of the propagation constants of both TE- and TM-polarized modes for all three waveguides. The results are compared with several published

TABLE 5.3

Comparisons of Effective Indices and Normalized Indices at λ = 1.55

Methods	Guide 1		Guide 2		Guide 3	
	n_{eff}	b	n_{eff}	b	n_{eff}	b
TE-Polarized Mode						
SVMM	3,887,148	0.4835	3,953,612	0.4391	4,368,918	0.3782
Sv-BPM[14]	388,711	0.4834	395,471	0.4405	436,805	0.3608
Helmholtz[15]	388,764	0.4839	395,560	0.4416	436,808	0.3614
SI[19]	38,874	0.4837	39,506	0.4354	43,688	0.3759
SV[25]	3,869,266	0.4656	3954	0.4401	4,368,112	0.3621
FD[26]	3,882,623	0.4789	3,952,147	0.4373	436,804	0.3611
TM-Polarized Mode						
SVMM	3,879,173	0.4755	390,647	0.3803	4,368,434	0.3685
Sv-BPM[14]	387,924	0.4756	390,693	0.3809	436,772	0.3543
Helmholtz[15]	387,990	0.4762	390,712	0.3811	346,772	0.3543
SI[19]	38,788	0.4752	39,032	0.3763	43,684	0.3669
SV[25]	3,867,447	0.4638	3,905,927	0.3796	4,367,719	0.3542
FD[26]	3,875,430	0.4718	3,905,701	0.3794	4,367,751	0.3549

results. The bolded entries of the table are results of our work. We can see from the tables that our results compare favorably with all the other published results.

The numerical results presented so far have indicated the order of accuracy of the SVMM programs. We can observe that the exemplar results are compared well with other published results.

5.5.3 Higher-Order Modes

In our earlier discussion, we indicated that the inverse power method can be used to work out the other eigenmodes of the waveguide. To illustrate that, we simulate the waveguide mode of the waveguide structure published by Rahman and Davies.[22] Table 5.4 outlines the parameters of the waveguide structure. Figures 5.29 and 5.30 show the fundamental mode and the leading asymmetric mode of the TE-polarized field.

The leading asymmetric mode of Figure 5.30 can be obtained with an initial eigenvalue that is close to the eigenvalue of the leading asymmetric mode. One way to acquire a good initial guess for an independent eigenvalue is by perturbing the last few significant digits of the last calculated eigenvalue. In our case, the eigenvalue of the fundamental mode (see Figure 5.29) was calculated to be 347.78889. We then proceeded to the calculation of the asymmetric mode with an initial guess of 346. Other eigenmodes can also be worked out in similar fashion. However, we need to remember that there is only a limited number of eigenmodes supported by a certain waveguide structure. A good indication that the particular eigenmode is physically not feasible is an effective index that is lower than that of the refractive index of the substrate, thus giving a negative value of the normalized index. This is illustrated in Figure 5.31.

The third-order mode distribution depicted in Figure 5.32 is acquired by further reducing the initial guess of the eigenvalues from 346 to 345. As a result, we get an effective index of 398, which is lower than the refractive index of the substrate, which is 40 in this case. This results in a normalized index b of -0.047. As shown in the contour plot, most of the field is radiated into the substrate of the waveguide. Similarly, Figure 5.32 shows the 3D plot of the third-order mode, which is a radiation mode that is not guided in the waveguide region.

This feature of SVMM that enables us to work out the higher-order modes is extremely important for finding out if the designed waveguide can support multi-mode operation. We will see in the chapter on circular optical waveguides how such a feature can be exploited in the design of a single-mode waveguide.

TABLE 5.4
Parameters of Rib Waveguide ($\lambda = 1.15$ µm)

Guide	n_g	n_s	d (µm)	h (µm)	w (µm)
Rahman and Davies[22]	44	40	0.5	0.5	3

Source: Rahman, B.M.A., and Davies, J.B., *IEE Proc. J*, 132, 349–355, 1985.

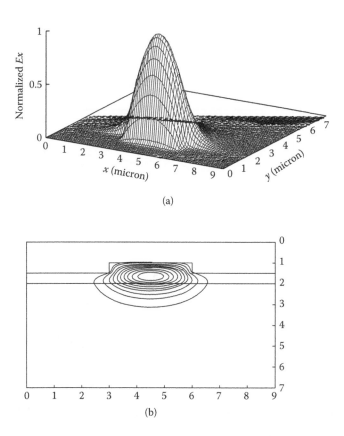

(a)

(b)

FIGURE 5.29 (a) Three-dimensional plot of fundamental mode of waveguide; (b) contour plot of the fundamental mode of the waveguide. (From Rahman, B.M.A., and Davies, J.B., *IEE Proc. J*, 132, 349–355, 1985.)

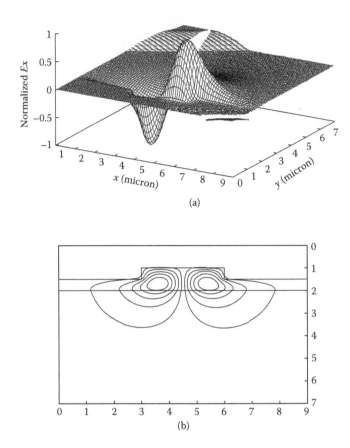

(a)

(b)

FIGURE 5.30 (a) Three-dimensional plot for the leading asymmetric mode of waveguide (calculated effective index = 4025302); (b) contour plot of leading asymmetric mode of waveguide. (From Rahman, B.M.A., and Davies, J.B., *IEE Proc. J*, 132, 349–355, 1985.)

5.6 CONCLUSIONS

In this chapter, the simplified approach for analytical study of channel waveguide, the 3D version, is described using the Marcatilli method and effective index techniques. Simplified analytical dispersion relations have been obtained for these 3D waveguides. An example design of a GeO_2-doped core rectangular channel waveguide is given.

Further, we have successfully developed numerical techniques based on a semivectorial finite difference analysis to solve the Helmholz equation. The numerical model that we have formulated can accurately and effectively model the guided modes in optical waveguides of arbitrary index profile distribution. A nonuniform mesh allocation scheme is employed in the formulation of the difference equations to free more computer memory for the computation of waveguide regions that bear greater significance. The accuracy of our computer program, SVMM, is assessed by computing the propagation constants and the effective indices of several rib

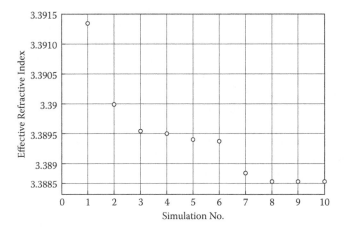

FIGURE 5.31   Refractive index variation with simulation number to obtain converged solution.

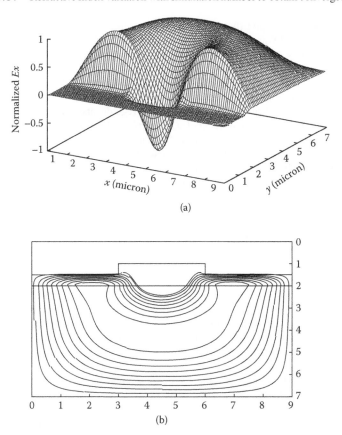

FIGURE 5.32   (a) Three-dimensional plot of the third-order mode, which is not supported by the waveguide structure (calculated effective index = 3980958, normalized index = −0.047314); (b) contour plot of radiated mode.

waveguides that have been known to be excellent benchmark waveguide structures. The results of our computation have compared favorably with other published
results.[14,15,19,25,26] We then continue to simulate the optical guided modes of diffused
optical waveguides in $LiNbO_3$. The computed mode size is consistent with published
experimental results. Our simulations, however, have shown the inadequacies of the
adopted diffusion model for its inability to model the diffused waveguide in a more
robust sense. It is suggested that further research be conducted for a more refined
and robust representation of the refractive index profile of the $Ti:LiNbO_3$-diffused
waveguide. Despite the shortcomings of the diffusion model that we have adopted,
we have demonstrated the potential of SVMM to be used as an analytical and design
tool for an integrated optical waveguide.

REFERENCES

1. M. Fukuma and J. Noda, "Optical Properties of Titanium-Diffused LiNbO3 Strip
   Waveguides and Their Coupling to Fiber Characteristics," *Appl. Opt.*, 20(4), 591–597,
   1980.
2. C. Bulmer et al., "High Efficiency Flip-Chip Coupling between Single Mode Fibers and
   LiNbO3 Channel Waveguides," *Appl. Phys. Lett.*, 37, 351–35, 1981.
3. V. Ramaswamy et al., "High Efficiency Single Mode Fiber to Ti:LiNbO3 Waveguide
   Coupling," *Elect. Lett.*, 10, 30–31. 1982.
4. R.C. Alferness et al., "Efficient Single Mode Fiber to Titanium Diffused Lithium Niobate
   Waveguide Coupling for l = 1.32μm," *IEEE J. Quant. Elect.*, QE-18, 1807–181, 1982.
5. E.A.J. Marcatilli, "Dielectric Rectangular Waveguide and Directional Couplers for
   Integrated Optics," *Bell Syst. Technol. J.*, 48, 2071–2102, 1969.
6. P.G. Suchoski and R.V. Ramaswamy, "Minimum Mode Size Low Loss Ti:LiNbO3
   Channel Waveguides for Efficient Modulator Operation at 1.3 μm," *IEEE J. Quant.
   Elect.*, QE-23(10), 1673–1679, 1987.
7. K.S Chiang, "Review of Numerical and Approximate Methods for the Modal Analysis
   of General Optical Dielectric Waveguides," *Opt. Quant. Elect.*, 26, S113–S134, 1994.
8. S.M. Saad, "Review of Numerical Methods for the Analysis of Arbitrary Shaped
   Microwave and Optical Dielectric Waveguides," *IEEE Trans. Microw. Th. Technol.*,
   MTT-33(10), 894–899, 1985.
9. S. Seki et al., "Two Dimensional Analysis of Optical Waveguides with a Non-Uniform
   Finite Difference Method," *IEE Proc. J Optoelect.*, 138(2), 123–127, 1991.
10. M.J. Robertson et al., "Semiconductor Waveguides: Analysis of Optical Propagation in
    Single Rib Structures and Directional Couplers," *IEE Proc. J Optoelect.*, 132(6), 336–
    342, 1985.
11. P. Lusse, P. Stuwe, et al., "Analysis of Vectorial Mode Fields in Optical Waveguides by
    a New Finite Difference Method," *IEEE J. Lightw. Technol.*, 12(11), 487–493, 1994.
12. M.S Stern, "Semivectorial Polarised *H* Field Solutions for Dielectric Waveguides with
    Arbitrary Index Profiles," *IEE Proc. J*, 135(5), 333–338, 1988.
13. W. Huang and H.A. Haus, "A Simple Variational Approach to Optical Rib Waveguides,"
    IEEE J. Lightw. Technol., 9(1), 56–61, 1991.
14. P.-L. Liu and B.-J. Li, "Semivectorial Beam Propagation Method for Analysing Polarised
    Modes of Rib Waveguides," *IEEE J. Quant. Elect.*, 28(4), 778–782, 1992.
15. P.-L. Liu and B.-J. Li, "Semivectorial Helmholtz Beam Propagation by Lanczos
    Reduction," *IEEE J. Quant. Elect.*, 29(8), 2385–2389, 1993.
16. T.M. Benson et al., "Rigorous Effective Index Method for Semiconductor Rib
    Waveguides," *IEE Proc. J*, 139(1), 67–70, 1992.

17. P.C. Kendall et al., "Advances in Rib Waveguide Analysis Using Weighted Index Method or the Method of Moments," *IEE Proc. J*, 137(1), 27–29, 1990.

18. S.V. Burke, "Spectral Index Method Applied to Rib and Strip-Loaded Directional Couplers," *IEE Proc. J*, 137(1), 7–10, 1990.

19. M.S. Stern, "Analysis of the Spectral Index Method for Vector Modes of Rib Waveguides," *IEE Proc. J*, 137(1), 21–26, 1990.

20. G. Ronald Hadley and R.E. Smith, "Full Vector Waveguide Modeling Using an Iterative Finite Difference Method with Transparent Boundary Conditions," *IEEE J. Lightw. Technol.*, 13(3), 465–469, 1995.

21. T.M. Benson et al., "Polarisation Correction Applied to Scalar Analysis of Semiconductor Rib Waveguides," *IEE Proc. J Optoelect.*, 139(1), 39–41, 1992.

22. B. M. A Rahman and J. B Davies, "Vectorial-H Finite Element Solution of GaAs/GaAlAs Rib Waveguides," *IEE Proc. J*, 132, 349–356, 1985.

23. C.M. Kim and R.V Ramaswamy, "Modelling of Graded Index Channel Waveguides Using Non-Uniform Finite Difference Method," *IEEE J. Lightw. Technol.*, 7(10), 1581–1589, 1989.

24. G.B. Hocker and W.K. Burns, "Mode Dispersion in Diffused Channel Waveguides by the Effective Index Method," *Appl. Opt.*, 16(1), 113–118.

25. M.S. Stern, "Semivectorial Polarised Finite Difference Method for Opitcal Waveguides with Arbitrary Index Profiles," *IEE Proc. J. Optoelect.*, 138, 185–190, 1990.

26. M.S. Stern, "Rayleigh Quotient Solution of Semivectorial Field Problems for Optical Waveguides with Arbitrary Index Profiles," *IEE Proc. J Optoelect.*, 138, 185–190, 1990.

27. K. Ogusu, "Numerical Analysis of the Rectangular Dielectric Waveguide and Its Modifications," IEEE Trans. Microw. Th. Tech., MTT-25(11), 874–885, 1977.

28. J. E. Goell, "A Circular-Harmonic Computer Analysis of Rectangular Dielectric Waveguide," *Bell. Syst. Techn. J.*, 48, 2133–2160, 1969.

29. R.L. Burden and J.D. Faires, *Numerical Analysis*, 4th ed., PWS-Kent Publishing Company, pp. 492–505.

30. W.H. Press et al., *Numerical Recipes—The Art of Scientific Computing*, Cambridge University Press, Cambridge, UK, pp. 377–379.

31. T.C. Oppe et al., *NSPCG User's Guide Version 1.0—A Package for Solving Large Sparse Linear Systems by Various Iterative Methods*, Center for Numerical Analysis, University of Texas, Austin.

32. S.K. Korotky, W.J. Minford, et al., "Mode Size and Method for Estimating the Propagation Constant of Single Mode Ti:LiNbO3 Strip Waveguides," *IEEE J. Quant. Elect.*, QE-18(10), 1796–1801, 1982.

33. M. Minakata, S. Shaito, and M. Shibata, "Two Dimensional Distribution of Refractive Index Changes in Ti Diffused LiNbO3 Waveguides," *J. Appl. Phys.*, 50(5), 3063–3067, 1979.

34. S. Fouchet et al., "Wavelength Dispersion of Ti Induced Refractive Index Change in LiNbO3 as a Function of Diffusion Parameters." *IEEE J. Lightw. Technol.*, LT-5(5), 700–708, 1987.

35. M. Minakata et al., "Precise Determination of Refractive Index Changes in Ti-Diffused LiNbO3 Optical Waveguides," *J. Appl. Phys*, 49(9), 4677–4682, 1978.

36. W.K. Burns et al., "Ti Diffusion in Ti:LiNbO3 Planar and Channel Optical Waveguides," *J. Appl. Phys.*, 50(10), 6175–6182, 1979.

37. M.D. Feit et al., "Comparison of Calculated and Measured Performance of Diffused Channel-Waveguide Couplers," *J. Opt. Soc. Am.*, 73(10), 1296–1304, 1983.

38. A. Sharma and P. Bindal, "Analysis of Diffused Planar and Channel Waveguides," *IEEE J. Quant. Elect.*, 29(1), 150–153, 1993.

39. E. Strake et al., "Guided Modes of Ti:LiNbO3 Channel Waveguides: A Novel Quasi Analytical Technique in Comparison with the Scalar Finite-Element Method," *IEEE J. Lightw. Technol.*, 6(6), 1126–1135, 1988.

40. M.R. Spiegel, *Mathematical Handbook of Formulas and Tables*, Shaum's Outline Series, McGraw Hill Book Company, New York, 1990.
41. G. Kotitz, "Properties of Lithium Niobate," EMIS data reviews series no. 5, INSPEC, The Institution of Electrical Engineers, London and New York, 1989.
42. N. Schulz et al., "Finite Difference Method without Spurious Solutions for the Hybrid-Mode Analysis of Diffused Channel Waveguides," *IEEE Trans. Microw. Th. Technol.*, 38(6), 722–729, 1990.
43. A. Sharma and P. Bindal, "An Accurate Variational Analysis of Single Mode Diffused Channel Waveguides," *Opt. Quant. Elect.*, 24, 1359–1371, 1992.
44. A. Sharma and P. Bindal, "Variational Analysis of Diffused Planar and Channel Waveguides and Directional Couplers," *J. Opt. Soc. Am. A*, 11(8), 2244–2248, 1994.
45. J. Ctyroky et al., "3-D Analysis of LiNbO3: Ti Channel Waveguides and Directional Couplers," *IEEE J. Quant. Elect.*, QE-20(4), 400–409, 1984.
46. M. Valli and A. Fioretti, "Fabrication of Good Quality Ti:LiNbO3 Planar Waveguides by Diffusion in Dry And wet O2 Atmospheres," *J. Modern Opt.*, 35(6), 885–890, 1988.
47. D.S. Smith et al., "Refractive Indices of Lithium Niobate," *Opt. Commun.*, 17(3), 332–335, 1976.
48. D.F. Nelson and R.M. Mikulyak, "Refractive Indices of Congruently Melting Lithium Niobate," *J. Appl. Phys.*, 45(8), 3688–3689, 1974.
49. K.T. Koai and P.-L. Liu, "Modelling of Ti:LiNbO3 Waveguide Devices: Part I—Directional Couplers," *IEEE J. Lightw. Technol.*, 7(3), 533–539, 1989.
50. G.B. Hocker and W.K. Burns, "Modes in Diffused Optical Waveguides of Arbitrary Index Profile," *IEEE J. Quant. Elect.*, QE-11(6), 270–276, 1975.
51. R.C. Alferness et al., "Characteristics of Ti Diffused Lithium Niobate Optical Directional Couplers," *Appl. Opt.*, 18(23), 4012–4016, 1979.

6 Optical Fibers

Single- and Few-Mode Structures and Guiding Properties

6.1 OPTICAL FIBERS: CIRCULAR OPTICAL WAVEGUIDES

6.1.1 GENERAL ASPECTS

Planar optical waveguides compose a guiding region, a slab embedded between a substrate and a superstrate having identical or different refractive indices. The lightwaves are guided by the confinement of the lightwaves with an oscillation solution. The number of oscillating solutions that satisfy the boundary constraints are the number of modes that can be guided. The guiding of lightwaves in an optical fiber is similar to that of the planar waveguide, except the lightwaves are guiding through a circular core embedded in a circular cladding layer.

Within the context of this book on guided wave photonics, optical fibers would be most relevant as circular optical waveguides. We should point out the following developments in optical fiber communication systems so that we are able to focus on modern optical systems that engineers have based on the fundamental understanding of the electromagnetic field theory:

- The step index and graded index multimode optical fibers find very limited applications in systems and networks for long-haul applications.
- The single-mode optical fibers have been achieved for very small differences in the refractive indices between the core and cladding regions. Thus the guiding in modern optical fibers for telecommunications is called weakling guiding. This development was intensively debated and agreed on by the optical fiber communications technology community during the late 1970s.
- The invention of optical amplification in rare-earth-doped single-mode optical fibers in the late 1980s has transformed the design and deployment of optical fiber communication systems and networks in the last decade and for the coming decades of the 21st century. The optical loss of the fiber and the optical components in the optical networks can be compensated for by using these fiber in-line optical amplifiers.
- Therefore, the pulse broadening of optical signals during the transmission and distribution in the networks has become much more important for system design engineers.

- Recently, due to several demonstrations of the use of digital signal processing of coherently received modulated lightwaves, multiple input multiple output (MIMO) techniques can be applied to enhance significantly the sensitivity of optical receivers, and thus the transmission distance and capacity of optical communication systems.[1] MIMO techniques would offer some possibilities of the use of different guided modes through a single fiber, for example, few-mode fibers that can support more than one mode but not too many, as in the case of multimode types. Thus, the conditions under which a circular optical waveguide can operate as a few-mode fiber are also described in this chapter.

Due to the above developments we shall focus the theoretical approach on the understanding of optical fibers toward the practical aspects for design optical fibers with minimum dispersion or for a specified dispersion factor. This can be carried out by, from practical measurements, the optical field distribution following a Gaussian distribution. Having known the field distribution, one would be able to obtain the propagation constant of the single-guided mode and, hence, the spot size of this mode—and thus the energy concentration inside the core of the optical fiber. Using the basic concept of optical dispersion by using the definition of group velocity and group delay, we would be able to derive the chromatic dispersion in single-mode optical fibers. After arming ourselves with the basic equations for dispersion, we would be able to embark on the design of optical fibers with a specified dispersion factor.

6.1.2 OPTICAL FIBER: GENERAL PROPERTIES

6.1.2.1 Geometrical Structures and Index Profile

An optical fiber consists of two concentric dielectric cylinders. The inner cylinder, or core, has a refractive index of $n(r)$ and radius a. The outer cylinder, or cladding, has an index n_2 with $n(r) > n_2$ and a larger outer radius. A core of about 4–9 μm and a cladding diameter of 125 μm are the typical values for a silica-based single-mode optical fiber. A schematic diagram of the structure of a circular optical fiber is shown in Figure 6.1. Figure 6.1(a) shows the core and cladding region of the circular fiber, while Figure 6.1(b) and (c) show the figure of the etched cross sections of a multimode and single mode, respectively. The silica fibers are etched in a hydroperoxide solution so that the core region doped with impurity would be etched faster than that of pure silica, and thus the exposure of the core region, as observed. Figure 6.2 shows the index profile and the structure of circular fibers. The refractive index profile can be step or graded.

The refractive index $n(r)$ of a circular optical waveguide is usually changed with radius r from the fiber axis ($r = 0$) and is expressed by

$$n^2(r) = n_2^2 + NA^2 s\left(\frac{r}{a}\right)$$ (6.1)

where NA is the numerical aperture at the core axis, while $s(r/a)$ represents the profile function that characterizes any profile shape ($s = 1$ at maximum) with a scaling parameter (usually the core radius).

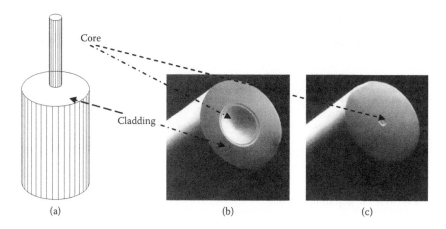

(a)   (b)   (c)

FIGURE 6.1 (a) Schematic diagram of the step index fiber: coordinate system, structure. The refractive index of the core is uniform and slightly larger than that of the cladding. For silica glass the refractive index of the core is about 1.478, and that of the cladding about 1.47, at the 1550 nm wavelength region. (b) Cross section of an etched fiber—multimode type—50 μm diameter. (c) Single-mode optical fiber-etched cross section.

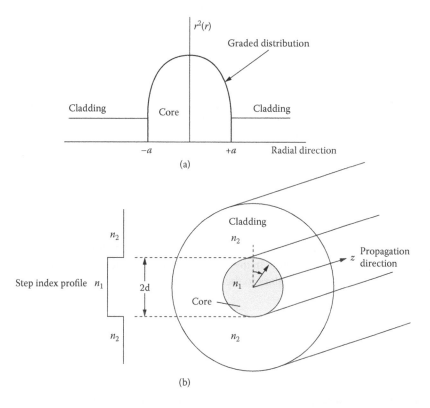

FIGURE 6.2 (a) Refractive index profile of a graded index profile. (b) Fiber cross section and step index profile with a as the radius of fiber.

6.1.2.1.1 Step Index Profile

In a step index profile the refractive index remains constant in the core region, thus

$$s\left(\frac{r}{a}\right) = 1 \quad \text{for } r \leq a \tag{6.2}$$

$$s\left(\frac{r}{a}\right) = 0 \quad \text{for } r > a \tag{6.3}$$

so we have for a step index profile

$$n^2(r) = n_1^2 \quad \text{for } r < a \tag{6.4}$$

and

$$n^2(r) = n_2^2 \quad \text{for } r > a \tag{6.5}$$

Exercise

Refer to the technical specification of the single-mode optical fiber Corning SMF-28. State whether the index profile of the fiber is a perfect step index profile or graded index profile. Is it true that the profile is a perfect step index distribution? If not, then what is the real manufactured profile?

6.1.2.1.2 Graded Index Profile

We consider hereafter the two most common types of graded index profiles: power-law index and the Gaussian profile.

6.1.2.1.2.1 Power-Law Index Profile
The core refractive index of an optical fiber usually follows a graded profile. In this case the refractive index rises gradually from the value n_2 of the cladding glass to the value at the fiber axis. Therefore, $s(r/a)$ can be expressed as

$$s\left(\frac{r}{a}\right) = \left\{1 - \left(\frac{r}{a}\right)^\alpha \quad \text{for } r \leq a \text{ and } = 0 \text{ for } r < a \tag{6.6}\right.$$

where α is the power exponent. Thus the index profile distribution $n(r)$ can be expressed in the usual way as (by using (6.6) and (6.2), by substituting $NA^2 = n_1^2 - n_2^2$)

$$
n^2(r) = \begin{cases} n_1^2 \left[1 - 2\Delta \left(\dfrac{r}{a} \right)^{\alpha} \right] & \text{for } r \le a \\[2ex] n_2^2 & \text{for } r > a \end{cases} \tag{6.7}
$$

where $\Delta = NA^2 / n_1^2$ is the relative refractive difference with a small difference between the cladding and the core regions. The profile shape given in (6.7) offers three special distributions:

- $\alpha = 1$: The profile function $s(r/a)$ is linear and is called a triangular profile.
- $\alpha = 2$: The profile is a quadratic function with respect to the radial distance and is called the parabolic profile.
- $\alpha = \infty$: The profile is a step type.

6.1.2.1.2.2 Gaussian Index Profile While in the Gaussian index profile, the refractive index changes gradually from the core center to a distance very far away from it, and $s(r)$ can be expressed as

$$
s\left(\frac{r}{a} \right) = e^{-\left(\frac{r}{a} \right)^2} \tag{6.8}
$$

6.1.3 FUNDAMENTAL MODE OF WEAKLY GUIDING FIBERS

The electric and magnetic fields $E(r,\phi,z)$ and $H(r,\phi,z)$ of the optical fibers in cylindrical coordinates can be found by solving Maxwell's equations. However, only the lower-order modes of ideal step index fibers are important for modern optical transmission systems. For single-mode optical fibers the relative refractive index difference $\Delta < 1\%$; that is, there is a very small difference between the refractive indices of the core and the cladding, and then optical waves are mildly confined and thus the optical field is gently guided. The electric and magnetic fields E and H can then take approximate solutions of the scalar wave equation in a cylindrical coordinate system (x, θ, ϕ) as

$$
\left[\frac{\delta^2}{\delta r^2} + \frac{1}{r} \frac{\delta}{\delta r} + k^2 n_j^2 \right] \varphi(r) = \beta^2 \varphi(r) \tag{6.9}
$$

where $n_j = n_1, n_2$, and $\varphi(r)$ is the spatial field distribution of the nearly transverse EM waves:

$$
\left[\frac{\delta^2}{\delta r^2} + \frac{1}{r} \frac{\delta}{\delta r} + k^2 n_j^2 \right] \varphi(r) = \beta^2 \varphi(r) \tag{6.10}
$$

where E_y, E_z, H_x, H_z are negligible and $\varepsilon = \varepsilon_0 n_2^2$ and $Z_0 = (\varepsilon_0 \mu_0)^{1/2}$ is the vacuum impedance. That is, the waves can be seen as a plane wave traveling down the fiber tube. This plane wave is reflected between the dielectric interfaces; in other words, it is trapped and guided along the core of the optical fiber.

6.1.3.1 Solutions of the Wave Equation for Step Index Fiber

The field spatial function $\varphi(r)$ would have the form of Bessel functions (from (6.9)):

$$\varphi(r) = A \frac{J_0(ur/a)}{J_0(u)}; \quad 0 < r < a: \text{ in core region} \tag{6.11}$$

$$\varphi(r) = A \frac{K_0(vr/a)}{K_0(v)}; \quad r > a: \text{ in cladding region} \tag{6.12}$$

where J_0 and K_0 are the Bessel functions of the first kind and a modification of the second kind, respectively, and u, v are defined as

$$\frac{u^2}{a^2} = k^2 n_1^2 - \beta^2 \tag{6.13a}$$

$$\frac{v^2}{a^2} = -k^2 n_2^2 + \beta^2 \tag{6.13b}$$

Following the Maxwell's equations relation, we find that E_z can take two possible solutions, which are orthogonal as

$$E_z = -\frac{A}{kan_2} \left(\begin{array}{c} \sin\phi \\ \cos\phi \end{array} \right\} \left| \begin{array}{cc} \dfrac{uJ_1\left(u\dfrac{r}{a}\right)}{J_0(u)} & \text{for } 0 \le r < a \\[4mm] \dfrac{vK_1\left(\dfrac{vr}{a}\right)}{K_0(v)} & \text{for } r > a \end{array} \right. \tag{6.14}$$

The terms u and v must satisfy simultaneously two equations:

$$u^2 + v^2 = V^2 = ka(n_1^2 - n_2^2)^{1/2} = kan_2(2\Delta)^{1/2} \tag{6.15}$$

and

$$u \frac{J_1(u)}{J_0(u)} = v \frac{K_1(v)}{K_0(v)} \tag{6.16}$$

where (6.16) is obtained by applying the boundary conditions at the interface $r = a$ (E_z is the tangential component and must be continuous at this dielectric interface). Equation (6.16) is commonly termed the eigenvalue equation of the wave equation for guiding. The solution of this equation would give specific discrete values of β, the propagation constants of the guided lightwaves.

6.1.3.2   Single- and Few-Mode Conditions

Equation (6.15) shows that the longitudinal field is in the order of $u/(kan_2)$ with respect to the transverse component. In practice $\Delta \ll 1$, and by using (6.15), we observe that this longitudinal component is negligible compared with the transverse component. Thus the guided mode is *transversely polarized*. The fundamental mode is then usually denominated as an LP_{01} mode (LP stands for linearly polarized), for which the field distribution is shown in Figure 6.5(a) and (b). The graphical representation of the eigenvalue equation (6.16) calculated as the variation of $b = \beta/k$ as the normalized propagation constant and the V-parameter is shown in Figure 6.6(d). There are two possible polarized modes, the horizontal and vertical polarizations, which are orthogonal to each other. These two polarized modes can be employed for transmission of different information channels. They are currently exploited, in the first two decades of the 21st century, in optical transmission systems employing polarization division multiplexed so as to offer a transmission bit rate of 100 Gb/s and beyond.[2,3] Furthermore, when the number of guided modes is higher than two polarized modes, they form a set of modes over which information channels can be simultaneously carried and *spatially* demultiplexed at the receiving end, so as to increase the transmission capacity, as illustrated in Figure 6.5(a) and (b)[4,5] Such few-mode fibers are employed in the most modern optical transmission system, whose schematic is shown in Figures 6.3 and 6.4. Obviously there must be a demultiplexing process to split the model patterns. A number of the guided modes supported by the fiber can be modulated and spatially multiplexed, then coupled to the fiber line for transmission to the other end. They are then spatially demultiplexed into individual modes and then fed into optical receivers. Note the two possible polarizations of the mode LP_{11}. The separated modes are coherently detected to give electronic signals that are then amplified and sampled and converted into digital forms. These digital signals are then processed in ultra-high-speed digital signal processors (DSPs) using the MIMO technique, which is well developed in wireless transmission technology.

The number of guided modes is determined by the number of intersecting points of the circle of radius V and the curves representing the eigenvalue solutions (6.16). Thus for a single-mode fiber the V-parameter must be less than 2.405, and for few-mode fiber this value is higher; for example, if $V = 2.8$, we have three intersecting points between the circle of radius V and three curves, and then the number of modes would be and their corresponding alternative polarized modes would be LP_{01}, LP_{11} as shown in Figure 6.5(b) and (c). For a single mode there are two polarized modes whose polarizations can be vertical or horizontal. Thus a single-mode fiber is not a monomode type but supports two polarized modes! The main issues are on the optical amplification gain for the transmission of modulated signals in such few-mode fibers. This remains to be the principal obstacle.

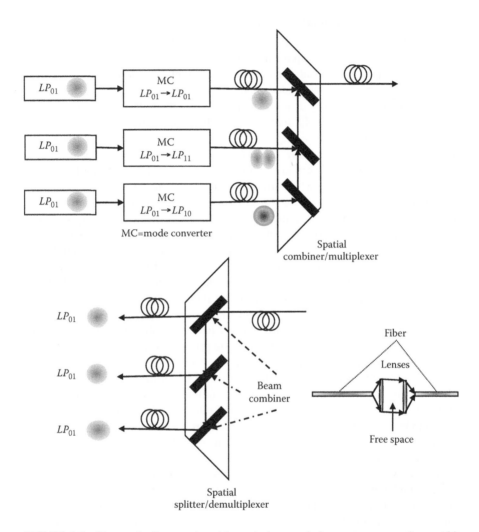

FIGURE 6.3 Few-mode fiber employed in optical transmission systems operating at 100 Gb/s and higher bit rates.

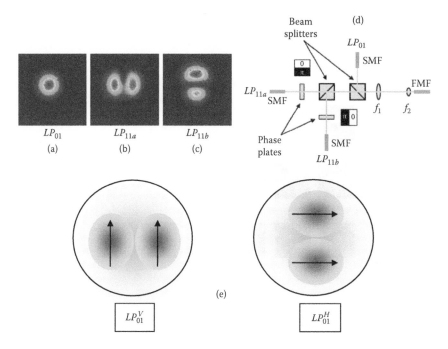

FIGURE 6.4 (a–c) Intensity profiles of the first few-order modes of a few-mode optical fiber employed for a 5×65 Gbps optical transmission system[6] and an optical system arrangement for the spatially demux and mux of modal channels. (d) Arrangements of the spatial multiplexer for coupling three modes into the transmission fiber. (e) Horizontal and vertical polarized modes $LP_{01}^{V,H}$; polarization directions are indicated by arrows.

We can illustrate the propagation of the fundamental mode and higher-order modes as in Figure 6.6(a) and (b). The rays of the modes can be axially straight or skewed and twisted around the principal axis of the fiber. Thus there are different propagation times between these modes. This property can be employed to compensate for the chromatic dispersion effect.[7] Figure 6.6 shows the graphical solution of the modes of optical fibers. In Figure 6.6(d) the regions of single operation, and then higher-order, second-order mode regions as determined by the value of the V-parameter, are indicated. Naturally due to manufacturing accuracy the mode regions would be variable from fiber to fiber.

Figure 6.7 shows the coherent mixing of a local oscillator laser and the signals, which can consist of two polarized channels and be modulated using quadrature amplitude modulation (QAM). A μ/2 phase shifter is required to split the real part, the in-phase component, and the quadrature one. Both polarization and optical phase shifting are employed so that the polarized channels can be transmitted in order to double the capacity of the transmission systems. The QAM can be used to obtain the best performance as well as to increase the spectral efficiency by high-order modulation. The optically processed signals are then detected by the photodetectors in pairs and connected back to back, the balanced detection, in which the mixing of the signals and local laser happens to recover the amplitude of the in-phase

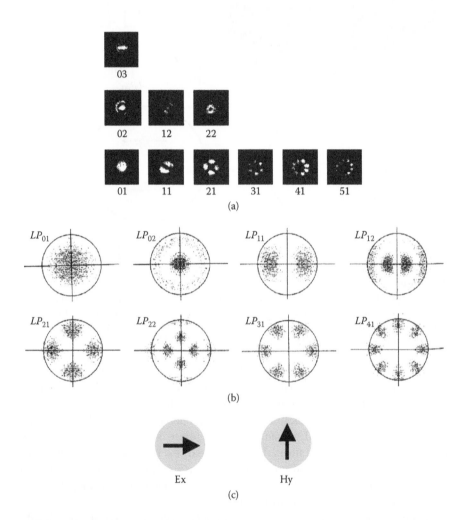

FIGURE 6.5 (a) Spectrum of guided modes in a few/multimode fiber; numbers indicate order of modes. (b) Calculated intensity distribution of LP-guided modes in step index optical fibers with $V = 7$. (c) Electric and magnetic field distribution of an LP_{01} mode polarized along Ox (H-mode) and Oy (V-mode) of the fundamental mode of a single-mode fiber.

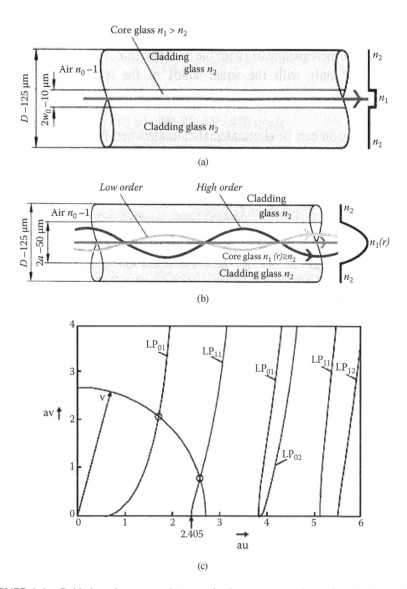

FIGURE 6.6 Guided modes as seen by a ray in the transverse plane of a circular optical fiber. (a) Ray model of lightwave propagating in the single-mode fiber. (b) Ray model of propagation of different modes guided in a few/multimode graded index fiber. (c) Graphical illustration of solutions for eigenvalues (propagation constant – wavenumber of optical fibers).

Continued

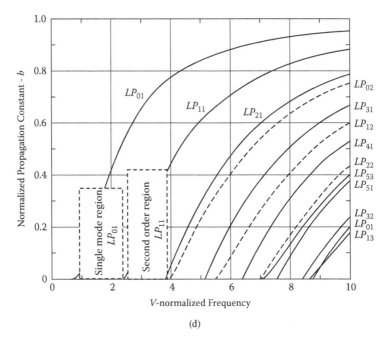

(d)

FIGURE 6.6 (*Continued*)   Guided modes as seen by a ray in the transverse plane of a circular optical fiber. (d) *b-V* characteristics of guided fibers.

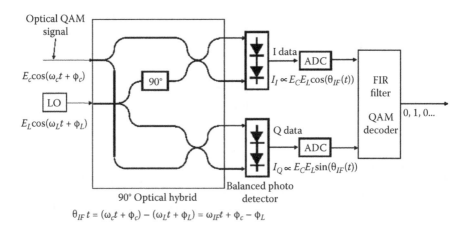

FIGURE 6.7 $\pi/2$ hybrid coupler for polarization demultiplexing and mixing with a local oscillator in a coherent receiver of a modern DSP-based optical receiver for detection of phase-modulated schemes.

and quadrature components, hence the magnitude and phase. These signals are then amplified and sampled to convert into a digital domain, so that they can be process digitally. This type of detection and processing in the digital domain represents the most modern technique to date for dual-polarization lightwave transmission and a coherent technique at an ultra-high bit rate of 100 Gb/s and beyond. When the frequency of the local laser equals that of the signals, we have homodyne detection; if not, then we have heterodyne detection or intradyne detection, depending on whether the frequency difference is outside or within the signal band.

6.1.3.3 Gaussian Approximation

6.1.3.3.1 Fundamental Mode Revisited

We note again that $\underset{\sim}{E}$ and $\underset{\sim}{H}$ are approximate solutions of the scalar wave equation, and the main properties of the fundamental mode of weakly guiding fibers can be observed as follows:

- The propagation constant β (along the z-direction) of the fundamental mode must lie between the core and cladding wavenumbers. This means the effective refractive index of the guided mode lies with the range of the cladding and core refractive indices.
- Accordingly, the fundamental mode must be nearly a transverse electromagnetic wave, as described by Equation (6.10).

$$\frac{2\pi n_2}{\lambda} < \beta < \frac{2\pi n_1}{\lambda} \tag{6.17}$$

- The spatial dependence $\varphi(r)$ is a solution of the scalar wave equation (6.9).

6.1.3.3.1.1 Gaussian Approximation
The main objectives are to find a good approximation for the field $\varphi(r)$ and the propagation constant β, which can be found though the eigenvalue equation and Bessel's solutions as shown in the previous section. It is desirable if we can approximate the field to good accuracy to obtain simple expressions and have a clearer understanding of light transmission on a single-mode optical fiber without going through graphical or numerical methods. Furthermore, experimental measurements and numerical solutions for step and power-law profiles show that $\varphi(r)$ is approximately Gaussian in appearance. We thus approximate the field of the fundamental mode as

$$\varphi(r) \cong Ae^{-\frac{1}{2}\left(\frac{r}{r_0}\right)^2} \tag{6.18}$$

where r_o is defined as the spot size, i.e., at which the intensity equals e^{-1} of its maximum. Thus by multiplying the wave equation (6.9) by $r\varphi(r)$, and using the identity,

$$r\varphi\frac{\delta^2\varphi}{\delta r^2} + \varphi\frac{\delta\varphi}{\delta r} = \frac{\delta}{\delta r}\left(r\varphi\frac{\delta\varphi}{\delta r}\right) - r\left(\frac{\delta\varphi}{\delta r}\right)^2 \tag{6.19}$$

then by integrating from 0 to infinitive and using

$$\left[r\Psi \frac{d\varphi}{dr} \right]_0^\infty = 0$$

we have

$$\beta^2 = \frac{\int_0^\infty \left[-\left(\frac{\delta\varphi}{\delta r}\right)^2 + k^2 n^2(r)\varphi^2 \right] r\,\delta r}{\int_0^\infty r\varphi^2\,\delta r} \tag{6.20}$$

The procedure to find the spot size is then followed by substituting $\varphi(r)$ (Gaussian), Equation (6.18) into (6.20), then differentiating and setting $\delta^2\beta/\delta r$ evaluated at to zero; that is, the propagation constant β of the fundamental mode *must* give the largest value of r_0.

Knowing r_0 and β, the fields E_x and H_y Equation (6.10) are fully specified.

6.1.3.3.1.1.1 Case 1: Step Index Fiber

Substituting the step index profile given by (6.10) and $\varphi(r)$ Equation (6.18) into (6.20) leads to an expression for β in terms of r_0 given by

$$V = NA \cdot k \cdot a = NA \frac{2\pi}{\lambda} a \tag{6.21}$$

The spot size is thus evaluated by setting

$$\frac{\delta^2\beta}{\delta r_0} = 0 \tag{6.22}$$

That is when the propagation constant is largest. For a step index core fiber, the spot size r_0 is given by

$$r_0^2 = \frac{a^2}{\ln V^2} \tag{6.23}$$

Substituting (6.23) into (6.21) we have

$$(a\beta)^2 = (akn_1)^2 - \ln V^2 - 1 \tag{6.24}$$

This expression is physically meaningful only when $V > 1$ (r_0 is positive).

6.1.3.3.1.1.2   Case 2: Gaussian Index Profile Fiber   Similarly, for the case of a Gaussian index profile, by following the procedures for a step index profile fiber we can obtain

$$(a\beta)^2 = (an_1 k)^2 - \left(\frac{a}{r_0}\right)^2 + \frac{V^2}{\left(\dfrac{a}{r_0} + 1\right)} \tag{6.25}$$

and

$$r_0^2 = \frac{a^2}{V - 1} \text{ by using } \frac{\delta^2 \beta}{\delta r_0} = 0 \tag{6.26}$$

that is, maximizing the propagation constant of the guided waves. The propagation constant is at maximum when the "light ray" is very close to the horizontal direction. Substituting (6.26) into (6.25), we have

$$(a\beta)^2 = (akn_1)^2 - 2V + 1 \tag{6.27}$$

Equation (6.27) is physically meaningful only when $V > 1$ ($r_0 > 0$).

It is obvious from (6.27) that the spot size of the optical fiber with a V-parameter of 1 is extremely large; hence, all energy is not concentrated in the core of the fiber but in the cladding region. It is very important that one must not design the optical fiber with a near unit value of the V-parameter. In practice we observe that the spot size is large but finite (observable). In fact, if $V < 1.5$, the spot size becomes large and the effect of this on the dispersion of the signal propagating through a single-mode fiber will be investigated in detail in the next chapter.

6.1.3.4   Cutoff Properties

Similar to the case of planar dielectric waveguides, from Figure 6.6 we observe that when we have $V < 2.405$, only the fundamental LP_{01} exists. Thus we have what is shown in Figure 6.8.

We note that for single-mode operation the V-parameter must be less than or equal to 2.405. However, in practice $V < 3$ can be acceptable for single-mode operation.

Indeed, the value 2.405 is the first zero of the Bessel function $J_0(u)$. In practice, one cannot really distinguish the V-value between 2.3 and 3.0; experimental observation also shows that the optical fiber can still support only one mode. Thus, designers usually take the value of V as 3.0 or less to design a single-mode optical fiber.

The V-parameter is inversely proportional with respect to the optical wavelength, which is directly related to the operating frequency. If an optical fiber is launched with lightwaves whose optical wavelength is smaller than the operating wavelength at which the optical fiber is single mode, then the optical fiber is supporting more than one mode. The optical fiber is said to be operating in a few regions and then a multimode region when the total number of modes reaches a few hundred.

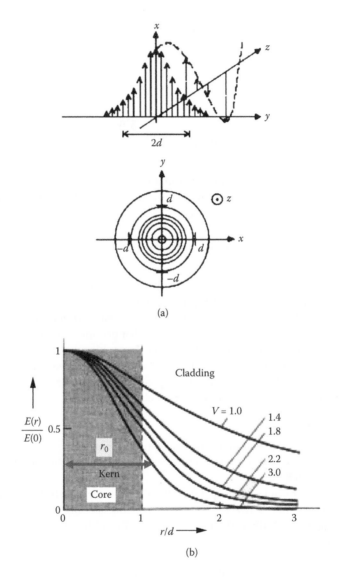

FIGURE 6.8 (a) Intensity distribution of the LP_{01} mode. (b) Variation of the spot size–field distribution with radial distance r with V as a parameter.

Thus one can define the cutoff wavelength for optical fibers as follows: the wavelength (λ_c) above which only the fundamental mode is guided in the fiber is called the cutoff wavelength λ_c. This cutoff wavelength can be found by using the V-parameter as $V_c = V_{at\ cut\ off} = 2.405$; thus,

$$\lambda_c = \frac{2\pi a NA}{V_c} \tag{6.28}$$

Exercise

An optical fiber has the following parameters: a core refractive index of 1.46, a relative refractive index difference of 0.3%, a cladding diameter of 125 µm, and a core diameter of 8.0 µm.

a. Find the fiber NA and hence the fiber acceptance angle.
b. What is the cutoff wavelength of this fiber?
c. What is the number of optical guided modes that can be supported if the optical fiber is excited with lightwaves of a wavelength of 810 nm?
d. If the cladding diameter is reduced to 50 and 20 µm, comment on the field distribution of the guided single mode.

In practice the fibers tend to be effectively single mode for larger values of V, say $V < 3$ for the step profile, because the higher order modes suffer radiation losses due to fiber imperfections. Thus if $V = 3$, from (6.15) we have $a < 3\lambda/2\ NA$; in this case $\lambda = 1$ µm and the numerical aperture NA must be very small (<<1) for radius a to have some reasonable dimension. Usually Δ is about 1% or less for standard single-mode optical fibers (SSMFs) employed in long-haul optical transmission systems so as to minimize the loss factor and the dispersion.

6.1.3.5 Power Distribution

The axial power density or intensity profile $S(r)$, the z-component of Poynting's vector, is given by

$$S(r) = \frac{1}{2} E_x H_y^* \tag{6.29}$$

Substituting (6.10) into (6.29) we have

$$S(r) = \frac{1}{2} \left(\frac{\varepsilon}{\mu} \right)^{1/2} e^{-\left(\frac{r}{r_0} \right)^2} \tag{6.30}$$

The total power is then given by

TABLE 6.1

Analytical Expressions for Total Optical Guided Power and Its Fractional Power Confined Inside the Core Region for Step Index and Gaussian Index Profiles

	Step Index Profile Fiber	Gaussian Profile Fiber
$S(r/a)$ for $V > 1$	$\dfrac{1}{2}\left(\dfrac{\varepsilon}{\mu}\right)^{1/2} e^{-\left(\frac{r}{a}\right)^2 \ln V^2}$	$\dfrac{1}{2}\left(\dfrac{\varepsilon}{\mu}\right)^{1/2} e^{-\left(\frac{r}{a}\right)(V-1)}$
Power P for $V > 1$	$\dfrac{1}{2}\left(\dfrac{\varepsilon}{\mu}\right)^{1/2} \dfrac{a^2}{\ln V^2}$	$\dfrac{1}{2}\left(\dfrac{\varepsilon}{\mu}\right)^{1/2} \dfrac{a^2}{V-1}$
$\eta(r)$ for $V > 1$	$1 - e^{-\left(\frac{r}{a}\right)^2 \ln V^2} = 1 - \dfrac{1}{V^2}$ for $r = a$	$1 - e^{-\left(\frac{r}{a}\right)^2 (V-1)}$

$$P = 2\pi \int_0^\infty rS(r)\,dr = \frac{1}{2}\left(\frac{\varepsilon}{\mu}\right)^{1/2} r_0^2 \qquad (6.31)$$

and hence the fraction of power $\eta(r)$ within 0 to r across the fiber cross section is given by

$$\eta(r) = \frac{\displaystyle\int_0^r rS(r)\,dr}{\displaystyle\int_0^\infty rS(r)\,dr\,r} = 1 - e^{-\left(\frac{r^2}{r_0^2}\right)} \qquad (6.32)$$

Table 6.1 gives the expressions for P and $\eta(r)$ of step index and Gaussian profile fibers (by substituting the appropriate values of r_0 into (6.31) and (6.32)).

As a rule of thumb—and experimentally confirmed—an optical fiber is best for the guided mode, and the optical power contained in the core is about 70–80% of the total power.

Exercise

Using Gaussian approximation for the intensity distribution of the fundamental mode of the single-mode optical fiber with $V = 2$, find the fraction of power in the core region with $a = 4\ \mu m$.

Exercise

Find the radius a for maximum confinement of light power, i.e., maximum mode spot size r_0, for step index and parabolic profile optical fibers.

6.1.3.6 Approximation of Spot Size r_0 of Step Index Single-Mode Fibers

As stated above, spot size r_0 would play a major role in determining the performance of single-mode fiber. It is useful if we can approximate the spot size as long as the fiber is operating over a certain wavelength. When a single-mode fiber is operating above the cutoff wavelength, a good approximation (greater than 96% accuracy) for r_0 is given by

$$\frac{r_0}{a} = 0.65 + 1.619V^{-3/2} + 2.879V^{-6} = 0.65 + 0.434\left(\frac{\lambda}{\lambda_c}\right)^{+3/2} + 0.0419\left(\frac{\lambda}{\lambda_c}\right)^{+6} \quad (6.33)$$

for $0.8 \le \dfrac{\lambda}{\lambda_c} \le 2.0$ single mode.

Exercise

What is the equivalent range for the V-parameter of Equation (6.33)? Inspect the b-V and $V^2(d^2(Vb)/dV^2)$ versus V and b; if possible, do a curve fitting to obtain the approximate relationship for r/a and V (MATLAB® procedure is recommended). In the chapter's appendix typical values of mode spot size or mode field diameter, approximately twice the spot size, are given for Corning SMF-28, the standard single-mode optical fiber and large effective area fibers, also single mode. Note that the value specified must vary with respect to the fluctuating value of the relative refractive index and the concentricity of the fiber core diameter.

Exercise

Refer to the technical specification of Corning SMF-28 and LEAF.

   a. State the core diameter of the fibers, the spot size, or mode field diameters of the fibers.
   b. Then estimate the effective areas of these fibers.
   c. What is the ratio of the effective area and the physical area of the cores of the fibers.

6.1.4 Equivalent Step Index Description

As we can observe, there are two possible orthogonally polarized modes, (E_x, H_y) and (E_y, H_x), that can be propagating simultaneously. The superposition of these modes can usually be approximated by a single *LP* mode. These modes' properties

Index Profiles:

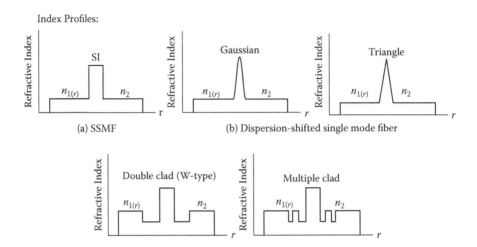

FIGURE 6.9   Index profiles of a number of modern fibers, e.g., dispersion-shifted single-mode fibers.

are well known for step index optical fibers, and analytical solutions are also readily available.

Unfortunately, practical single-mode optical fibers never have a perfect step index profile due to the variation of the dopant diffusion and polarization. These non-step index fibers can be approximated, under some special conditions, by the *equivalent step index* (ESI) profile technique.

A number of index profiles of modern single-mode fibers, e.g., non-zero-dispersion-shifted fibers, are shown in Figure 6.9. The ESI profile is determined by approximating the fundamental mode electric field spatial distribution $\varphi(r)$ by a Gaussian function as described in Section 6.1.3.1. The electric field can thus be totally specified by the e^{-1} width of this function or *mode spot size* (r_0). Alternatively, the term *mode field diameter (MFD)* is also used and equivalent to twice the size of the mode spot size r_0.

6.1.4.1   Definitions of ESI Parameters

The ESI description can be used to design a single-mode fiber with a graded index, W-type or segmented core profiles (under some limitations). These non-step index profiles can be described by ESI parameters denoted in the following: V_e is the effective or equivalent V-parameter, a_e is the ESI core radius, λ_{ec} is the ESI cutoff wavelength, and Δ_e is the equivalent relative index difference. These parameters are related to two moments M_0, M_1 defined as

$$M_n = \int_0^\infty [n^2(r) - n^2(a)] r^n \, dr \qquad (6.34)$$

for $n = 1, 2$. The effective V_e-parameter and core radius are given by

$$V_e^2 = 2k^2 \int_0^\infty [n^2(r) - n^2(a)] r \, dr \qquad (6.35)$$

$$V_e^2 = 2k^2 M_1 \quad \text{and} \quad a_e = 2\frac{M_1}{M_0} \qquad (6.36)$$

It follows from (6.35) and (6.36) that the parameters λ_{ec} and Δ_e can be found by setting

$$V_e^2 = 2k^2 a_e^2 n_1^2 \Delta_e \qquad (6.37)$$

and $V_e = 2.405$, the cutoff condition for step index fibers. Therefore the cutoff wavelength for an ESI profile fiber is

$$\lambda_{ec} = \frac{2\pi\sqrt{2M_1}}{2.405} \qquad (6.38)$$

It is noteworthy that V_e as given in (6.37) is equivalent to the mode *volume*. Physically, the significance of V_e can be compared to the *average density* of a disk with a local density equal to $[n^2(r) - n^2(a)]$.

6.1.4.2 Accuracy and Limits

The ESI approximation is generally accurate to within 2%, at least over the wavelength range $0.8 < \lambda/\lambda_c < 1.5$. For most practical purposes this range is the operating wavelength to minimize the dispersion property of single-mode optical fibers.

6.1.4.3 Examples of ESI Techniques

6.1.4.3.1 Graded Index Fiber

These index profiles of graded fiber are given by Equation (6.7). We thus have

$$n^2(r) - n^2(a) = s(r/a) = 1 - \left(\frac{r}{a}\right)^\alpha \qquad (6.39)$$

Substituting (6.39) into (6.35) gives

$$\frac{V_e}{V} = \left(\frac{\alpha}{\alpha+2}\right)^{1/2} \qquad (6.40)$$

where we have used $V = k \cdot a \cdot NA$ as the *V*-parameter of a step index fiber with the core index at the fiber axis of n_1. Then, we have

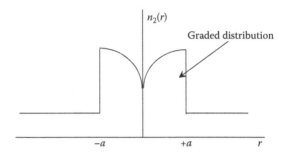

FIGURE 6.10 Refractive index profile of a graded index fiber with a central dip. This is a typical profile of manufactured fiber if good collapsing of the fiber preform is not achieved.

$$\lambda_{ec} = \frac{V}{2.405}\left(\frac{\alpha}{\alpha+2}\right)^{1/2} \tag{6.41}$$

Exercise

Given a single-mode optical fiber with a triangular profile index distribution whose equivalent V-parameter is equal to 2 at 1550 nm wavelength, what is the V-parameter value at the center of the core of the fiber? If the diameters of the cores of the two fibers are kept identical, then what is the ratio of the refractive indices at the core center of the fibers? Repeat for a parabolic profile.

6.1.4.3.2 Graded Index Fiber with a Central Dip

The fiber index profile with a central dip and grade gradually increases to the outer cladding, as shown in Figure 6.10.

Similar to (6.6) for a graded index fiber with a maximum index at the core axis, we have

$$S(r/a) = 1 - \gamma(1-x)^{\alpha} \quad \text{for } o < r < a \tag{6.42}$$

where γ is the depth and $0 < \gamma < 1$. When $r = 0$, we have a step index profile, and when $r = 1$, we have the central axis refractive index equal to the cladding index.

Using (6.34) and (6.35), V_e can be easily found and given by

$$\frac{V_e^2}{V^2} = 1 - \frac{2\gamma}{(\alpha+1)(\alpha=2)} \tag{6.43}$$

6.1.4.4 General Method

The general technique to find the ESI parameters for optical fibers can be started by raising the stationary expression in (20) for expressing β of the actual fiber compared to its equivalent propagation constant β_e as

$$\beta^2 = \beta_e^2 + k^2 \frac{\displaystyle\int_0^\infty [n^2(r) - n_e^2(r)] r \varphi^2(r)\, dr}{\displaystyle\int_0^\infty r \varphi^2(r)\, dr} \tag{6.44}$$

where $n_e^2(r)$ is the equivalent counterpart of $n(r)$ when the fiber is expressed in its equivalent step form. The field expression $\psi(r)$ is assumed (in fact to be obtained) to be similar for the actual fiber and its step equivalence. Once the field $\varphi(r)$ can be replaced by the approximate field shape, we can find V_e and a_e that minimize $\beta^2 - \beta_e^2$ given in (6.44). Generally these parameters are functions of both V and a, and thus it is impossible to get one ESI technique that is applicable to a wide range of wavelengths, and it is required that we apply complicated numerical calculations.

6.2 SPECIAL FIBERS

Besides optical fibers for transmission of signals, there are a number of special types of fibers that are essential for optical systems, for applications in either telecommunication systems or sensing devices.

In fiberoptics, a polarization-maintaining optical fiber (PMF or PM fiber) is an optical fiber in which the polarization of linearly polarized lightwaves launched into the fiber is maintained during propagation, with little or no cross-coupling of optical power between the polarization modes. Such fiber is used in special applications where preserving polarization is essential. Several different designs of PM fiber are used. Most work by inducing stress in the core via a noncircular cladding cross section or via rods of another material included within the cladding. Several different shapes of rod are used, and the resulting fiber is sold under brand names such as PANDA and Bow Tie. The differences in performance between these types of fibers are subtle. Some of the differences are explained in the article "PANDA-Style Fibers Move beyond Telecom."[8]

Typical cross sections of PMFs or high-birefringence fibers are shown in Figure 6.11. From the guided mode point of view, we can observe that the refractive index of the fiber would take two different values along the horizontal and vertical directions due to the directional effects of stress (Figure 6.11(a)–(c)) or the geometrical profile (Figure 6.11(d)). The difference in refractive indices in these directions can be tailored so that only one mode is guided and dominant with the polarization in the direction along which the single-mode condition is satisfied, as described in Section 6.1.3. In Figure 6.11(a), the core of the fiber is circular and different stresses are applied to the core under the drawing process or by inserting a silica circular tube into the preform during the drawing process (see drawing process in Section 6.4). This fiber cross section figure looks like the face of a panda, and so the name PANDA fiber, while in Figure 6.11(b), the core refractive index distribution looks like a bow tie and hence the name bow-tie fiber.

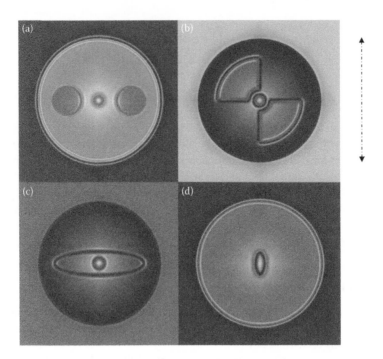

FIGURE 6.11   Cross section of polarization-maintaining fibers: (a) PANDA, (b) Bow Tie, (c) elliptical clad, and (d) elliptical core.

Polarization-maintaining optical fibers are used in special applications, such as in fiber optic sensing, interferometry, and quantum key distribution.[9,10] They are also commonly used in telecommunications for the connection between a source laser and a modulator, since the modulator requires polarized light as input. They are rarely used for long-distance transmission, because PMF is expensive and has much higher attenuation than SMF.

PMF does not polarize light like a polarizer does. Rather, PM fiber maintains the existing polarization of linearly polarized light that is launched into the fiber with the correct orientation. If the polarization of the input light is not aligned with the stress direction in the fiber, the output will vary between linear and circular polarization (and generally be elliptically polarized). The exact polarization will then be sensitive to variations in temperature and stress in the fiber. The output of a PMF is typically characterized by its polarization extinction ratio (PER)—the ratio of correctly to incorrectly polarized light, expressed in decibels. The quality of PMF patchcords and pigtails can be characterized with a PER meter. These fibers are sometime called HiBi fibers or high-birefringent fibers. The axes of a PMF are termed as fast and slow (see Figure 6.12), indicating the propagation wavenumber β_x, β_y of high and low values, as these propagation constants are determined in vacuum, divided by the effective index of the guided polarized mode. Hence, along the fast axis the polarized lightwaves see a smaller value of refractive index and vice versa.

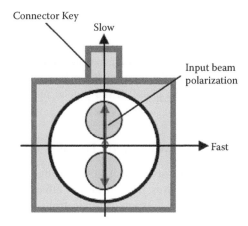

FIGURE 6.12   Alignment of fiber axis and input polarization of lightwaves: fast and slow axes.

The principles of the design of these polarization-maintaining fibers can be similar to the effective index method described in Section 5.3 of Chapter 5 for three-dimensional optical waveguides, except that the cross section of these fibers is circular with birefringence in the refractive index along the horizontal and vertical axes or elliptical geometry.

6.3   NONLINEAR OPTICAL EFFECTS

In this section the nonlinear effects on the guided lightwaves propagating through a long length of optical fibers, the single-mode type, are described. Unfortunately when the intensity of the optically modulated waves reaches a certain level, the refractive index of the core changes and, hence, the phase of the waves—thus dispersion. Besides this phase change, there are also other effects, such as frequency doubling and wavelength conversion. Although the nonlinear coefficient of the fiber silica is small, the length of propagation is very long, and so the accumulation of these phase changes is quite significant, and it is very much visible when the width of the pulses is short at ultra-high speed, e.g., 40 Gb/s.

These nonlinear effects play important roles in the transmission of optical pulses along single-mode optical fibers as distortion due to the modification of the phase of the lightwaves. The nonlinear effects can be classified into three types: the effects that change the refractive index of the guided medium due to the intensity of the pulse, the self-phase modulation; the scattering of the lightwave to other frequency-shifted optical waves when the intensity reaches a certain threshold, Brillouin and Raman scattering (SBS, SRS) phenomena; and the mixing of optical waves to generate a fourth wave, the degenerate four-wave mixing (FWM). Besides these nonlinear effects there is also the photorefractive effect, which is due to the change of the refractive index of silica due to the intensity of ultraviolet optical waves. This phenomenon is used to fabricate grating whose spacing between dark and bright regions satisfies the Bragg diffraction condition. These are fiber Bragg gratings (FBGs) and are used as optical filters and dispersion compensators when the spacing varies or is chirped.

In modern coherent optical communication systems incorporating digital signal processors at the coherent receiver, the compensation can be done in the electronic domain, and back propagation of the lightwaves can be implemented to reverse the nonlinear effects imposed on the phase of the guided mode using the frequency domain transfer function.[11–13]

6.3.1   NONLINEAR SELF-PHASE MODULATION EFFECTS

All optical transparent materials are subject to the change of the refractive index with the intensity of the optical waves, the optical Kerr effect. This physical phenomenon originates from the harmonic responses of electrons of optical fields, leading to the change of the material susceptibility. The modified refractive index $n_{1,2}^{K}$ of the core and cladding regions of the silica-based material can be written as

$$n_{1,2}^{K} = n_{1,2} + \overline{n}_2 \frac{P}{A_{eff}} \tag{6.45}$$

where n_2 is the nonlinear index coefficient of the guided medium; the average typical value of n_2 is about 2.6×10^{-20} m²/W. P is the average optical power of the pulse sequence, and A_{eff} is the effective area of the guided mode, which is the e^{-1}-value of the intensity distribution of the guided field. The nonlinear index changes with the doping materials in the core. Although the nonlinear index coefficient is very small, the effective area is also very small, about 50–70 μm², and the length of the fiber under the propagation of optical signals is very long and the accumulated phase change quite substantial. This leads to the self-phase modulation (SPM) and cross-phase modulation (XPM) effects in the optical channels.

6.3.2   SELF-PHASE MODULATION

The origin of the SPM is due to the phase variation of the guided lightwaves exerted by the intensity of its own power or field accumulated along the propagation path, which is quite long, possibly a few hundred to thousands of kilometers. Under a linear approximation we can write the modified propagation constant of the guided linearly polarized mode in a single-mode optical fiber as

$$\beta^{K} = \beta + k_0 \overline{n}_2 \frac{P}{A_{eff}} = \beta + \gamma P \tag{6.46}$$

where

$$\gamma = \frac{2\pi \overline{n}_2}{\lambda A_{eff}}$$

where n_2 and γ are the nonlinear coefficient and parameter of the guided medium, respectively, taking effective values of 2.3×10^{-23} m^{-2} and from 1 to 5 (kmW)$^{-1}$, depending on the effective area of the guided mode and the operating wavelength. Thus, the smaller the mode spot size or mode field diameter, the larger the nonlinear SPM effect. For a dispersion-compensating fiber the effective area is about 15 μm^2, while for SSMF and nonzero-dispersion-shafted fibers (NZ-DSF), the effective area ranges from 50 to 80 μm. Thus the nonlinear threshold power of dispersion compensating fiber (DCF) is much lower than that of SSMF and NZ-DSF. The maximum launched power into DCF would be limited at about 0 dBm or 1.0 mW in order to avoid a nonlinear distortion effect, while it is about 5 dBm for SSMF.

The accumulated nonlinear phase changes due to the nonlinear Kerr effect over the propagation length L is given by

$$\phi_{NL} = \int_0^L (\beta^K - \beta)dz = \int_0^L \gamma P(z)dz = \gamma P_{in}L_{eff} \quad \text{with } P(z) = P_{in}e^{-\alpha z} \quad (6.47)$$

defined as the representation of the attenuation of the optical signals along the propagation direction z. In order to consider that the nonlinear SPM effect is small compared with the linear chromatic dispersion effect, one can set $\phi_{nL} = 1$ or $\phi_{NL} = 0.1$ rad, and the effective length of the propagating fiber is set at $L_{eff} = 1/\alpha$, with optical losses equalized by cascaded optical amplification subsystems. Then, the maximum input power to be launched into the fiber can be set at

$$P_{in} < \frac{0.1\alpha}{\gamma N_A} \quad (6.48)$$

For $\gamma = 2(\text{W} \cdot \text{km})^{-1}$ and $N_A = 10$, $\alpha = 0.2$ dB/km (or 0.0434×0.2 km^{-1}), and then $P_{in} < 2.2$ mW or about 3 dBm. Similarly, this threshold level is about 0 dBm for DCF with an effective area about at the 1.550 μm spectral region. In practice, due to the randomness of the arrival 1 and 0, this nonlinear threshold input power can be set at about 10 dBm as the total average power of all wavelength multiplexed channels of WDM transmission systems, launched into the fiber link.

6.3.3 CROSS-PHASE MODULATION

The change of the refractive index of the guided medium as a function of the intensity of the optical signals can also lead to the phase of optical channels in different spectral regions close to that of the original channel. This is XPM effects, which are critical in wavelength division multiplexed (WDM) channels, and even more critical in dense WDM when the frequency spacing between channels is 50 GHz or even narrower, the cross-interference between channels generating unwanted noises in the optical domain, and thence to the detected electronic signals at the receiver. In such systems the nonlinear phase shift of a particular channel depends not only on its power but also on that of other multiplexed channels. The phase shift of the ith channel can be written as[14]

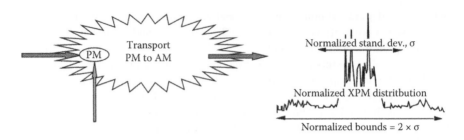

FIGURE 6.13 Illustration of XPM effects—phase modulation (PM) conversion to amplitude modulation (AM) and, hence, interference between adjacent channels.

$$\phi^i_{NL} = \gamma L_{eff} \left(P^i_{in} + 2\sum_{j\neq i}^M P_j \right) \quad \text{with } M = \text{number of multiplexed channels} \quad (6.49)$$

The factor of 2 in (6.49) is due to the bipolar effects of the susceptibility of silica materials, and the total phase noises are integrated over both sides of the channel spectrum. The XPM thus depends on the bit pattern and the randomness of the synchronous arrival at 1. It is hard to estimate analytically, so numerical simulations would normally be employed to obtain the XPM distortion effects using the nonlinear Schrodinger wave propagation equation involving the signal envelopes of all channels. The evolution of slow-varying complex envelopes $A(z, t)$ of optical pulses along a single-mode optical fiber is governed by the nonlinear Schrodinger equation (NLSE)[7]:

$$\frac{\partial A(z,t)}{\partial z} + \frac{\alpha}{2} A(z,t) + \beta_1 \frac{\partial A(z,t)}{\partial t} + \frac{j}{2}\beta_2 \frac{\partial^2 A(z,t)}{\partial t^2} - \frac{1}{6}\beta_3 \frac{\partial^3 A(z,t)}{\partial t^3}$$
$$= -j\gamma |A(z,t)|^2 A(z,t) \tag{6.50}$$

where z is the spatial longitudinal coordinate, α accounts for fiber attenuation, β_1 indicates DGD, β_2 and β_3 represent second- and third-order factors of fiber chromatic dispersion (CD), and γ is the nonlinear coefficient. This equation would be derived from Maxwell's equations under external perturbation.

The phase modulation due to nonlinear phase effects is then converted to amplitude modulation and the cross talk to other adjacent channels. This is shown in Figure 6.13.

6.3.4 STIMULATED SCATTERING EFFECTS

Scattering of lightwaves by the impurities can happen due to the absorption and vibration of the electrons and dislocation of molecules in silica-based materials. The back scattering and absorption are commonly known as Raleigh scattering losses in fiber propagation in whose phenomena the frequency of the optical carrier does not change. Other scattering processes in which the frequency of the lightwave carrier is shifted to other frequency regions are commonly known as inelastic scattering, the Raman scattering, and Brillouin scattering. In all cases the scattering of photons to a lower energy level photon with an energy difference between these levels falls with the energy of phonons. Optical phonons result from the electronic vibration for Raman scattering, while acoustic phonons or mechanical vibration of the linkage between molecules leads to Brillouin scattering. At high power, when the intensity exceeds a certain threshold, the number of scattered photons is exponentially grown, and then the phenomena are a simulated process. Thus the phenomena can be called stimulated Brillouin scattering (SBS) and stimulated Raman scattering (SRS). SRS and SBS were first observed in the 1970s.[15–17]

6.3.4.1 Stimulated Brillouin Scattering

Brillouin scattering comes from the compression of silica materials in the presence of an electric field, the electrostriction effect. Under the pumping of an oscillating electric field of frequency f_p, an acoustic wave of frequency F_a is generated. Spontaneous scattering is an energy transfer from the pump wave to the acoustic wave and then a phase matching to transfer a frequency-shifted optical wave of frequency as a sum of the optical signal waves and the acoustic wave. This acoustic wave frequency shift is around 11 GHz with a bandwidth of around 50 to 100 MHz (due to the gain coefficient of the SBS), and a beating envelope would be modulating the optical signals. Thus, jittering of the received signals at the receiver would be formed and, hence, the closure of the eye diagram in the time domain.

Once the acoustic wave is generated, it beats with the signal waves to generate the sideband components. This beating beam acts as a source and further transfers the signal beam energy into the acoustic wave energy and amplifies this wave to generate further jittering effects. The Brillouin scattering process can be expressed by the following coupled equations:[14]

$$\frac{dI_p}{dz} = -g_B I_p I_s - \alpha_p I_p$$

$$-\frac{dI_s}{dz} = +g_B I_p I_s - \alpha_s I_s$$

(6.51)

The SBS gain g_B is frequency dependent with a gain bandwidth of around 50 to 100 MHz for a pump wavelength at around 1.550 μm. For silica fiber g_B is about 5e-11 mW^{-1}. The threshold power for the generation of SBS can be estimated (using Equation (6.51)) as

$$g_B P_{th\_SBS} \frac{L_{eff}}{A_{eff}} \approx 21 \quad \text{with the effective length } L_{eff} = \frac{1 - e^{-\alpha L}}{\alpha}$$

(6.52)

where

I_p = intensity of pump beam

I_s = intensity of signal beam

g_B = Brillouin scattering gain coefficient

α_s, α_p = losses of signal and pump waves

For the standard single-mode optical fiber (SSMF), this SBS power threshold is about 1.0 mW. Once the launched power exceeds this power threshold level the beam energy is reflected back. Thus, the average launched power is usually limited to a few dBm due to this low threshold power level.

6.3.4.2 Stimulated Raman Scattering

Stimulated Raman scattering (SRS) occurs in silica-based fiber when a pump laser source is launched into the guided medium, and the scattering light from the molecules and dopants in the core region are shifted to a higher energy level and then jump down to a lower energy level—hence, the amplification of photons in this level. Thus a transfer of energy from different frequency and energy level photons occurs. The stimulated emission happens when the pump energy level reaches above the threshold level. The pump intensity and signal beam intensity are coupled via the following equations:

$$\frac{dI_p}{dz} = -g_R I_p I_s - \alpha_p I_p$$

$$-\frac{dI_s}{dz} = +g_R I_p I_s - \alpha_s I_s$$

(6.53)

where

I_p = intensity of pump beam

I_s = intensity of signal beam

g_R = Raman scattering gain coefficient

α_s, α_p = losses of signal and pump waves

The spectrum of the Raman gain depends on the decay lifetime of the excited electronic vibration state. The decay time is in the range of 1 ns, and the Raman (gain) bandwidth is about 1 GHz. In single-mode optical fibers the bandwidth of the Raman gain is about 10 THz. The pump beam wavelength is usually about 100 nm below the amplification wavelength region. Thus, in order to extend the gain spectra, a number of pump sources of different wavelengths are used. Polarization multiplexing of these beams is also used to reduce the effective power launched in the fiber to avoid the damage of the fiber. The threshold for stimulated Raman gain is given by

$$g_R P_{th\_SRS} \frac{L_{eff}}{A_{eff}} \approx 16 \quad \text{with the effective length } L_{eff} = \frac{1 - e^{-\alpha L}}{\alpha}$$

(6.54)

or $\approx 1/\alpha$ for long length

For SSMF with an effective area of 50 μm^2, $g_R \sim 1e - 13$ m/W, and then the threshold power is about 570 mW near the C-band spectral region. This would require at least two pump laser sources, which should be polarization multiplexed. The SRS is used frequently in modern optical communication systems, especially when no undersea optical amplification is required; the distributed amplification of SRS offers significant advantages compared with lumped amplifiers such as an erbium-doped fiber amplifier (EDFA). The broadband-gain and low-gain ripple of SRS is also another advantage for DWDM transmission.

6.3.4.3 Four-Wave Mixing Effects

Four-wave mixing (FWM) is considered a scattering process in which three photons are mixed to generate the fourth wave. This happens when the momentum of the four waves satisfies a phase matching condition, that is, the condition of maximum power transfer. Figure 6.14 illustrates the mixing of different wavelength channels to generate interchannel cross talk. The phase matching can be represented by a relationship between the propagation constant along the z-direction in a single-mode optical fiber as

$$\beta(\omega_1) + \beta(\omega_2) - \beta(\omega_3) - \beta(\omega_4) = \Delta(\omega) \tag{6.55}$$

where ω_1, ω_2, ω_3, ω_4 are the frequencies of the first to fourth waves and Δ is the phase mismatching parameter. If the channels are equally spaced with a frequency spacing of Ω, as in DWDM optical transmission, then we have $\omega_1 = \omega_2$; $\omega_3 = \omega_1 + \Omega$; $\omega_4 = \omega_1 - \Omega$. We can use the Taylor series expansion around the propagation constant at the center frequency of the guide carrier β_0; then we can obtain[14]

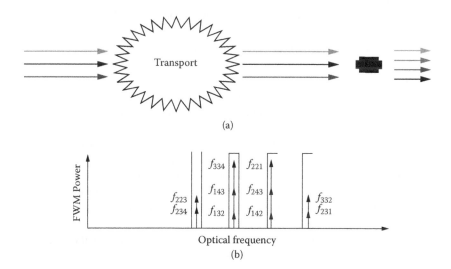

(a)

(b)

FIGURE 6.14 Illustration of FWM of optical channels: (a) Momentum vectors of channels; (b) frequencies resulting from mixing of different channels.

$$\Delta(\omega) = \beta_2 \Omega^2 \tag{6.56}$$

The phase matching is thus optimized when β_2 is zero, indicating that in the region where there is no dispersion FWM is biggest, and hence, there is maximum interchannel cross talk. This is the reason dispersion-shifted fiber is not commonly used when the zero dispersion wavelength falls in the spectral region of operation of a channel. In modern transmission fiber the zero-dispersion wavelength is shifted to outside the C-band, say 1.510 μm, so that there is a small dispersion factor at 1550 nm, and the C-band ranges from 2 to 6 ps/nm.km, for example, Corning LEAF or nonzero-dispersion-shifted fibers (NZ-DSFs). This small amount of dispersion is sufficient to avoid the FWM with a channel spacing of 100 or 50 GHz.

The XPM signal is proportional to instantaneous signal power. Its distribution is bounded <5 channels and otherwise effectively unbounded. Thus, the link budgets include XPM evaluated at maximum outer bounds.

6.4   OPTICAL FIBER MANUFACTURING AND CABLING

This section is devoted to a brief description of the manufacturing of optical fibers and the cabling of several fibers for optical communication systems. The manufacturing techniques and cabling process affect the transmission and physical properties of the fibers. We focus on the aspects for a general understanding of optical transmission systems.

As we have described in previous sections, the SSMF structure is a cylindrical core with a refractive index slightly higher than that of the cladding region. For optical communications operating in the 1300 and 1700 nm wavelength regions, the silica material is the base material. A pure silica tube is the starting structure, and a combination of silica, germanium dioxide (GeO_2), and P_2O_5 is then deposited inside the tube. Other dopants, such as B_2O_3 and flouride, can also be used to reduce the refractive index of some small regions of the core. These are the segmented core and W-type fibers, which are described in the next chapter.

Once the deposition of the impurities is done (see Figure 6.15) the tube is collapsed to produce silica preforms as shown in Figure 6.15(a). Also shown in this figure is a schematic of the fiber-drawing machine and fiber-drawing tower. The refractive index of the fiber preform is also shown in this figure, as noted in its caption, and its details are shown in Figure 6.15(b) and (c) and Figure 6.16. The fiber preform is necessary and fabricated by starting with a pure silica tube rotating in a chemical vapor chamber containing the composition of silica and doping impurities for forming the core region. After the deposition of the core material, the silica tube is then collapsed into the preform. This preform is then placed in a drawing tower as shown in Figure 6.17, heated by a microwave section to melting, and drawn into a circular optical waveguide fiber with a control feedback subsystem to ensure the uniformity of the core of the fiber. In addition, the drawn fiber may be spun during the drawing process to obtain uniformity in the fiber core ellipticity to minimize the polarization mode dispersion (PMD), which will be treated in the next chapter. This PMD is very critical for a modern optical fiber system operating at ultra-speed. Figure 6.18 shows the scenarios of

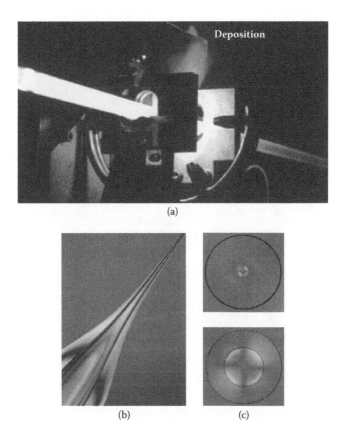

(a)

(b)   (c)

FIGURE 6.15 Schematic of a fiber deposition and fabrication of a fiber preform. (a) Deposition of core material and collapsing. (b) Fiber preform, before drawing into fiber strands. (c) Cross section of fiber preform with refractive index profile exactly the same as the fiber index profile of single-mode (upper) and multimode (lower) types.

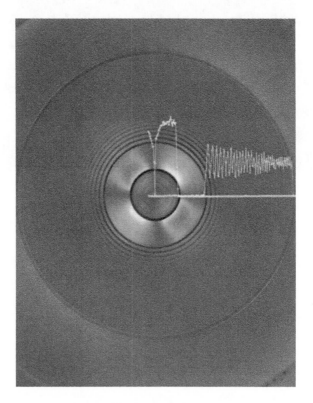

FIGURE 6.16   Refractive index profile across a single-mode fiber preform. (Note: Non-step-like profile—so why modeled as a step index structure?)

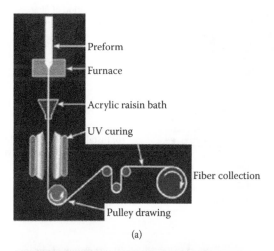

Preform

Furnace

Acrylic raisin bath

UV curing

Fiber collection

Pulley drawing

(a)

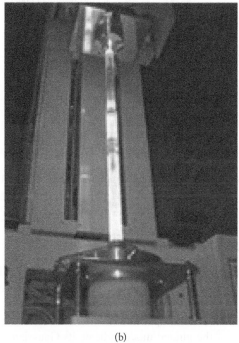

(b)

FIGURE 6.17 (a) Schematic of fiber-drawing machine. (b) Picture of fiber microwave furnace and diameter monitoring and feedback control.

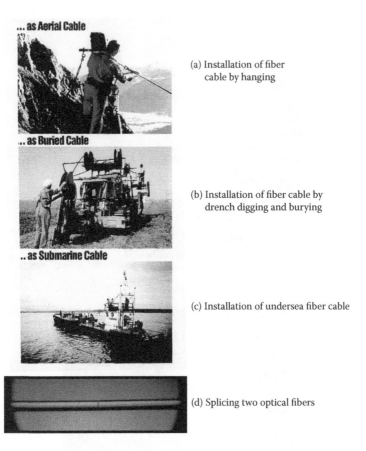

(a) Installation of fiber
    cable by hanging

(b) Installation of fiber cable by
    drench digging and burying

(c) Installation of undersea fiber cable

(d) Splicing two optical fibers

FIGURE 6.18   Installation of fiber cables at different terrains and undersea.

installation of fiber cables by aerial hanging, ploughing into the ground and undersea. The PMD is most serious for the aerial environment due to the randomness of wind direction and speeds of the wind and thence the random vibration of the cables.

6.5   CONCLUDING REMARKS

This chapter introduces the fundamental concepts of waveguiding in the circular optical waveguide or optical fibers, including approaches from the Maxwell equations and the wave equations leading to guided conditions for the modes subject to boundary conditions. However, experimental observations and measurement of the intensity distribution of the guided mode indicate its Gaussian distribution. It is an engineering approach to substitute this known solution to obtain essential parameters for the guided mode. This condition is even more important when the difference in the refractive indices of the core and cladding region is very small and the mode is guided gently, the weakly guiding phenomena.

The basic properties of the fiber structure, its profile, the spot size, the cutoff wavelength, and the Gaussian approximation, are described. The Gaussian approximation makes the understanding of the optical guided mode simple. It also allows us to obtain directly the optical mode distribution and thus several other approximations required to obtain the simplest form of important parameters of single-mode optical fibers.

Once the basic properties of single-mode optical fibers are found, they form the basic set of parameters so that optical fibers whose effective index profiles are non-step can be found based on the ESI technique, which converts the parameters to an equivalent step-like profile and hence other optical properties.

Only structural and wave properties of lightwave signals traveling in optical fibers are presented here. As optical communication systems engineers, we have to understand and develop techniques for analyzing and identifying the transmission of digital and analog signals through optical fibers, that is, the attenuation and broadening of optical signals after transmission through a medium, namely, attenuation and broadening via dispersion of lightwave pulses. These topics will be treated in the next chapter, where the theory of electromagnetism has been transformed into the signal propagation and the distortion mechanism due to interference effects of different spectral/frequency component lightwaves under modulation traveling at different speeds due to the confinement of the optical waveguides and the material refractive index dependence on wavelength. Furthermore, examples of the design of optical fibers are given in the next chapter, especially on how to consider the flattening and compensating of dispersion characteristics.

When the time-dependent and nonlinear parameters along the propagation direction are included in the wave equation, it becomes a nonlinear type, and in order to illustrate the behavior of the guided mode along the fiber transmission line, it can be considered a cascade of several optical lenses so that the nonlinear effects can be operated in the frequency domain. The numerical technique involved in this propagation will be described in Chapter 7.

6.6 PROBLEMS

6.6.1 PROBLEM 1

An optical fiber with a step index profile, a core diameter of 62.5 μm, and a numerical aperture of 0.2 at a wavelength of 1550 nm is used for signal distribution and transmission in a local area network.

   a. What is the V-parameter of this optical fiber?
   b. How many guided modes would it support? Can you comment on this number regarding the velocities of lightwaves?
   c. Select a cladding diameter. Give reasons for your selection.
   d. Find the maximum acceptance angle of this fiber and estimate the coupling loss of a laser source with a uniform radiation cone of 30°.

6.6.2 PROBLEM 2

An optical fiber has the following parameters: index profile = step-like, core diameter = 9.0 μm, numerical aperture = 0.11, and cladding refractive index = 1.4844.

a. Find the normalized frequency of the fiber at 1550 nm wavelength.
b. Is the fiber operating in the single-mode or multimode region at 1550 nm? If it is in the single-mode region, estimate its mode field diameter and spot size. Sketch its field and intensity distribution across the fiber cross section.
c. Find the cutoff wavelength of the fiber. If lightwaves of wavelengths smaller than the cutoff wavelength are launched into the fiber, is the fiber still operating in the single-mode region?

6.6.3 PROBLEM 3

A single-mode step index optical fiber has the following parameters: core diameter = 8.0 μm, cladding diameter = 0.125 mm, core refractive index = 1.460, and relative index difference = 0.2% at 1550 nm.

a. Confirm that the fiber can be operating in the single-mode regime at 1550 nm wavelength.
b. Find the fiber cutoff wavelength. If the refractive index difference fluctuates within 20% due to the manufacturing of the fiber, what is the cutoff region of this fiber?
c. What is the fiber mode field diameter if it is operating at 1.550 μm wavelength?

6.6.4 PROBLEM 4

For the optical fiber in Problem 3, if the refractive index profile is parabolic ($\alpha = 2$) or triangular ($\alpha = 1$) with the given numerical aperture at the central axis, repeat (a), (b), and (c).

6.6.5 PROBLEM 5

Single-mode optical fibers produced by Corning (see Appendix 6.1) have typical characteristics as per the technical data sheet.

a. Using the fiber physical characteristics and technical data on its numerical aperture, confirm the fiber functional characteristics such as the cutoff wavelength range.
b. If this fiber is launched with a 0.850 μm laser, how many modes would it support? Sketch the fields for LP_{01} and LP_{11} modes.
c. If lightwaves at 1.550 μm travel over 10 km of this fiber, calculate the travel time of the waves.
d. Estimate the fiber core diameter at 1.310 μm wavelength.
e. If the same spot size of (d) is required for the fiber to operate at 1.550 μm, can you advise the manufacturer on any change of the fiber physical parameters?

6.6.6 PROBLEM 6

a. The optical fiber in Problem 3 is used in an optical fiber transmission system with a laser source operating at 1.310 μm having an output power of

1.0 mW. The fiber length is 50 km. An optical receiver can detect an average optical power of 0.1 μW. Is it possible to detect the optical power at the end of the fiber length?

b. Referring to the technical data of the standard optical fiber, estimate the spreading of the optical pulse after transmitting through the 50 km length fiber if the source has an optical line width of 2.0 nm.

6.6.7 PROBLEM 7

A step index optical fiber is used for an optical communication system operating at 1.310 μm and having a core radius of 25 μm and refractive indices in the core and cladding regions of 1.460 and 1.4550, respectively.

a. What is the numerical aperture of the fiber?
b. Estimate the number of guided modes.

6.6.8 PROBLEM 8

a. Show that for a graded index fiber having a core refractive index

$$n^2(r) = n_2^2 \left[1 + 2\Delta s \left(\frac{r}{a} \right) \right]$$

with $s(r/a) = 1 - (r/a)^\alpha$, the acceptance angle $\alpha(r)$ is given by $\sin\alpha(r) = [n^2(r) - n_2^2]^{1/2}$.

b. If the optical fiber has a parabolic profile shape, show that

$$\sin\alpha(r) = NA \sqrt{1 - \left(\frac{r}{a} \right)^{1/2}}$$

where $NA = n_2(2\Delta)^{1/2}(1 + \Delta)$.

c. A parabolic graded index silica optical fiber has a cladding refractive index of 1.460 and a relative index difference at the core axis of 1%. Find the maximum acceptance angle at the core axis of the fiber. Plot $\sin\alpha(r)$ as a function of r. What is the acceptance angle of the fiber at the core and cladding interface? Comment on the launch of a laser source into this fiber.

6.6.9 PROBLEM 9

a. For a single-mode optical fiber having a graded index central dip, that is, $s(r/a) = 1 - (1 - r/a)\alpha$, the ESI parameters of V and the radius are given by

$$\frac{V_e}{V} = \left[1 - \frac{2\gamma}{(\alpha+1)(\alpha+2)}\right]^{1/2} \quad \text{and} \quad \frac{a_e}{a} = \frac{(\alpha+1)(\alpha+2)(\alpha+3)-6\gamma}{(\alpha+1)(\alpha+2)(\alpha+3)-2\gamma}$$

where $V = ka(2\Delta)^{1/2}$.

b. The fiber has a physical core radius of 8.0 µm, a maximum relative index difference of 0.3%, and a cladding refractive index of 1.460. Find its ESI parameters for the normalized frequency and radius at 1.550 µm wavelength. Find also its ESI cutoff wavelength and its mode field diameter at this wavelength.

6.6.10 PROBLEM 10

A silica single-mode optical fiber with a mode spot size of 9 µm is launched with an optical data signal sequence of an average power of 10 mW.

a. Assuming the nonlinear coefficient n_2 is $2.3e^{-20}$ m/W, estimate the nonlinear factor γ—then the nonlinear phase expected after propagating through a 100 km span of this fiber, given that the attenuation factor of the fiber is 0.2 dB/km. *Hint*: You need to find the effective length of the optical sequence traveling distance.

b. If the optical data sequence is modulated using the QPSK scheme, sketch the constellation of the data sequence before and after propagating through the fiber.

c. Repeat (a) and (b) for the Corning SMF-28 and LEAF as given in the chapter's appendix.

APPENDIX 6.1: TECHNICAL SPECIFICATION OF CORNING SINGLE-MODE OPTICAL FIBERS

Corning® SMF-28™ Optical Fiber
Product Information

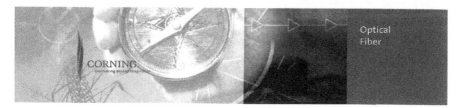

PI1036
Issued: April 2001
Supercedes: March 2001
ISO 9001 Registered

Corning® Single-Mode Optical Fiber

The Standard For Performance

Corning® SMF-28™ single-mode optical fiber has set the standard for value and performance for telephony, cable television, submarine, and utility network applications. Widely used in the transmission of voice, data, and/or video services, SMF-28 fiber is manufactured to the most demanding specifications in the industry. SMF-28 fiber meets or exceeds ITU-T Recommendation G.652, TIA/EIA-492CAAA, IEC Publication 60793-2 and GR-20-CORE requirements.

Taking advantage of today's high-capacity, low-cost transmission components developed for the 1310 nm window, SMF-28 fiber features low dispersion and is optimized for use in the 1310 nm wavelength region. SMF-28 fiber also can be used effectively with TDM and WDM systems operating in the 1550 nm wavelength region.

Features and Benefits

• Versatility in 1310 nm and 1550 nm applications.

• Outstanding geometrical properties for low splice loss and high splice yields.

• OVD manufacturing reliability and product consistency.

• Optimized for use in loose tube, ribbon, and other common cable designs.

The Sales Leader

Corning SMF-28 fiber is the world's best selling fiber. In 2000, SMF-28 fiber was deployed in over 45 countries around the world. All types of network providers count on this fiber to support network expansion into the 21st Century.

Environmental Specifications

Environmental Test Condition	Induced Attenuation (dB/km)	
	1310 nm	1550 nm
Temperature Dependence −60°C to +85°C*	≤0.05	≤0.05
Temperature-Humidity Cycling −10°C to + 85°C*, up to 98% RH	≤0.05	≤0.05
Water Immersion, 23° ± 2°C*	≤0.05	≤0.05
Heat Aging, 85°C ± 2°C*	≤0.05	≤0.05

*Reference temperature = +23°C

Operating Temperature Range

−60°C to + 85°C

Dimensional Specifications

Standard Length (km/reel): 2.2 − 50.4*
*Longer spliced lengths available at a premium.

Glass Geometry

Fiber Curl: ≥ 4.0 m radius of curvature
Cladding Diameter: 125.0 ± 1.0 μm
Core-Clad Concentricity: ≤ 0.5 μm
Cladding Noncircularity: ≤ 1.0%

Defined as: $\left[1 - \dfrac{\text{Min. Cladding Diameter}}{\text{Max. Cladding Diameter}}\right] \times 100$

Coating Geometry

Coating Diameter: 245 ± 5 μm
Coating–Cladding Concentricity: <12 μm

Mechanical Specifications

Proof Test

The entire fiber length is subjected to a tensile proof stress ≥ 100 kpsi (0.7 GN/m²*).
*Higher proof test levels available at a premium.

Performance Characterizations
Characterized parameters are typical values.

Core Diameter: 8.2 μm

Numerical Aperture: 0.14

NA is measured at the one percent power level of a one-dimensional far-field scan at 1310 nm.

Zero Dispersion Wavelength (λ_0): 1313 nm

Zero Dispersion Slope (S_0): 0.086 ps/(nm²·km)

Refractive Index Difference: 0.36%

Effective Group Index of Refraction, (N_{eff} @ nominal MFD):

1.4677 at 1310 nm
1.4682 at 1550 nm

Fatigue Resistance Parameter (n_d): 20

Coating Strip Force:

Dry: 0.6 lbs. (3N)
Wet, 14-day room temperature: 0.6 lbs. (3N)

Rayleigh Backscatter Coefficient (for 1 ns pulse width):

1310 nm: −77 dB
1550 nm: −82 dB

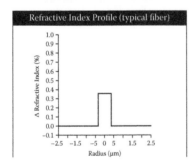

Refractive Index Profile (typical fiber)

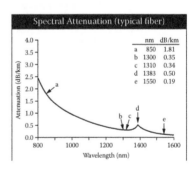

Spectral Attenuation (typical fiber)

nm	dB/km
a	850	1.81
b	1300	0.35
c	1310	0.34
d	1383	0.50
e	1550	0.19

Corning® LEAF® Optical Fiber
Product Information

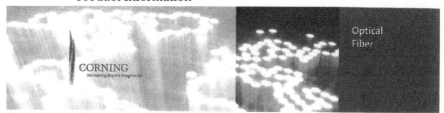

PI1107
Issued: May 2001
Supercedes: April 2001
ISO 9001 Registered

A powerful Network Needs:
Backbone by LEAF Fiber.

With the ever-accelerating race for bandwidth, network designers are challenged to build a network for the present that will also maximize future technologies. Deploy the fiber that revolutionized network technology and gives you room to move. Break the bandwidth barrier with a fiber so technologically advanced it gives you the optical backbone you need for today's and tomorrow's networks – Corning® LEAF® optical fiber.

Find out what the world's most powerful networks have in common: Backbone by LEAF fiber.

The Large Effective Area Advantage

LEAF fiber's large effective area (A_{eff}) offers higher power-handling capability, improved optical signal-to-noise ratio, longer amplifier spacing, and maximum dense wavelength division multiplexing (DWDM) channel plan flexibility compared with other nonzero dispersion-shifted fibers (NZ-DSFs).

Fiber with a large A_{eff} also provides a critical performance advantage – the ability to uniformly reduce all nonlinear effects (Figure 1). Nonlinear effects represent the greatest performance limitations in today's multichannel DWDM systems.

The Next Generation

In addition to outperforming other NZ-DSFs in the conventional band (C-Band: 1530–1565 nm), LEAF fiber facilitates the next technological development in fiber–optic networks – the migration to the long band (L-Band: 1565–1625 nm). In both C-Band and L-Band operation, LEAF fiber has demonstrated greater ability to handle more channels by reducing nonlinear effects such as four-wave mixing, self-phase modulation and cross-phase modulation in multichannel DWDM transmission.

Reduce Network Costs

With its increased optical reach advantage, LEAF fiber requires fewer amplifiers and regenerators, and therefore provides immediate and long-term cost savings. LEAF fiber is also compatible with installed base fibers and photonic components. In fact, LEAF fiber's slightly larger mode-field diameter improves its splicing performance, especially when connecting to standard single-mode fiber such as Corning® SMF-28™ fiber. And, as with all Corning optical fiber, LEAF fiber's geometry package is the best in the industry. With LEAF fiber, it is easy and economical to increase the information-carrying capacity of your network.

Figure 1

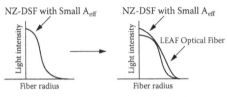

NZ-DSF with Small A_{eff} NZ-DSF with Small A_{eff}

LEAF fiber's larger A_{eff} increases the area where the light can propagate, thereby reducing nonlinear effects.

Fiber for Today and Tomorrow

While LEAF fiber is exceptionally suited to operate with already installed 2.5 Gbps systems, it is techno-economically optimized for today's high-channel-count 10 Gbps systems and provides the ability to upgrade in the future to tomorrow's high bit systems. Additionally, LEAF fiber's unparalleled specifications on polarization mode dispersion (PMD) allow fiber installed today to operate at data rates higher than 10 Gbps. The combination of LEAF fiber's large A_{eff} and its demonstrated Raman upgradeability allows transmission engineers to design and build networks advantaged over other fiber plants. As the world's most advanced NZ-DSF, LEAF fiber is ready for future technology when your network is.

LEAF Fiber – All about Value

With LEAF fiber's proven large A_{eff} advantage, the industry's best geometry package, and inherent future-proof design, LEAF fiber continues to be the fiber of choice for today's high-capacity and tomorrow's all-optical networks. Network providers on the cutting edge have embraced large A_{eff} technology as the fiber "backbone" for high-data-rate networks now and in the future.

Technology Awards

Corning Incorporated has received multiple industry awards for its patented LEAF optical fiber. Independent panels of experts have chosen LEAF fiber based on its technical merits for the following awards:

Annual Technology Award from *Fiberoptic Product News*

Commercial Technology Achievement Award for Fiber-Optics from *Laser Focus World Magazine*

Circle of Excellence Award from *Photonics Spectra Magazine*

R&D 100 Award from *R&D Magazine*

Coating

Corning fiber is protected for long-term performance and reliability by the CPC™ coating system. Corning's enhanced, dual acrylate CPC coatings provide excellent fiber protection and are easy to work with. CPC coatings are designed to be mechanically stripped and have an outside diameter of 245 μm. CPC coatings are optimized for use in many single- and multi-fiber cable designs, including loose tube, ribbon, slotted core, and tight buffer cables.

Optical Specifications

Attenuation

≤0.25 dB/km at 1550 nm

≤0.25 dB/km at 1625 nm

No point discontinuity greater than 0.10 dB at 1550 nm

Attenuation at 1383 ± 3 nm shall not exceed 1.0 dB/km

Attenuation vs Wavelength		
Range (nm)	Ref. λ (nm)	Max Increase α (dB/km)
1525–1575	1550	0.05
1625	1550	0.05

The attenuation in a given wavelength range does not exceed the attenuation of the reference wavelength (λ) by more than the value α. In all cases, a maximum attenuation of ≤0.25 dB/km applies at 1550 nm and 1625 nm.

Dispersion Calculation

$$\text{Dispersion} = D(\lambda) = \left(\frac{D(1565\text{ nm}) - D(1530\text{ nm})}{35} \, ^*(\lambda - 1565)\right) + D(1565\text{ nm})$$

λ = Operating wavelength up to 1565

$$\text{Dispersion} = D(\lambda) = \left(\frac{D(1625\text{ nm}) - D(1565\text{ nm})}{60} \, ^*(\lambda - 1625)\right) + D(1625\text{ nm})$$

λ = Operating wavelength from 1565–1625

Special selections of LEAF fiber attributes are available upon request.

Ordering Information

To order Corning® LEAF® optical fiber, contact your sales representative, or call the Optical Fiber Customer Service Department at **910-395-7659** (North America) and **607-974-7174** (International). Please specify the following parameters when ordering.

Fiber Type: Corning® LEAF® Nonzero Dispersion Shifted Single-Mode Fiber

Fiber Attenuation Cell: dB/km

Fiber Quantity: kms

Other: (Requested ship date, etc.)

Corning Incorporated
www.corning.com/opticalfiber

One Riverfront Plaza
Corning, NY 14831
U.S.A.

Phone: 800-525-2524 (U.S. and Canada)
607-786-8125 (International)

Fax: 800-539-3632 (U.S. and Canada)
607-786-8344 (International)

Email: info@corningfiber.com

Europe

Berkeley Square House
Berkeley Square
London W1X 5PE
U.K.

Phone: +800 2800 4800 (U.K.*, Ireland, France, Germany,
The Netherlands, Spain and Sweden)
*Callers from U.K. dial (00) before the phone number

+800 781 516 (Italy)

+44 7000 280 480 (All other countries)

Fax: +44 7000 250 450

Email: europe@corningfiber.com

Asia-Pacific

Australia
Phone: 1-800-148-690
Fax: 1-800-148-568

Indonesia
Phone: 001-803-015-721-1261
Fax: 001-803-015-721-1262

Malaysia
Phone: 1-800-80-3156
Fax: 1-800-80-3155

Philippines

Singapore
Phone: 800-1300-955
Fax: 800-1300-956

Thailand
Phone: 001-800-1-3-721-1263
Fax: 001-800-1-3-721-1264

Central & Latin America

Brazil
Phone: 000817-762-4732
Fax: 000817-762-4996

Mexico
Phone: 001-800-235-1719
Fax: 001-800-339-1472

Venezuela
Phone: 800-1-4418
Fax: 800-1-4419

Greater China

Beijing
Phone: (86) 10-6505-5066
Fax: (86) 10-6505-5077

Hong Kong
Phone: (852) 2807-2723
Fax: (852) 2807-2152

Shanghai
Phone: (86) 21-6361-0826 ext. 107
Fax: (86) 21-6361-0827

Taiwan
Phone: (886) 2-2716-0338
Fax: (886) 2-2716-0339

E-mail: luyc@corning.com

REFERENCES

1. C. Xia, N. Bai, I. Ozdur, X. Zhou, and G. Li, "Supermodes for Optical Transmission," *Opt. Express*, 19(17), 16653, 2011.
2. A.H. Gnauck, P.J. Winzer, A. Konczykowska et al., "Generation and Transmission of 21.4-Gbaud PDM 64-QAM Using a Novel High-Power DAC Driving a Single I/Q Modulator," *IEEE J. Lightw. Technol.*, 30(4), 2012.
3. M. Nakazawa, "Giant Leaps in Optical Communication Technologies Towards 2030 and Beyond," Presentation at ECOC 2010 Plenary Talk, Torino, September 20, 2010.
4. A. Safaai-Jazi and J.C. McKeeman, "Synthesis of Intensity Patterns in Few-Mode Optical Fibers," *IEEE J. Lightw. Technol.*, 9(9), 1047, 1991.
5. M. Salsi, C. Koebele, D. Sperti et al., "Transmission at 2x100Gb/s, over Two Modes of 40km-Long Prototype Few-Mode Fiber, Using LCOS Based Mode Multiplexer and Demultiplexer," Presentation at Optical Fiber Conference, OFC 2011, and National Fiber Optic Engineers Conference (NFOEC), Los Angeles, CA, March 6, 2011, Postdeadline Session B (PDPB).
6. S. Randel, R. Ryf, A. Sierra et al., "6×56-Gb/s Mode-Division Multiplexed Transmission over 33-km Few-Mode Fiber Enabled by 6×6 MIMO Equalization," *Opt. Express*, 19(17), 16697, 2011.
7. C.D. Poole, J.M. Wiesenfeld, D.J. DiGiovanni, and A.M. Vengsarkar, "Optical Fiber-Based Dispersion Compensation Using Higher Order Modes Near Cutoff," *IEEE J. Lightw. Technol.*, 12(10), 1994.
8. A. Carter and B. Samson, "PANDA-Style Fibers Move beyond Telecom," *Laser Focus World*, August 2008.
9. C.H. Bennett, F. Bessette, G. Brassard, L. Salvail, and J. Smolin, "Experimental Quantum Cryptography," *J. Cryptol.*, 5(1), 3–28, 1992.
10. A.R. Dixon, Z.L. Yuan, J.F. Dynes, A.W. Sharpe, and A.J. Shields, *Opt. Express*, 16(23), 18790–18979.
11. L.N. Binh, L. Ling, and L.C. Li, "Volterra Series Transfer Function in Optical Transmission and Nonlinear Compensation," in *Nonlinear Optical Systems*, ed. L.N. Binh and D.V. Liet, CRC Press, Boca Raton, FL, 2012, Chapter 10.
12. A. Mecozzi, C.B. Clausen, and M. Shtaif, "Analysis of Intrachannel Nonlinear Effects in Highly Dispersive Optical Pulse Transmission," *IEEE Photon. Technol. Lett.*, 12, 392–394, 2000.
13. K. Kikuchi, "Coherent Detection of Phase-Shift Keying Signals Using Digital Carrier-Phase Estimation," Presentation at Proceedings of IEEE Conference on Optical Fiber Communications, Institute of Electrical and Electronics Engineers, Anaheim, CA, 2006, Paper OTuI4.
14. G.P. Agrawal, *Fiber Optic Communications Systems*, 3rd ed., J. Wiley, New York, 60–67, 2002.
15. R.H. Stolen, E.P. Ippen, and A.R. Tynes, *Appl. Phys. Lett.*, 20, 62, 1972.
16. E.P. Ippen and R.H. Stolen, *Appl. Phys. Lett.*, 21, 539, 1972.
17. R.G. Smith, *Appl. Opt.*, 11, 2489, 1972.

7 Optical Fiber Operational Parameters

Operational parameters of any electronic or photonic devices are specified for operations such as amplification, switching, or transmission, so they are required in both time and frequency domains related to the speed of the system operation. This chapter describes the mechanism and properties of lightwave modulated signals when propagating over a distance of optical fibers, the single-mode optical fibers (SMFs) only, in particular standard SMF (SSMF) and nonzero-dispersion-shifted fibers (NZ-DSFs), and dispersion-compensating fibers. Attenuation and dispersion effects, the two principal impairments in the design of transmission systems, are described in detail.

7.1 INTRODUCTORY REMARKS

In Chapter 8 the basic structure and fundamental aspects of lightwaves propagating in planar optical waveguides are treated. The single-mode optical fibers are the basic structure for standard communication transmission systems and were introduced in Chapter 6. This chapter deals with the transmission of optical signals over optical fibers, mainly the loss and spreading of optical signals transmitted through optical fibers, namely, the attenuation and dispersion effects. These two parameters influence significantly the transmission distance and bit rate, or the high or low speed of the data rate, which can be transmitted over the guided wave medium.

Attenuation and dispersion are the two most important effects that play major parts in optical fiber transmission systems. The attenuation of an optical signal would limit the availability of optical power along the transmission path, and for very low attenuation, dispersion limits the repeater spacing below what would be possible from the attenuation factor.

The fiber loss has been reduced from 100 dB/km (i.e., transmission possible over only a few meters) at 1300 nm in 1970 to about 0.25 and 0.15 dB/km, which is very close to a theoretically possible transparent limit and transmission of over several hundred kilometers of fibers, for 1300 and 1550 nm wavelength regions, respectively, in 1980.

The dispersion and pulse broadening of optical fibers have also been reduced due to "smart design" of optical fiber structures. In early 1970 we saw a remarkable development of theories for the understanding of lightwave guiding in optical fibers of multimode types. The breakthrough in the reduction of loss in optical fibers and the ability to manufacture optical fibers with a very small core diameter leads to the design of single optical fibers.[1] The remarkable theoretical development of optical waveguiding in a weakly guiding (i.e., a very small difference between the core and cladding regions) fiber structure leads to a plane-wave-like transmission of light-waves. Further, the availability of narrow linewidth lasers allows systems engineers to design and implement several high-speed long-distance fiber optic communication systems.

The attenuation that arises from intrinsic material properties and from waveguide properties is described, and a general attenuation coefficient is derived. The chromatic dispersion for SM fiber in a linear limit means that we assume the optical power launched into the fiber to be less than the threshold for nonlinear effects is then treated. The effects of optical waveguide parameters on the dispersion factors are analyzed. The balance of the opposite-signed dispersion factors between material and waveguides is analyzed so that a minimum dispersion factor can be designed for optical fibers with dispersion-compensated or -shifted characteristics achieved.

7.2   SIGNAL ATTENUATION IN OPTICAL FIBERS

Optical loss in optical fibers is one of the two main fundamental limiting factors, as it reduces the average optical power reaching the receiver. The optical loss is the sum of three major components: intrinsic loss, microbending loss, and splicing loss.

7.2.1   Intrinsic or Material Attenuation

Intrinsic loss consists mainly of absorption loss due to OH impurities and Rayleigh scattering loss. The intrinsic absorption is a function of λ^{-6}. Thus in silica fibers, the longer the operating wavelength, the lower the loss is. However, it also depends on the transparency of the optical materials that are used to form the optical fibers. For silica fiber the optical material loss is low over the wavelength range 0.8 to 1.8 μm. Over this wavelength range there are three optical windows that optical communications are utilizing. The first window over the central wavelength 810 nm is about 20.0 nm bandwidth over the central wavelength. The second and third windows most commonly used in present optical communications are over 1300 and 1550 nm with a range of 80 and 40 nm, respectively. The intrinsic losses are about 0.3 and 0.15 dB/km at 1550 and 1300 nm regions, respectively.

This is a few hundred thousand times improvement over the original transmission of signal over 5.0 m with a loss of about 60 dB/km. Most communication fiber systems are operating at 1300 nm due to the minimum dispersion at this range. For power-hungry systems optical or extra-long systems should operate at 1550 nm.

7.2.2   Absorption

The absorption loss in silica glass is composed mainly of ultraviolet (UV) and infrared (IR) absorption tales of pure silica. The IR absorption tale of pure silica has been shown to be due to the vibration of the basic tetrahedron, and thus strong resonances occurs around 8 to 13 μm with a loss of about 10^{-10} dB/Km. This loss is shown in the curve IR of Figure 7.1. Overtones and combinations of these vibrations lead to various absorption peaks in the low-wavelength range, as shown by curve UV.

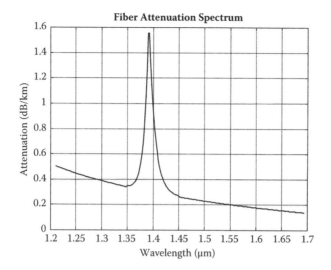

FIGURE 7.1 Attenuation of optical signals as a function of wavelength. The minimum loss at wavelength: at $\lambda = 1.3$ µm about 0.3 dB/km and at $\lambda = 1.5$ µm about 0.13 dB/km. For cabled fibers the attenuation factor at 1550 nm is 0.25 dB/km.

Various impurities that also lead to spurious absorption effects in the wavelength range of interest (1.2–1.6 µm) are transition metal ions and water in the form of OH ions. These sources of absorption have been practically reduced in recent years.

7.2.3 RAYLEIGH SCATTERING

The Raleigh scattering loss, L_R, which is due to microscopic nonhomogeneities of the material, shows a λ^{-4} dependence and is given by

$$L_R = (0.75 + 4.5D)\lambda^{-6} \text{ dB/Km} \tag{7.1}$$

where Δ is the relative index difference as defined above and λ is the wavelength in µm. Thus, to minimize the loss Δ should be made as low as possible.

7.2.4 WAVEGUIDE LOSS

The losses due to waveguide structure arise from power leakage, bending, microbending of the fiber axis, and defects and joints between fibers. The power leakage is significant only for depressed cladding fibers.

7.2.5 BENDING LOSS

When a fiber is bent the plane-wave fronts associated with the guided mode are pivoted at the center of curvature, and their longitudinal velocity along the fiber axis increases with the distance from the center of curvature. As the fiber is bent further over a critical curve, the phase velocity exceeds that of a plane wave in the cladding and radiation occurs.

The bend loss L_B for a radius R (radius of curvature) is given by

$$L_B = -10 \log_{10}\left(1 - 890\frac{r_0^6}{\lambda^4 R^2}\right) \text{ for silica} \qquad (7.2)$$

7.2.6 MICROBENDING LOSS

Microbending loss results from power coupling from the guided fundamental mode of the fiber to radiation modes. This coupling takes place when the fiber axis is bent randomly in a high spatial frequency. Such bending can occur during packing of the fiber during the cabling process.

The microbending loss of an SMF is a function of the fundamental mode spot size r_0. Fibers with large spot size are extremely sensitive to microbending. It is therefore desirable to design the fiber to have as small a spot size as possible to minimize bending loss. The microbending loss can be expressed by the relation

$$L_m = 2.15 \times 10^{-4} r_0^6 \lambda^{-4} L_{mm} \qquad (7.3)$$

where L_{mm} is the microbending loss of a 50 μm core multimode fiber having a numerical aperture (NA) of 0.2.

7.2.7 JOINT OR SPLICE LOSS

Ultimately the fibers will have to be spliced together to form the final transmission link. With fiber cable that averages 0.4–0.6 dB/km, splice loss in excess of 0.2 dB/splice drastically reduces the nonrepeated distance that can be achieved. It is therefore extremely important that the fiber be designed such that splicing loss be minimized.

Splice loss is mainly due to axial misalignment of the fiber core, as shown in Figure 7.2.

Splicing techniques, which rely on aligning the outside surface of the fibers, require extremely tight tolerances on the core to outside surface concentricity. Offsets on the order of 1 μm can produce significant splice loss. This loss is given by

$$L_s = \frac{10}{\ln 10}\left(\frac{d}{r_0}\right)^2 \text{ dB} \qquad (7.4)$$

where d is the *axial* misalignment of the fiber cores. It is obvious that minimizing optical loss involves making trade-offs between the different sources of loss. It is advantageous to have a large spot size to minimize both Raleigh and splicing losses, whereas minimizing bending and microbending losses requires a small spot size. In

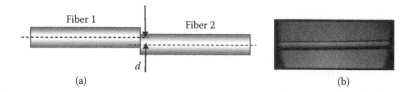

FIGURE 7.2 (a) Misalignment in splicing of two optical fibers, generating losses; (b) aligned spliced fibers.

addition, as will be described in the next section, the spot size plays a significant role in the chromatic dispersion properties of single-mode fibers.

7.2.8 ATTENUATION COEFFICIENT

Under general conditions of power attenuation inside an optical fiber, the attenuation coefficient of the optical power P can be expressed as

$$\frac{dP}{dz} = -\alpha P \tag{7.5}$$

where α is the attenuation coefficient in linear scale. This attenuation coefficient can include all effects of power loss when signals are transmitted through the optical fibers.

Considering optical signals with an average optical power entering at the input of the fiber length L, P_{in}, and output optical power, P_{out}, we have that P_{in} and P_{out} are related to the attenuation coefficient α as

$$P_{out} = P_{in}e^{(-\alpha L)} \tag{7.6}$$

It is customary to express α in dB/km by using the relation

$$\alpha(\text{dB/km}) = -\frac{10}{L}\log_{10}\left(\frac{P_{out}}{P_{in}}\right) = 4.343\alpha \tag{7.7}$$

with L the distance in kilometers. SSMFs with a small Δ would have a loss/attenuation coefficient of about 0.2 dB/km; i.e., the purity of the silica would be very high. Such a purity of a bar of silica would allow us to see through a 1 km thick glass bar without distortion. The attenuation curve for silica glass is shown in Figure 7.1.

7.3 SIGNAL DISTORTION IN OPTICAL FIBERS

7.3.1 BASICS ON GROUP VELOCITY

Consider a monochromatic field given by

$$E_x = A\cos(\omega t - \beta z) \tag{7.8}$$

where A is the wave amplitude/envelope carried by the optical carrier, ω is the radial frequency of the lightwave carrier, and β is the propagation constant of the light-waves guided along the z-direction at the operating wavelength. If we set $(\omega t - \beta z)$ constant, then the wave phase velocity is defined as the differential variation with respect to time along the propagation direction, given by

$$v_p = \frac{dz}{dt} = \frac{\omega}{\beta} \tag{7.9}$$

Now consider the propagating wave, which is a superimposition of two mono-chromatic fields of frequencies $\omega + \delta\omega$ and $\omega - \delta\omega$ of

$$E_{x1} = A\cos[(\omega + \delta\omega)t - (\beta + \delta\beta)z)] \tag{7.10}$$

$$E_{x2} = A\cos[(\omega - \delta\omega)t - (\beta - \delta\beta)z)] \tag{7.11}$$

The total field is then given by

$$E_x = E_{x1} + E_{x2} = 2A\cos(\omega t - \beta z)\cos(\delta\omega t - \delta\beta z) \tag{7.12}$$

If $\omega \gg \delta\omega$, then $\cos(\omega t - \beta z)$ varies much faster than $\cos(\delta\omega t - \delta\beta z)$. By setting the constant we can define the group velocity as

$$v_g = \frac{d\omega}{d\beta} \rightarrow v_g^{-1} = \frac{d\beta}{d\omega} \tag{7.13}$$

The group delay t_g per unit length (setting L at 1.0 km) is thus given as

$$t_g = \frac{L(\text{of } 1\text{ Km})}{v_g} = \frac{d\beta}{d\omega} \tag{7.14}$$

We are interested in the group velocity because when the lightwave carrier is modulated so as to carry information, the spectrum of the lightwaves is broadened from a single line of the single carrier. This broadening of the spectrum consists of several frequency components that travel down the optical fiber with different veloci-ties because the refractive index of the fiber and the propagation constant vary with respect to frequency. Thus, interference between these components would generate the broadening of the original modulating pulse before propagating. This broadening would then limit the operation speed of the transmission systems.

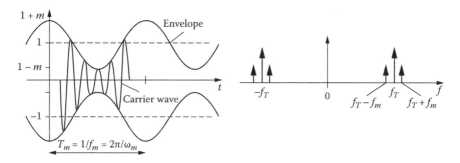

FIGURE 7.3 Modulated signal in time domain and frequency domain.

The pulse spread $\Delta\tau$ per unit length due to a group delay of light sources of spectral width σ_λ (i.e., the full width at half maximum (FWHM) of the optical spectrum of the light source) is

$$\Delta\tau = \frac{dt_g}{d\lambda}\sigma_\lambda \tag{7.15}$$

Then, the spread of the group delay is due to the spread of the source wavelength in ps/km. The linewidth of the light source makes a significant difference in the distortion of the optical signal transmitted through a long length of single-mode optical fiber. The narrower the source linewidth, the less dispersed the optical pulses are. A typical linewidth of Fabry–Perot semiconductor lasers is about 1 to 2.0 nm, while the distributed feedback (DFB) laser would exhibit a linewidth of 100 MHz. To how many nanometers is this 100 MHz optical frequency equivalent?

In the case where the laser source has a very narrow spectral width, such as the external laser cavity whose bandwidth is only 100 MHz, the spectral width would take the value of the 3 dB bandwidth of the modulated spectrum, which is equivalent to that in the baseband of the signal. This is illustrated in Figure 7.3.

The variation of the phases of different spectral components can also be illustrated, as phasors, in Figures 7.4 and 7.5. Due to this phase difference or delay times between components, interference of these waves would thus create dispersion or broadening of the pulse in the time domain.

Optical signal traveling along a fiber becomes increasingly distorted. This distortion is a consequence of *intermodal* delay effects and *intramodal* dispersion. Intermodal delay effects are significant in multimode optical fibers due to each mode having a different value of group velocity at a specific frequency. While intermodal dispersion is pulse spreading that occurs within a single mode, it is the result of the group velocity being a function of the wavelength λ and is therefore referred to as *chromatic dispersion*.

Two main causes of intermodal dispersion are

- Material dispersion arising from the variation of the refractive index of the fiber core and cladding, $n(\lambda)$, as a function of wavelengths. This causes a wavelength dependence of the group velocity of any guided mode.

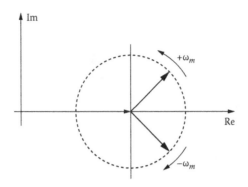

FIGURE 7.4 Vector phasor diagram representing the complex envelope. The phasor rotates at an optical angular frequency, and the optical carrier phase is indicated by the angle of the vectors in this plane.

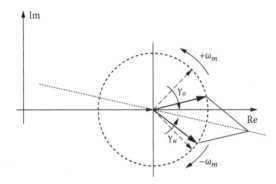

FIGURE 7.5 Magnitude of complex envelope when not sinusoidal or affected by nonlinear distortion effects.

- Waveguide dispersion occurring due to the dependence of the propagation constant $\beta(\lambda)$ of the guided mode as a function of wavelength λ and core radius a and the refractive index difference.

The group velocity associated with the fundamental mode is thus frequency dependent. As a result, different spectral components of the light pulse travel at different group velocities; this phenomenon is then referred to as the *group velocity dispersion* (GVD), intramodal dispersion, or material dispersion and waveguide dispersion.

7.3.2 Group Velocity Dispersion

7.3.2.1 Material Dispersion

The refractive index of silica varies as a function of wavelength, as depicted in Figure 7.6. The refractive index is plotted over the wavelength region of 1.0 to 2.0 μm, which is the lowest loss range for silica fiber and, thus, the operating region for optical communication systems. This region is commonly known as the 1300 and 1550 nm window.

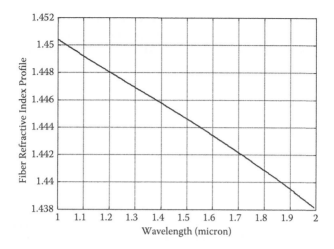

FIGURE 7.6 Variation in the refractive index as a function of optical wavelength of silica fiber.

The propagation constant β of the fundamental mode guided in an optical fiber can be written as

$$\beta(\lambda) = \frac{2\pi n(\lambda)}{\lambda}$$ (7.16)

The group delay t_{gm} per unit length, by using (7.14), can be obtained for material dependence as

$$t_{gm} = \frac{d\beta}{d\omega}$$ (7.17)

Now use

$$d\omega = d\left(\frac{2\pi c}{\lambda}\right) = -\frac{2\pi c}{\lambda^2}d\lambda$$ (7.18)

Then (7.17) becomes

$$t_{gm} = -\frac{\lambda^2}{2\pi c}\frac{d\beta}{d\lambda}$$ (7.19)

Substituting (7.16) into (17.9) we have

$$t_{gm} = \frac{1}{c}\left[n(\lambda) - \frac{\lambda dn(\lambda)}{d\lambda}\right]$$ (7.20)

Thus the pulse dispersion per unit length $\Delta\tau_m/\Delta\lambda$ due to material can be found by differentiating (7.20) and summing up all spectral components of the lightwave group of a source having a root mean square (RMS) spectral width σ_λ as

$$\Delta\tau_m = -\frac{\lambda}{c}\frac{d^2n}{d\lambda^2}\sigma_\lambda \tag{7.21}$$

If setting $\Delta\tau_m = M(\lambda)\sigma_\lambda$, we obtain

$$M(\lambda) = -\frac{\lambda}{c}\frac{d^2n}{d\lambda^2} \tag{7.22}$$

$M(\lambda)$ is defined as the material dispersion factor or material dispersion parameter; its unit is commonly expressed in ps/(nm.km), or equivalently in SI units of 10^{-6} s/m^2.

If the refractive index can be expressed as a function of the optical wavelength, the material dispersion can be estimated. In fact, in practice optical material engineers have to characterize all optical properties of new materials. The refractive index $n(\lambda)$ can usually be expressed in Sellmeier's dispersion formula as

$$n^2(\lambda) = 1 + \sum_k \frac{G_k\lambda^2}{(\lambda^2 - \lambda_k^2)} \tag{7.23}$$

where G_k are Sellmeier's constants and k is an integer normally taken in a range of $k = 1, 2, 3$. In the late 1970s several silica-based glass materials were manufactured and their properties measured. The refractive indices are usually expressed using Sellmeier's coefficients. These coefficients for several optical fiber materials are given in Table 7.1.

By using curve fitting, the refractive index of pure silica, $n(\lambda)$, can be expressed as

$$n(\lambda) = c_1 + c_2\lambda^2 + c_3\lambda^{-2} \tag{7.24}$$

where $c_1 = 1.45084$, $c_2 = -0.00343$ μm^{-2}, and $c_3 = 0.00292$ μm^2. From Table 7.1 and (7.24), we can use (7.22) to determine the material dispersion factor for certain wavelength range.

For the doped core of the optical fiber, Sellmeier's expression (7.23) can be approximated by using the curve-fitting technique to approximate it to the form given in (7.24). The material dispersion factor $M(\lambda)$ becomes zero at wavelengths around 1280 nm and about -10 ps/(nm.km) at 1550 nm, as calculated in Figure 7.7. However, the attenuation at 1350 nm is about 0.4 dB/km, compared to 0.2 dB/km at 1550 nm, as shown in Table 7.1. So from the operation of optical transmission systems, at the operating wavelength of 1300 nm, optical modulated signals would suffer almost no dispersion and, hence, no distortion, but they are limited by the high attenuation factor. On the other hand, operating in the 1550 nm wavelength, the loss is low but the modulated signals would suffer distortion due to pulse broadening by

TABLE 7.1

Sellmeier's Coefficients for Several Optical Fiber Silica-Based Materials with Germanium Doped in the Core Region

Sellmeier's Constants	Germanium Concentration, C (mol%)			
	0 (pure silica)	3.1	5.8	7.9
G_1	0.6961663	0.7028554	0.7088876	0.7136824
G_2	0.4079426	0.4146307	0.4206803	0.4254807
G_3	0.8974794	0.8974540	0.8956551	0.8964226
λ_1	0.0684043	0.0727723	0.0609053	0.0617167
λ_2	0.1162414	0.1143085	0.1254514	0.1270814
λ_3	9.896161	9.896161	9.896162	9.896161

the dispersion phenomena. This dilemma has been considered and was solved in early 1990 when fiber-based optical amplification was achieved using erbium-doped silica core fiber and the design of zero-dispersion fiber with the zero-dispersion wavelength shifted to the vicinity of 1550 nm. The detection of this kind of optical communication system is direct, and noises contributed by these optical amplifiers are high—hence, limiting the transmission distance. In the 21st century coherent detection associated with digital signal processing has enabled the reduction of optical amplifiers installed along the transmission link and compensation of dispersion conducted in the electronic digital domain. These advanced techniques have allowed the best performance for optical transmission systems.

Communication engineers prefer to deploy optical communication systems at wavelength ranges over which the loss is at a minimum, as well as the dispersion factor. Unfortunately, this cannot be simultaneously satisfied. Over the last three decades of advancement of transmission systems, in the 1980s of the last century, we were employing sources at 1310 nm when the dispersion was almost zero and repeating the channels every 40 km in the electronic domain. Until the 1990s, when optical amplification could be employed using Er-doped fiber amplifier at 1550 nm, all optical systems were switched to this infrared region. The fiber dispersion could then be compensated by designing dispersion-compensating fiber (DCF) to insert after the transmission fiber or dispersion-shifted fibers whose zero-dispersion wavelength, λ_{ZD}, could be tailored to be in the lowest attenuation region, as shown in Figure 7.7. However, due to the low values of material dispersion, the waveguide dispersion factor requires fibers of diameter sufficiently small—hence, the nonlinear distortion and high attenuation factor due to bending. Additional optical amplifiers must be employed to compensate for this high DCF loss, and further optical noises are added to the overall transmitted signals, thus limiting the transmission distance. These noises can be eliminated by techniques that can compensate the dispersion in either the optical or electronic domain. In the optical domain, with devices such as fiber Bragg gratings, few mode fibers can be used to delay different frequency components opposite of those that exerted on the spectrum of the signals. We will address the design of such

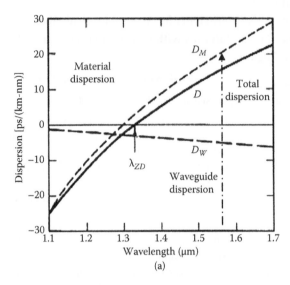

(a)

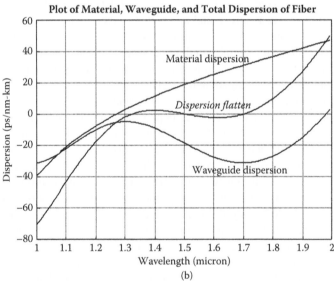

(b)

FIGURE 7.7 Chromatic dispersion factor of (a) SSMF; (b) dispersion-flattened fiber: plotted curves representing the material dispersion factor as a function of optical wavelength for silica-based optical fiber (dispersion flatten curve) with a zero-dispersion wavelength at 1290 nm. This curve is generated as an example. For standard single-mode optical fibers that are currently installed throughout the world, the total dispersion is around +17 ps/(nm.km) at 1550 nm and almost zero at 1310 nm (see also Figure 7.8). Estimate the waveguide dispersion curve for the standard single-mode (SM) optical fiber at around 1300 and 1550 nm windows.

advanced optical fibers in the next few sections, after the description of the waveguide dispersion phenomena and associated mathematical expressions.

A system set up to measure the dispersion property is shown in Figure 7.8. A tunable laser source with a very narrow line width is modulated by an external optical integrated modulator, the RF signals of which come from the phase detector at the output of the fiber line. The wavelength of the laser is scanned across the spectral region of interests. The differential group delay (DGD) is then obtained by finding the ration between the phase and the difference in wavelength. From this DGD we can derive the dispersion factor.

Likewise signals can be coherently detected, amplified, and then compensated in a digital electronic domain. This technique is currently most preferable, as it is programmable provided that high-speed analog-to-digital converters (ADCs) are available. ADCs at a sampling rate of 65 GSa/s are now commercially available, and an ultra-high sampling digital oscilloscope has also been developed, allowing optical communication systems operating in the 100 Gb/s and beyond, even at 1 Tbps capacity using superchannels.

7.3.2.2 Waveguide Dispersion

The effect of waveguide dispersion can be approximated by assuming that the refractive index of the material is independent of wavelength. Let us now consider the group delay, i.e., the time required for a mode to travel along a fiber of length L. This kind of dispersion depends strongly on Δ and V-parameters. To obtain the results of fiber parameters, we define a normalized propagation constant b as

$$b = \frac{\dfrac{\beta^2}{k^2} - n_2^2}{n_1^2 - n_2^2} \tag{7.25}$$

for small Δ. We note that β/k is in fact the *effective refractive index* of the guided optical mode propagating along the optical fiber; that is, the guided waves traveling the axial direction of the fiber see it, or slow down, as a medium with a refractive index of an equivalent effective index.

In case the fiber is a weakly guided waveguide, that is, the refractive index difference between the core and cladding regions is very small, the effective refractive index taking a value of significantly close to that of the core or cladding index, the normalized propagation constant (7.25), can be approximated by

$$b \cong \frac{\dfrac{\beta}{k} - n_2}{n_1 - n_2} \tag{7.26}$$

Solving (7.26) for β, we have

$$\beta = n_2 k (b\Delta + 1) \tag{7.27}$$

The group delay due to waveguide dispersion is then given by (per unit length)

$$t_{wg} = \frac{d\beta}{d\omega} = \frac{1}{c}\frac{d\beta}{dk} \tag{7.28}$$

$$t_{wg} = \frac{1}{c}\left[n_1 + n_2\Delta\frac{d(bk)}{dk}\right] = \frac{1}{c}\left[n_1 + n_2\Delta\frac{d(bk)}{dk}\right] = \frac{1}{c}\left[n_1 + n_2\Delta\frac{d(bV)}{dV}\right] \tag{7.29}$$

Equation (7.29) can be obtained from (7.28) by using the expression of the *V-parameter*. Thus, the pulse temporal spreading $\Delta\tau_{wg}$ due to the waveguide dispersion per unit length by a source having an optical bandwidth (or linewidth σ_λ) is given by

$$\Delta\tau_{wg} = \frac{dt_{gw}}{d\lambda}\sigma_\lambda = -\frac{n_2\Delta}{c\lambda}V\frac{d^2(Vb)}{dV^2}\sigma_\lambda \tag{7.30}$$

and the waveguide dispersion factor or waveguide dispersion parameter (similar to the material dispersion factor) is then defined as

$$D(\lambda) = -\frac{n_2(\lambda)\Delta}{c\lambda}V\frac{d^2(Vb)}{dV^2} \tag{7.31}$$

in units of ps/(nm.km). In the range of $0.9 < \lambda/\lambda_c < 2.6$, the dimensionless factor $V(d^2(Vb)/dV^2)$ can be approximated (to <5% error) by

$$V\frac{d^2(Vb)}{dV^2} \cong 0.080 + 0.549(2.834 - V)^2 \tag{7.32}$$

or alternatively, using the definition of cutoff wavelength and the expression of the *V*-parameters, we obtain

$$V\frac{d^2(Vb)}{dV^2} \cong 0.080 + 3.175\left(1.178 - \frac{\lambda_c}{\lambda}\right)^2 \tag{7.33}$$

Exercise

Prove the equivalence of Equations (7.28) to (7.33). Note that the sign assignment of the material and waveguide dispersion factors must be the same. Otherwise, a negative and positive of these dispersion factors would create confusion. Can you explain what would happen to the pulse if it is transmitted through an optical fiber having a total negative dispersion factor?

From (7.33) and (7.32) we can calculate the waveguide dispersion factor and hence the pulse dispersion factor for a particular source spectral width, σ_λ. It is noted that the dispersion considered in this chapter is for step index fiber only. For grade index fiber, equivalent step index (ESI) parameters must be found and the chromatic dispersion can then be calculated.

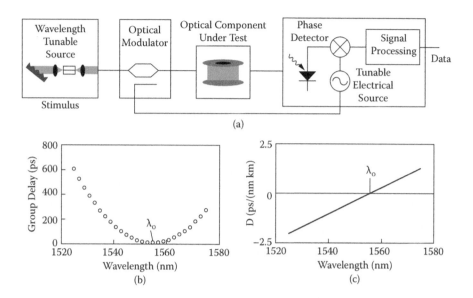

FIGURE 7.8   (a) Chromatic dispersion measurement of two-port optical device; (b) relative group delay versus wavelength; (c) dispersion parameter versus wavelength.

7.3.2.3   Alternative Expression for Waveguide Dispersion Parameter

Alternatively, the waveguide dispersion parameter can be expressed as a function of the propagation constant β by using $\omega = 2\pi c/\lambda$ and (7.33); then the waveguide dispersion factor can be written as

$$D(\lambda) = -\frac{2\pi c}{\lambda^2}\beta_2 = -\frac{2\pi c}{\lambda^2}\frac{d\beta^2}{d\omega^2} \tag{7.34a}$$

Thus, the waveguide dispersion factor is directly related to the second-order derivative of the propagation constant with respect to the optical radial frequency.

   An example for the design of an optical fiber operating in the single-mode region is given in Figure 7.7. The cladding material is pure silica. Shown in this figure are the curves of the material dispersion, waveguide dispersion, and total dispersion for a single-mode optical fiber with a nonuniform refractive index profile in the core.

7.3.2.4   Higher-Order Dispersion

We observe also from Figure 7.6 that the bandwidth length product of the optical fiber can be extended to infinity if the system is operating at the wavelength such that the total dispersion is zero. However, the dispersive effects do not disappear completely at this zero-dispersion wavelength. Optical pulses still experience broadening due to higher-order dispersion effects. It is easily imagined that the total dispersion factor cannot be made zero "flattening" over the optical spectrum. This is higher-order dispersion, which governs by the slope of the total dispersion curve, defined as the dispersion slope

$$S(\lambda) = \frac{d(M+D)}{d\lambda}; \; S(\lambda)$$

can thus be expressed as

$$S(\lambda) = \left(\frac{2\pi c}{\lambda^2}\right)^2 \frac{d^3\beta}{d\lambda^3} + \left(\frac{4\pi c}{\lambda^3}\right)\frac{d^2\beta}{d\lambda^2} \qquad (7.34b)$$

$S(\lambda)$ is also known as the differential-dispersion parameter or dispersion slope and commonly specified by fiber manufacturer, as given in Appendix 7.1 of this chapter for standard single-mode optical fiber (SSMF) and large effective area fiber (LEAF).

7.3.2.5 Polarization Mode Dispersion

Figure 7.9 illustrates the conceptual dynamics of the difference of the two polarized fields of the guided fundamental mode of a single-mode fiber, the linearly polarized mode LP_{01}, due to the birefringence of the fiber created by external effects during the installation of fibers or during the drawing process from a fiber-drawing tower. The propagation or transmission length is very long, possibly longer than a few thousand kilometers; the variation of the fiber core birefringence and concentricity would lead to different propagation velocities and delay time between these polarized

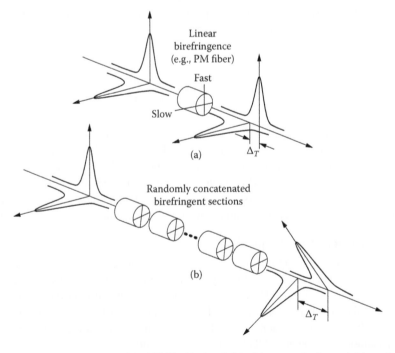

FIGURE 7.9 Conceptual model of PMD: (a) simple birefringence device and (b) randomly concatenated birefringence.

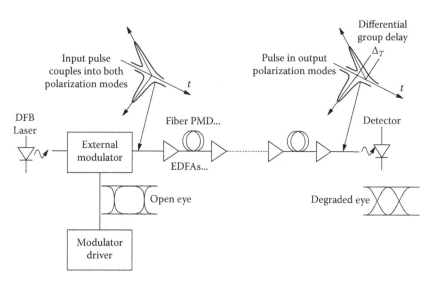

FIGURE 7.10   Effect of PMD in a digital optical communication system, degradation of the received eye diagram.

components—hence, the term *polarization mode dispersion* (PMD). The PMD pulse is broadening due to the delay times of different spectral components of a modulated pulse or pulse sequence. Figure 7.10 illustrates the degradation of a pulse sequence in terms of an observed eye diagram in optical fiber communication systems. Years ago this PMD effect was not critical for bit rates less than 2.5 Gb/s, but since the end of the last century the bit rate has been increased to 10 Gb/s and presently 100 G, and even beyond to 400 GSymbols/s, and so the PMD is even more serious.

The delay between two PSPs is normally negligibly small at 10 Gb/s. However, at high bit rate and in ultra-long-haul transmission, PMD severely degrades the system performance.[2–5] The instantaneous value of DGD ($\Delta\tau$) varies along the fiber and follows a Maxwellian distribution[3,6,7] (see Figure 7.11).

The Maxwellian distribution is governed by the following expression:

$$f(\Delta\tau) = \frac{32(\Delta\tau)^2}{\pi^2 \langle \Delta\tau \rangle^3} \exp\left\{ -\frac{4(\Delta\tau)^2}{\pi \langle \Delta\tau \rangle^2} \right\} \Delta\tau \geq 0 \qquad (7.35)$$

The mean DGD value $\langle \Delta\tau \rangle$ is commonly termed fiber PMD and provided in the fiber specifications. For example, for SSMF the PMD is about $0.1\ \mathrm{ps/ps/\sqrt{km}}$, and for old type about $0.5\ \mathrm{ps/\sqrt{km}}$, which is quite substantial. Optical signals traveling over 100 km, a 5 ps PMD, would be added to the total pulse broadening. The following expression gives an estimate of the maximum transmission limit L_{max} due to the PMD effect:

$$L_{\max} = \frac{0.02}{\langle \Delta\tau \rangle^2 \cdot R_b^2} \qquad (7.36)$$

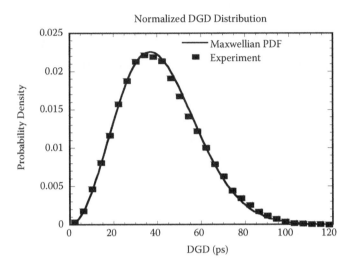

FIGURE 7.11   Maxwellian distribution of PMD random process. (PDF = probability density function.)

where R_b is the bit rate. Based on (7.36), L_{max} for both old vintage fiber and contemporary fibers can be obtained as follows:

- $\langle \Delta\tau \rangle$ = 1 ps/km (old fiber vintages): For bit rate of R = 40 Gb/s, the maximum distance L_{max}= 12.5 km; for R = 10 Gb/s, L_{max}= 200 km.

- $\langle \Delta\tau \rangle$ = 0.1 ps/km (contemporary fiber for modern optical systems): For bit rate R = 40 Gb/s, the maximum transmission distance L_{max}= 1250 km; for R = 10 Gb/s, L_{max}= 20,000 km under no chromatic dispersion effect.

The PMD effects can, however, be compensated for in the electronic domain in a coherent transmission and digital signal processor (DSP)-based optical receiver.[8]

Question

Inspect the technical specifications of Corning SMF-28 and LEAF fibers given in the Appendix 6.1 and extract the values of the PMD. Explain the difference between the values of the fibers. What is the standard value allowable for PMD in modern fibers?

7.3.3   TRANSMISSION BIT RATE AND THE DISPERSION FACTOR

The effect of dispersion on the system bit rate B_r is obvious and can be estimated by using the criterion

$$B_r \cdot \Delta t < 1 \qquad (7.37)$$

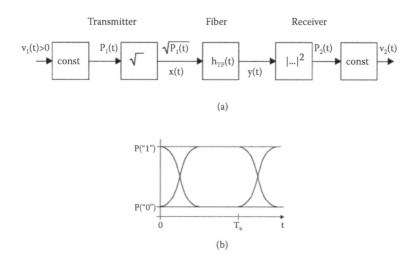

(a)

(b)

FIGURE 7.12 (a) Schematic of an optical transmission system operating under direct detection and equivalent transfer functions of subsystems; (b) typical ideal eye diagrams for non–return-to-zero amplitude shift keying (NRZ-ASK) modulation format.

where $\Delta\tau$ is the total pulse broadening. With the fiber length, the total dispersion $D_T(\lambda) = M(\lambda) + D(\lambda)$, and a source linewidth σ_λ, the criterion becomes

$$B_r \cdot L \cdot |D_T|\sigma_\lambda \leq 1 \qquad (7.38)$$

For a total dispersion factor of 1 ps/(nm · km) and a semiconductor laser of linewidth of 2–4 nm, the bit rate length product cannot exceed 100 Gb/s-km. That is, if a 100 km transmission distance is used, then the bit rate cannot be higher than 1.0 Gb/s.

Figure 7.12 shows a schematic diagram of subsystems in cascade of an optical transmission system, including transmitter, fiber, and receiver, which is a direct detection type. Under direct detection the photodetector can be represented by square-law detection, that is, the square of the absolute value of the optical field amplitude. This is due to the fact that a photodetection diode absorbs optical power in the sensitive spectrum of the material of the detector, and then electrons and holes are generated and collected at the electrodes based on biasing voltage dropped across the electrodes.

7.3.4 EFFECTS OF MODE HOPPING

So far we have assumed that the source center emission wavelength is unaffected by the modulation. In fact, when a short current pulse is applied to a semiconductor laser, its center emission wavelength may hop from one mode to its neighbor at a longer wavelength. In the case where a multilongitudinal mode laser is used, this hopping effect is negligible; however, it is very significant for a single longitudinal mode laser. Currently external cavity lasers can offer a very narrow linewidth of about 100 kHz and high stability without any mode hopping effects. These lasers are employed in digital coherent transmission systems operating at 100 Gbps and beyond bit rates.

7.4 ADVANCED OPTICAL FIBERS: DISPERSION-SHIFTED, -FLATTENED, AND -COMPENSATED OPTICAL FIBERS

In the beginning of the 1980s, there was great interest to reduce the total dispersion $[M(\lambda) + D(\lambda)]$ of single-mode optical fiber at 1550 nm, where the loss is lowest for silica fiber. There were two significant trends; one was to reduce the linewidth and stabilize the laser center wavelength, and the other was to reduce the dispersion at this wavelength. The fibers designed for long-haul transmission systems usually exhibit a near-zero dispersion at a certain spectral window. These are dispersion-shifted fibers; that is, at this wavelength we prefer to have the total dispersion = $[M(\lambda) + D(\lambda)] \sim 0$. The material dispersion factor $M(\lambda)$ is natural and slightly affected by variation of doping material and concentration. However, the waveguide dispersion factor $D(\lambda)$, and hence the total dispersion $D_T(\lambda)$, can be tailored by designing appropriate refractive index profiles and geometrical structure to balance the material dispersion effects. Note that the dispersion factors due to material and waveguide take algebraic values; thus, they can be designed to take opposite values to cancel each other. See Appendices 1 and 2 (Sections 7.6 and 7.7) for MATLAB® files as examples for the design of these fibers.

An advanced optical fiber design technique can offer the design of dispersion-flattened fibers where the dispersion factor is flat over the wavelength range from 1300 to 1600 nm by tailoring the refractive index profile of the core of optical fibers in such distribution as the W-profile, the segmented profile, and multilayer core structure.

Question

What is the principal phenomenon for an optical fiber so that the dispersion characteristic is flattened over the wavelength range 1300 to 1550 nm?

Another type of optical fiber that would be required for compensating the dispersion effect of an optical signal after transmission over a length of optical fiber is the dispersion-compensated fiber, whose dispersion factor is many times larger than that of the standard communication fiber with an opposite sign. This can be designed by setting the total dispersion to the required compensated dispersion, and the waveguide dispersion can be found over the required operating range. Then optical fiber structures can be tailored. A sample design of an advanced optical fiber, the non-zero-dispersion-shifted type, is given in Appendix 7.2.

7.5 PROPAGATION OF OPTICAL SIGNALS IN OPTICAL FIBER TRANSMISSION LINE: SPLIT-STEP FOURIER METHOD

The simulation of optical communication becomes more important nowadays due to the costs of setting up a test platform for operation at every speed, on the order of more than 40 and 100 Gbps and beyond. Under simulation platforms the propagation of an optical data sequence over a long line of optically amplified fiber is essential. Thus numerical techniques must be resorted to to achieve accurate and reasonably fast results. It is noted that the length of the data sequence would be on the order of several billion bits if a bit error rate (BER) of 10^{-9} is desired. This section gives an introduction to the numerical technique, the split-step Fourier method, to solve the nonlinear Schrodinger equation so that it can be integrated into the simulation platform for optical transmission systems.

7.5.1 SYMMETRICAL SPLIT-STEP FOURIER METHOD (SSFM)

The evolution of slow-varying complex envelopes $A(z, t)$ of optical pulses along a single-mode optical fiber is governed by nonlinear Schrodinger equation (NLSE):

$$\frac{\partial A(z,t)}{\partial z} + \frac{\alpha}{2} A(z,t) + \beta_1 \frac{\partial A(z,t)}{\partial t} + \frac{j}{2} \beta_2 \frac{\partial^2 A(z,t)}{\partial t^2} - \frac{1}{6} \beta_3 \frac{\partial^3 A(z,t)}{\partial t^3} \tag{7.39}$$

$$= -j\gamma |A(z,t)|^2 A(z,t)$$

where z is the spatial longitudinal coordinate, α accounts for fiber attenuation, β_1 indicates DGD, β_2 and β_3 represent second- and third-order dispersion factors of fiber chromatic dispersion (CD), and γ is the nonlinear coefficient. In a single-channel transmission, Equation (7.39) includes the following effects: fiber attenuation, fiber CD and PMD, dispersion slope, and self-phase modulation (SPM) nonlinearity. Fluctuation of optical intensity caused by the Gordon–Mollenauer effect is also included in this equation.

The solution of NLSE and hence the modeling of pulse propagation along a single-mode optical fiber is solved numerically by using the split-step Fourier method (SSFM). In SSFM, fiber length is divided into a large number of small segments δz. In practice, fiber dispersion and nonlinearity are mutually interactive at any distance along the fiber. However, these mutual effects are small within δz, and thus, effects of fiber dispersion and fiber nonlinearity over δz are assumed to be statistically independent of each other. As a result, SSFM can separately define two operators: (1) the linear operator that involves fiber attenuation and fiber dispersion effects and (2) the nonlinearity operator that takes into account fiber nonlinearities. These linear and nonlinear operators are formulated as follows:

$$\hat{D} = -\frac{j\beta_2}{2} \frac{\partial^2}{\partial T^2} + \frac{\beta_3}{6} \frac{\partial^3}{\partial T^3} - \frac{\alpha}{2} \tag{7.40}$$

$$\hat{N} = j\gamma |A|^2$$

where $j = \sqrt{-1}$, A replaces $A(z, t)$ for simpler notation, and $T = t - z/v_g$ is the reference time frame moving at the group velocity. Equation (7.40) can be rewritten in a shorter form, given by

$$\frac{\partial A}{\partial z} = (\hat{D} + \hat{N})A \tag{7.41}$$

and the complex amplitudes of optical pulses propagating from z to $z + \delta z$ are calculated using the following approximation:

$$A(z+h,T) \approx \exp(h\hat{D})\exp(h\hat{N})A(z,T) \qquad (7.42)$$

Equation (7.42) is accurate to the second order of the step size δz. The accuracy of SSFM can be improved by including the effect of fiber nonlinearity in the middle of the segment rather than at the segment boundary (see Figure 7.13). This modified SSFM is known as the symmetric SSFM.

Equation (7.73) can now be modified as

$$A(z+\delta z,T) \approx \exp\left(\frac{\delta z}{2}\hat{D}\right)\exp\left(\int_{z}^{z+\delta z}\hat{N}(z')dz'\right)\exp\left(\frac{\delta z}{2}\hat{D}\right)A(z,T) \qquad (7.43)$$

This method is accurate to the third order of the step size δz. In symmetric SSFM, the optical pulse propagates along a fiber segment δz in two stages. First, the optical pulse propagates through the linear operator that has a step of $\delta z/2$, in which the fiber attenuation and dispersion effects are taken into account. Then, the fiber nonlinearity is calculated in the middle of the segment. After that, the pulse propagates through the second half of the linear operator. The process continues repetitively in consecutive segments of size δz until the end of the fiber. It should be highlighted that the linear operator is computed in the frequency domain, while the nonlinear operator is calculated in the time domain.

7.5.2 MATLAB® Program and MATLAB Simulink® Models of the SSFM

A MATLAB program is given below. This program performs the propagation of the optical signals along the optical fiber transmission distance, as shown in Figure 7.14. This program must be included in the folder storing the MATLAB Simulink model. In this folder an initialization program (see Appendix 8.2) must also be included to set the data and parameters required for the Simulink model and subroutines. (See Figure 7.15.) Furthermore, the SSMF including Raman gain amplification effects is given in Appendix 8.1, a modification of the MATLAB file given below to include the Raman scattering effect.

7.5.2.1 SSFM MATLAB Program

```
function output = ssprop_matlabfunction_modified(input)
nt = input(1);
u0 = input(2:nt+1);
dt = input(nt+2);
dz = input(nt+3);
nz = input(nt+4);
alpha_indB = input(nt+5);
betap = input(nt+6:nt+9);
```

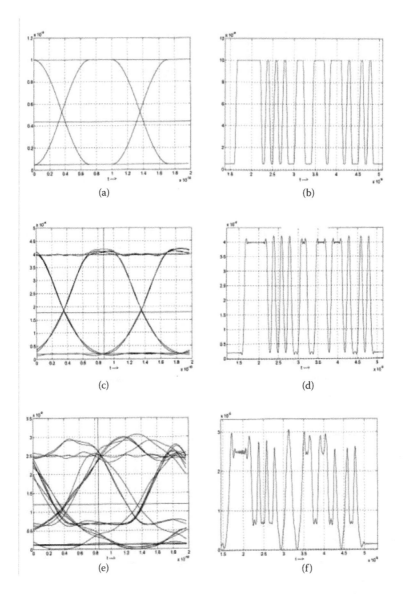

FIGURE 7.13 Eye diagram and time sequence of random signals at 10 Gb/s transmission over standard SMF after (a and b) 0 km (i.e., at the transmitter), (c and d) 20 km, and (e and f) 80 km distance.

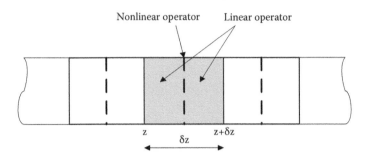

FIGURE 7.14   Schematic illustration of symmetric SSFM.

```
gamma = input(nt+10);
P_non_thres = input(nt+11)
maxiter = input(nt+12);
tol = input(nt+13);
  tic;
%tmp = cputime;
% This section solves the NLSE for pulse propagation in an
optical fiber using the
% SSFM
% The following effects are included: group velocity
dispersion
% (GVD),higher order dispersion, loss, and self-phase
modulation (gamma).
% USAGE
% u1 = ssprop(u0,dt,dz,nz,alpha,betap,gamma);
% u1 = ssprop(u0,dt,dz,nz,alpha,betap,gamma,maxiter);
% u1 = ssprop(u0,dt,dz,nz,alpha,betap,gamma,maxiter,tol);
% INPUT
% u0 - starting field amplitude (vector)
% dt - time step - [in ps]
% dz - propagation stepsize - [in km]
% nz - number of steps to take, ie, ztotal = dz*nz
% alpha - power loss coefficient [in dB/km], need to convert
to linear to
% have P = P0*exp(-alpha*z)
% betap - dispersion polynomial coefs, [beta_0... beta_m] [in
ps^(m-1)/km]
% gamma - nonlinearity coefficient [in (km^-1.W^-1)]
% maxiter - max number of iterations (default = 4)
% tol - convergence tolerance (default = 1e-5)
%% OUTPUT
%% u1 - field at the output
% Convert alpha_indB to alpha in linear scale
%- - - - - - - -

alpha = log(10)*alpha_indB/10;    % alpha (1/km)

%- - - - - - - -
```

```
ntt = length(u0);
w = 2*pi*[(0:ntt/2-1),(-ntt/2:-1)]'/(dt*nt);
%w = 2*pi*[(ntt/2:ntt-1),(1:ntt/2)]'/(dt*ntt);
clear halfstep
  halfstep = -alpha/2;
for ii = 0:length(betap)-1;
  halfstep = halfstep - j*betap(ii+1)*(w.^ii)/factorial(ii);
end

clear LinearOperator
% Linear Operator in Split Step method
LinearOperator = halfstep;
% pause
halfstep = exp(halfstep*dz/2);
%
u1 = u0;
ufft = fft(u0);
%
% Nonlinear operator will be added if the peak power is
greater than the
% Nonlinear threshold
iz = 0;
while (iz < nz) & (max((abs(u1).^2 + abs(u0).^2)) > P_non_
thres)
  iz = iz+1;
  uhalf = ifft(halfstep.*ufft);
  for ii = 1:maxiter,
    uv = uhalf.* exp(-j*gamma*(abs(u1).^2 +
    abs(u0).^2)*dz/2);
    ufft = halfstep.*fft(uv);
    uv = ifft(ufft);

    %fprintf('You are using SSFM\n');

    if (max(uv-u1)/max(u1) < tol)
      u1 = uv;
      break;
    else
      u1 = uv;
    end
  end
  if (ii = = maxiter)
    warning(sprintf('Failed to converge to%f in%d
    iterations',...
      tol,maxiter));
  end

  u0 = u1;

end
```

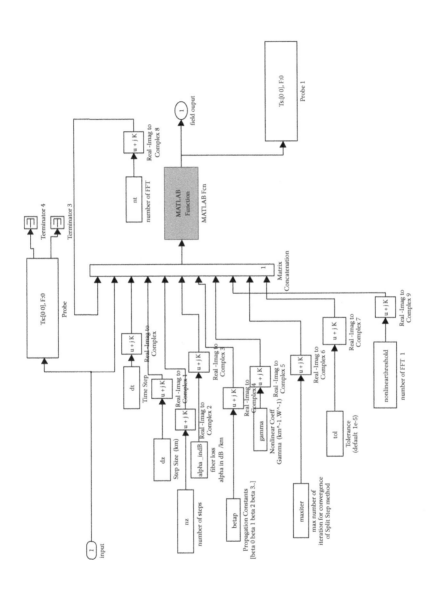

FIGURE 7.15 MATLAB® Simulink® model of the SSMF. MATLAB function includes the MATLAB program. Other inputs are data to pass to the MATLAB program SSMF, usually defined in the initialization file (given in Appendix 8.2).

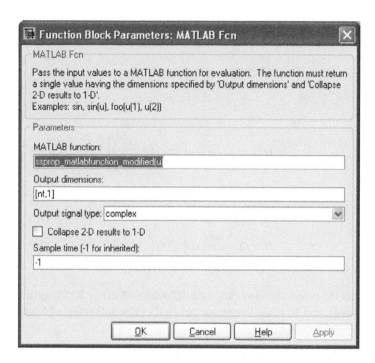

FIGURE 7.16 Screenshot of MATLAB Simulink model under the mask of the MATLAB function of the SSMF algorithm.

```
if (iz < nz) & (max((abs(u1).^2 + abs(u0).^2)) < P_non_thres)

% u1 = u1.*rectwin(ntt);
   ufft = = fft(u1);
   ufft = ufft.*exp(LinearOperator*(nz-iz)*dz);
   u1 = ifft(ufft);
%fprintf('Implementing Linear Transfer Function of the Fiber
Propagation');
end
   toc;
output = u1;
```

7.5.2.2 MATLAB Simulink Model

The MATLAB program (subroutine) is incorporated in the MALAB Simulink model for signal propagation. Under the mask of the block MATLAB function [ssprop _ matlabfunction _ modified(input)] is the inclusion of the MATLAB program subroutine given above and illustrated in Figure 7.16.

7.5.2.3 Modeling of Polarization Mode Dispersion (PMD)

First-order PMD can be implemented by modeling the optical fiber as two separate paths representing the propagation of two PSPs. The symmetrical SSFM can be implemented on each polarized transmission path, and then their outputs are superimposed to form the output optical field of the propagated signals. The transfer function to represent the first-order PMD is given by

$$H(f) = H^+(f) + H^-(f) \qquad (7.44)$$

where

$$H^+(f) = \sqrt{\kappa} \exp\left[j2\pi f\left(-\frac{\Delta\tau}{2} \right) \right] \qquad (7.45)$$

and

$$H^-(f) = \sqrt{\kappa} \exp\left[j2\pi f\left(-\frac{\Delta\tau}{2} \right) \right] \qquad (7.46)$$

in which κ is the power splitting ratio ($\kappa = 1/2$ when a 3 dB or 50:50 optical coupler/splitter is used), and $\Delta\tau$ is the instantaneous DGD value following a Maxwell distribution (refer to Equation (7.35)).[9,10]

7.5.2.4   Optimization of Symmetrical SSFM

7.5.2.4.1   Optimization of Computational Time

A huge amount of time is spent in symmetric SSFM for FFT and IFFT operations, in particular when fiber nonlinear effects are involved. In practice, when optical pulses propagate toward the end of a fiber span, the pulse intensity is greatly attenuated due to the fiber attenuation. As a result, fiber nonlinear effects are negligible for the rest of that fiber span, and hence the transmission is operating in a linear domain in this range. In this research, a technique to configure symmetric SSFM is proposed in order to reduce the computational time. If the peak power of an optical pulse is lower than the nonlinear threshold of the transmission fiber, for example, around −4 dBm, symmetrical SSFM is switched to a linear mode operation. This linear mode involves only fiber dispersions and fiber attenuation, and its low-pass equivalent transfer function for the optical fiber is

$$H(\omega) = \exp\left\{ -j\left[\frac{1}{2}\beta_2\omega^2 + \frac{1}{6}\beta_3\omega^3 \right] \right\} \qquad (7.47)$$

If β_3 is not considered in this fiber transfer function, which is normally the case due to its negligible effects on 40 Gb/s and lower bit rate transmission systems, the above transfer function has a parabolic phase profile.[9,10]

7.5.2.4.2   Mitigation of Windowing Effect and Waveform Discontinuity

In symmetric SSFM, mathematical operations of FFT and IFFT play very significant roles. However, due to a finite window length required for FFT and IFFT operations,

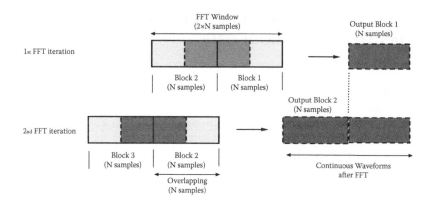

FIGURE 7.17 Proposed technique for mitigating windowing effect and waveform discontinuity caused by FFT/IFFT operations.

these operations normally introduce overshooting at two boundary regions of the FFT window, commonly known as the windowing effect of FFT. In addition, since the FFT operation is a block-based process, there exists the issue of waveform discontinuity; i.e., the right-most sample of the current output block does not start at the same position as the left-most sample of the previous output block. The windowing effect and the waveform discontinuity problems are resolved with the following technique. Referring to illustrations given in Figure 7.17, the actual window length for FFT/IFFT operations consists of two blocks of samples—hence, the 2N sample length. The output, however, is a truncated version with the length of one block (N samples) and output samples taken in the middle of the two input blocks. The next FFT window overlaps the previous one by one block of N samples.

A sample MATLAB program is given in Appendix 8.1, describing the SSMF process, which can be integrated into a MATLAB Simulink model whose initialization MATLAB file is also shown in Appendix 8.2. A number of Simulink models integrating SSMF can also be found in Binh.[11,12]

7.5.3 REMARKS

The attenuation and dispersion of optical signals transmitted through silica optical fibers are described. Attenuation can be reduced by using the optical wavelength in the longer wavelength range. For example, for silica fiber the preferred wavelength is 1.55 μm. However, natural forces are not kind to us, and a dispersion factor of about 18 ps/(nm.km) generates pulse broadening for signal transmitted at this wavelength in a circular fiber.

Longer-wavelength carriers can be used in the mid-infrared range of about 2.5 to 5 μm. At this wavelength range different kinds of glasses must be used, such as chalcogenite type or fluoride type. Another technique presently used to compensate for the dispersion effect is to reduce the linewidth of the lasers, or to use equalizing techniques, such as spectrum inversion at the transmitter ends or at the center of the transmission length by optical filtering at the receiving end. Alternatively, the optical fiber can be tailored to achieve dispersion-shifted or -flattened characteristics.

APPENDIX 7.1: PROGRAM LISTINGS FOR DESIGN
OF STANDARD SINGLE-MODE FIBER

```
% totdisp_SMF.m
%
% MatLab script for calculating total dispersion for Non-Zero
Dispersion Shifted
% Fiber (NSDSF). The script plots material dispersion, the
waveguide dispersion, and
% total dispersion for the designed fiber.
% Optical Fiber Design
lambda = [1.1:0.01:1.700]*1e-6;% setting spectral region
G1 = 0.7028554;            %Sellmeier's coefficients for
germanium: doped silica
G2 = 0.4146307;            % (concentration B in table)
G3 = 0.8974540;
lambda1 = 0.0727723e-6;    %Wavelengths for germanium doped
silica
lambda2 = 0.1143085e-6;
lambda3 = 9.896161e-6;


c = 299792458;      %Speed of light
pi = 3.1415926;     %Greek letter pi
a = 4.1e-6;         %Core radius
delta = 0.003;      %Greek letter delta (ref. index difference
                    %between core and cladding)

% Calculating the refractive index
%- - - - - - - - - - - - - - - - - - -

npow2oflambda = 1 + (G1.*lambda.^2./(lambda.^2.-
lambda1*lambda1))...
   + (G2.*lambda.^2./(lambda.^2.-lambda2*lambda2))...
   + (G3.*lambda.^2./(lambda.^2.-lambda3*lambda3));

noflambda = sqrt(npow2oflambda);

pointer = find(lambda = =1.550e-6);
n1 = noflambda(pointer)    %Refractive index in the core

% Calculating the material dispersion
%- - - - - - - - - - - - - - - - - - -

t1 = diff(noflambda);
t2 = diff(lambda);
t3 = t1./t2;

t4 = diff(t3);
t5 = diff(lambda);
t5 = adjmat(t5);
```

```
lambda = adjmat(lambda);
lambda = adjmat(lambda);

%Material dispersion
Matdisp = - (lambda./c).* (t4./t5);

% Converting to ps/nm.km
Matdisp = Matdisp.*1e6;
figure 1)
clf
hold

xlabel('nm')
ylabel('ps/nm.km')
title('Standard Single Mode Fiber')
plot(lambda, Matdisp, '.-')
grid on

% Calculating waveguide dispersion
%- - - - - - - - - - - - - - - - -

V = (2 * pi * a * n1 * sqrt(2 * delta))./(lambda);

Dlambda1 = - (n1 * delta)./(c * lambda);
%plot(lambda,Dlambda2)

Dlambda2 = 0.080 + 0.549 * (2.834 - V).^2;
%plot(lambda,Dlambda2)

Dlambda = Dlambda1.* Dlambda2;

% Converting to ps/nm.km
Dlambda = Dlambda.*1e6;
plot(lambda,Dlambda, '-')

% Calculating total dispersion
%- - - - - - - - - - - - - - - - -

TotDisp = Matdisp + Dlambda;
plot(lambda, TotDisp, ':')

legend('Material Dispersion', 'Waveguide Dispersion', 'Total
Dispersion', 0)

% Finding the dispersion at 1460, 1550 and 1625 nm
%- - - - - - - - - - - - - - - - - - - - - - - - -

pointer = find(lambda = =1.460e-6);
Disp1460 = TotDisp(pointer)
pointer = find(lambda = =1.550e-6);
Disp1550 = TotDisp(pointer)
```

```
pointer = find(lambda = =1.6250e-6);
Disp1625 = TotDisp(pointer)
% refind_SMF.m
%
% MatLab script for calculating of possible values for
% refractive index in the core, n1, and cladding, n2,
% its corresponding relative refractive index.
%
% PROJECT DESIGN: Optical Fiber Design
%


lambda = [1.1:0.01:1.700]*1e-6;

n1 = [1.0487:0.001:1.8587];  %Refractive index of the core
n2 = [1.0435:0.001:1.8535];  %Refractive index of the cladding

% Calculating the refractive index for the index profile
%- - - - - - - - - - - - - - - - - - - - - - - - - - - - -

delta = (n1 - n2)./n1;
deltap = delta * 100;

plot(n1, deltap)
grid on

xlabel('n1')
ylabel('Delta - Relative Refractive Index(%)')
title('Refractive Index')
```

APPENDIX 7.2: PROGRAM LISTINGS OF THE DESIGN OF NON-ZERO-DISPERSION-SHIFTED FIBER

```
% totdisp_NZDSF.m
%
% MatLab script for calculating total dispersion for
% Non-Zero Dispersion Shifted Fiber. The script plots
% the material dispersion, the waveguide dispersion, and
% the total dispersion of the designed fiber.
%
% PROJECT 1: Optical Fiber Design
%


lambda = [1.1:0.001:1.700]*1e-6;

G1 = 0.7028554;            %Sellmeier's coefficients for
                          germanium
G2 = 0.4146307;            %doped silica (concentration B in
                          table)
G3 = 0.8974540;
```

```
lambda1 = 0.0727723e-6;  %Wavelengths for germanium doped
                          silica
lambda2 = 0.1143085e-6;
lambda3 = 9.896161e-6;

c = 299792458;       %Speed of light
pi = 3.1415926;      %Greek letter pi
a = 2.4e-6;          %Core radius
delta = 0.0043;      %Greek letter delta (ref. index difference
                     %between core and cladding)

% Calculating the refractive index
%- - - - - - - - - - - - - - - - - -

npow2oflambda = 1 + (G1.*lambda.^2./(lambda.^2.-
lambda1*lambda1))...
   + (G2.*lambda.^2./(lambda.^2.-lambda2*lambda2))...
   + (G3.*lambda.^2./(lambda.^2.-lambda3*lambda3));

noflambda = sqrt(npow2oflambda);

pointer = find(lambda = =1.550e-6);
n1 = noflambda(pointer) %Refractive index in the core @1550nm

% Calculating the material dispersion
%- - - - - - - - - - - - - - - - - -

t1 = diff(noflambda);
t2 = diff(lambda);
t3 = t1./t2;

t4 = diff(t3);
t5 = diff(lambda);
t5 = adjmat(t5);

lambda = adjmat(lambda);
lambda = adjmat(lambda);

%Material dispersion
Matdisp = - (lambda./c).* (t4./t5);

% Converting to ps/nm.km
Matdisp = Matdisp.*1e6;
figure 1)
clf
hold

xlabel('nm')
ylabel('ps/nm.km')
title('Non-Zero Dispersion Shifted Fiber')
```

```
plot(lambda, Matdisp, '.')
grid on

% Calculating waveguide dispersion
%— — — — — — — — — — — — — — — -

V = (2*pi*a*n1*sqrt(2*delta))./(lambda);

pointer = find(lambda = =1.550e-6);
V1550 = V(pointer)

Dlambda1 = - (n1 * delta)./(c * lambda);
Dlambda2 = 0.080 + 0.549 * (2.834 - V).^2;
Dlambda = Dlambda1.* Dlambda2;

% Converting to ps/nm.km
Dlambda = Dlambda.*1e6;
plot(lambda,Dlambda, '-')

% Calculating total dispersion
%— — — — — — — — — — — — — — — -

TotDisp = Matdisp + Dlambda;
plot(lambda, TotDisp, '+')

legend('Material Dispersion', 'Waveguide Dispersion', 'Total
Dispersion', 0)

% Finding the dispersion at 1460, 1550 and 1625 nm
%— — — — — — — — — — — — — — — — — — — — — — — —

pointer = find(lambda = =1.460e-6);
Disp1460 = TotDisp(pointer)
pointer = find(lambda = =1.550e-6);
Disp1550 = TotDisp(pointer)
pointer = find(lambda = =1.6250e-6);
Disp1625 = TotDisp(pointer)
% refine_NZDSF.m
%
% MatLab script for calculating of possible values for
% refractive index in the core, n1, and cladding, n2,
% its corresponding relative refractive index.
%

lambda = [1.1:0.01:1.700]*1e-6;
n1 = [1.0487:0.001:1.8587]; %Refractive index of the core
n2 = [1.0324:0.001:1.8424]; %Refractive index of the cladding

% Calculating the refractive index for the index profile
%— — — — — — — — — — — — — — — — — — — — — — — — — — -
```

```
delta = (n1 - n2)./n1;
deltap = delta * 100;

plot(n1, deltap)
grid on
xlabel('n1')
ylabel('Delta - Relative Refractive Index(%)')
title('Refractive Index')
```

7.6 PROBLEMS

7.6.1 PROBLEM 1

What is the wavelength range of infrared light, ultraviolet light, and far-infrared light? What are the approximate wavelengths of the colors in the color band of resistors? Are they corresponding to the color of the rainbow?

7.6.2 PROBLEM 2

A GeO_2-doped silica-based optical fiber has the following parameters:

- Step index profile
- Refractive index difference at the core of 0.5%
- Core diameter of 9.0 μm

a. Calculate the refractive index of the fiber core and cladding at 1.310 and 1.55 μm wavelengths.
b. What is the estimate loss of this fiber at the above wavelengths?
c. What are the V-parameters of the fiber at these wavelengths?
d. What are the material dispersion and waveguide dispersion factors at these wavelengths? Hence, the total dispersion factors?
e. This fiber is to be used in optical systems of bit rates of 2.2 Gb/s. What is the maximum fiber length that the signal can be transmitted without suffering the allowable signal degradation?

7.6.3 PROBLEM 3

a. Give a brief account of the pros and cons for optical fiber communication systems operating at 810, 1300, and 1550 nm wavelength regions.
b. Why does silica optical fiber become very lossy at the 1400 nm wavelength region?
c. What are the typical optical fiber losses at the above wavelength regions and, hence, the typical cable losses?

7.6.4 PROBLEM 4

a. Show that the material dispersion factor is zero at the wavelength given by

$$\lambda^4 = -\frac{3c_3}{c_2}$$

b. The coefficients c_1, c_2, and c_3 for pure and GeO_2-doped silica fiber are

Coefficient c	Pure Silica	7.9% GeO_2-Doped Silica
c_1	1.45084	1.46286
c_2 in μm^{-2}	–0.00334	–0.00331
c_3 in μm^2	0.00292	0.00320

Find the zero-dispersion wavelengths due to the material of these fibers.

c. Derive an expression for the group delay per km unit length. Plot this group delay versus wavelength for part (b).

d. Find the transit time difference of lightwaves propagating through the fiber emitted by light sources centered at 810 and 1550 nm with a linewidth of 10 nm.

7.6.5 PROBLEM 5

Using the approximate expression for the normalized propagation constant b as a function of V, derive the group velocity delay due to the waveguide and the dispersion factor due to the waveguide as a function of V.

7.6.6 PROBLEM 6

Using the data of the SMF optical fiber manufactured by Corning, calculate

a. The material dispersion factor
b. The waveguide dispersion factor

at 1330 and 1550 nm.

7.6.7 PROBLEM 7

A silica optical fiber has a cladding refractive index of 1.4680 at 1550 nm and a relative refractive index of 0.5%. The fiber core diameter is 8.2 μm with an accuracy of 0.5 μm.

a. Estimate its V-parameter at the above operating wavelength. Hence confirm whether the fiber is single mode or weakly multimode.

b. If it is single mode at the operating wavelength, estimate its mode spot size and mode field diameter and the cutoff wavelength range of this fiber.

c. Using the Sellmeier's coefficients for pure silica, estimate the material dispersion of this fiber. Calculate the waveguide dispersion factor and thence the total dispersion factor of the fiber.

d. This fiber is used for signal transmission over several optically amplified spans. Sketch the transmission system, including all transmission subsystems and the transmission fibers and dispersion-compensating fibers.

e. Estimate the dispersion factor of the dispersion-compensating fibers so that only 15 km of DCF is required for 100 km of transmission fibers.

7.6.8 PROBLEM 8

Using the dispersion factor of the DCF obtained in problem 7, and the material dispersion in that question:

a. Find the waveguide dispersion required for achieving the total dispersion of this DCF fiber.

b. For it to operate as a single-mode type, use the approximation expression of the waveguide parameter to find the required core radius and refractive index difference.

7.6.9 PROBLEM 9

Give reasons why the dispersion of the non-zero-dispersion-shifted fibers must not be zero in the transmission band 1530 to 1620 nm. State the dispersion factor of the Corning LEAF fiber.

For the Corning LEAF fibers, what is the effective area of the fiber? Using the nonlinear threshold of the SMF-28 of 3 mW, estimate the nonlinear threshold of the LEAF fibers.

7.6.10 PROBLEM 10

Write a brief paragraph about the phenomenon of:

• The polarization mode dispersion effect in single-mode optical fibers
• Nonlinear dispersion effects due to the self-phase modulation effect

Estimate the total PMD of the SMF-28 and LEAF fibers over a transmission span of 100 km—thence the total PMD over 10 spans.

Search the Internet for an method of PMD compensation over long-haul transmission.

7.6.11 PROBLEM 11

For the SMF-28 estimate the total pulse broadening, including the chromatic dispersion (CD) and PMD, for an NRZ 10 Gb/s data channel transmitted over a single span of 100 km.

Reestimate this broadening for a 40 Gb/s transmission bit rate.

7.6.12 PROBLEM 12

I. Describe briefly the attenuation or loss of silica fibers and the total chromatic dispersion effect of the guided mode as a function of operating wavelength of lightwaves guided in a weakly guided single-mode optical fiber for long-haul optical communication systems.

II. Give a brief account of the optical transmission loss of silica fibers for optical communication systems as a function of the operating wavelength in the C-band, L-band, and S-band wavelength regions.

III. A circular optical fiber is manufactured with the cladding region of pure silica with a refractive index of 1.486 at 1550 nm wavelength, and the core refractive index difference is 0.003. Design an optical fiber so that it is single moded at the operating wavelength of 1550 nm with a total dispersion factor $D_T(\lambda)$ of not higher than 0.2 ps/(nm-km). The following parameters should be determined for the fiber: the fiber core diameter, the fiber cutoff wavelength, and its total dispersion factor.

The following materials can be used for the design of the fiber:

The material and waveguide dispersion factors $D_M(\lambda)$ and $D_W(\lambda)$ for a single-mode step index circular optical waveguide are given by

$$D_M = 122\left(1 - \frac{\lambda_{ZD}}{\lambda}\right) \text{ ps/(nm.km)}$$

where $\lambda_{ZD} = 1290$ nm for pure silica with a low doping concentration of impurities.

$$D_W = -\frac{n_2\Delta}{c\lambda}\left[V\frac{d^2(Vb)}{dV^2}\right] \text{ s/m}^2$$

where the normalized propagation constant b of a step index circular optical fiber can be approximately given by

$$b(V) = \left(1.1428 - \frac{0.9960}{V}\right)^2 \quad \text{for } 1.4 < V < 2.2$$

V is the normalized frequency parameter and $c = 3 \times 10^8$ m/s is the velocity of light in free space.

7.6.13 PROBLEM 13: FIBER DESIGN MINI-PROJECT

Design dispersion-shifted single-mode optical fibers at 1550 nm. Groups of students of three or four can be formed and requested to select particular combinations of doped material with different Sellmeier's coefficients as core and cladding materials for the design.

7.6.13.1 Design Project Objectives

To design the geometrical and index profile of silica optical fibers to meet certain dispersion properties as required. Further, a set of fiber performance with dispersion as the main factor must be investigated as a function of the fiber core radius and the relative refractive index difference.

7.6.13.1.1 System Applications

The designed fiber must be incorporated with a dispersion-compensated fiber—in this case the standard single-mode optical fiber (SMF) whose specifications are given in lectures. If the total length of the transmission is 10,000 km, specify the length of your designed fiber and the SMF so that the average fiber dispersion is 0.01 ps/nm.km. It can be assumed that the spacing of optical amplifiers is 100 km.

a. Design specifications: A number of types of single-mode optical fibers are required to be designed for optical communication systems in long- and short-haul transmission. Tables 7.2 through 7.4 show the characteristics of the required fibers:

TABLE 7.2
Materials for Core or Cladding Regions

Fiber No.	Optical Fibers Profile	Material Type Core	Systems Requirement	Other Requirements
		See tables below for Sellmeier's constants		Maximum dispersion in the wavelength range 1510 to 1590 nm in ps/(nm.km)
1	Triangular	A–J	Dispersion-shifted wavelength at 1550 nm	<1.5
2	Parabolic	A–J	Dispersion-shifted wavelength at 1520 nm	<1.5
3	Triangular	A–J	Dispersion-shifted wavelength at 1560 nm	<2
4	Parabolic	A–J	Dispersion-shifted wavelength at 1530	<3
5	Triangular	A–J	Dispersion-compensated wavelength at 1550 nm	>0.5
6	Parabolic	A–J	Dispersion-compensated wavelength at 1540 nm	>1.0
9	Triple-clad	A–J	Dispersion flattened over 1500 to 1590 nm	Not higher than 0.2 ps/(nm.km)
10	Triangular in core and segmented in cladding	A–J	Dispersion flattened over 1500 to 1590 nm	Not higher than 0.2 ps/(nm.km)

TABLE 7.3

Sellmeier's Coefficients for Several Optical Fiber Silica-Based Materials with Germanium Doped in the Core Region

Sellmeier's Constants	Germanium Concentration, C (mol%)			
Types	A	B	C	D
	0	3.1%	5.8%	7.9%
G_1	0.6961663	0.7028554	0.7088876	0.7136824
G_2	0.4079426	0.4146307	0.4206803	0.4254807
G_3	0.8974794	0.8974540	0.8956551	0.8964226
λ_1	0.6840432	0.0727723	0.0609053	0.0617167
λ_2	0.1162414	0.1143085	0.1254514	0.1270814
λ_3	9.896161	9.896161	9.896162	9.896161

Note: Wavelength in μm and G_k in μm^{-2}.

Source: After S. Kobayashi, Shibata, S., and Izawa, T., *Proc. International Conference on Integrated Optics and Optical Fiber Communication*, Tokyo, Japan, 1977, p. 309.

Practical limits for the fiber core radius and relative index difference must be taken into account. A set of curves must be obtained with the core radius or the relative index difference as a parameter. Make sure that the material dispersion factors are correctly modeled.

b. Software environment: The preferred package is MATLAB or MATLAB 4.2 or 5.1 for Windows. A sample of the design is given below.

7.6.13.2 Assessment of Design Assignments

A major design assignment on optical fibers for communication systems is counted for 10% of the total marks allocated for the optical systems part. Twenty percent is awarded to design groups selecting index profile types 9 or 10.

The design assignment is specified for different groups of students. The maximum number of members of each group is two. A higher number of group members can only be accepted in exceptional circumstances, and in this case the complexity of the design assignment is increased accordingly.

When the assignment is submitted, candidates may be requested for an individual oral presentation. The written design submitted by each group is awarded equally for group members and counted for only 40% of the total mark of the design assignment; the other 60% is awarded for the oral presentation.

a. Sample design program: This is a sample program for the design written in MATLAB to give a guideline for the estimation of material and waveguide dispersion as well as total dispersion and pulse broadening. Conditions for

TABLE 7.4

Sellmeier's Coefficients for Several Optical Fiber Silica-Based Materials with Germanium Doped in the Core Region

Sellmeier's Constants	Concentration Composition			
Types	E	F	G	H
	Quenched SiO_2	13.5 GeO_2:86.5 SiO_2	9.1 P_2O_5:90.0 SiO_2	13.3 B_2O_3:86.7 SiO_2
G_1	0.696750	0.711040	0.695790	0.690618
G_2	0.408218	0.408218	0.452497	0.401996
G_3	0.890815	0.704048	0.712513	0.898817
λ_1	0.069066	0.064270	0.061568	0.061900
λ_2	0.115662	0.129408	0.119921	0.123662
λ_3	9.900559	9.425478	8.656641	9.098960

Sellmeier's Constants	Concentration Composition			
Types	I	J	K	L
	1.0 F:99.0 SiO_2	16.2 Na_2O:32.5 $B2O_3$:50.6 SiO_2		
G_1	0.691116	0.796468		
G_2	0.399166	0.497614		
G_3	0.890423	0.358924		
λ_1	0.068227	0.094359		
λ_2	0.116460	0.093386		
λ_3	9.993707	5.999652		

Note: Wavelength in μm and G_k in $μm^{-2}$.
Source: After J.W. Fleming, *Elect. Lett.*, 14, 326–332, 1978.

a single-mode structure and total dispersion at a particular wavelength or wavelength range must be satisfied. Also, in Appendices 7.1 and 7.2 sample MATLAB files are given for the design of single-mode optical fibers and non-zero-dispersion-shifted fibers.

```
%— — — — — OPTICAL Fiber Design LABORATORY 1— — — — — — %
% Sample Design of Optical Fibers having Different Doping
Profile
% Copy right reserved ° LN Binh, 2012
% clear after each run.
Clear all
Clc;
c = 2.997925e8;% velocity of light in vacuum

%setting Sellmeiers constants
G1 = 0.711040;
```

```
G2 = 0.408218;
G3 = 0.704048;

lambda1 = 0.064270e-6;
lambda2 = 0.129408e-6;
lambda3 = 9.425478e-6;

% select a reasonable value of core radius and relative
refractive index
a = 4.1e-6;
delta = 0.0025;

start = input('Enter lambda start point (nm)-  -: ');
finish = input('Enter lambda end point (nm)- -  -: ');
resolution = input('Enter lambda resolution (nm)- -  : ');
disp('');

lambda = start*1e-9;
lambdavector(1,1) = lambda;

for(i = 1:(((finish-start)/resolution)+1))

   n1squared = 1+((G1*power(lambda,2))/(power(lambda,2)-power(l
   ambda1,2)))+((G2*power(lambda,2))/(power(lambda,2)-power(lam
   bda2,2)))+((G3*power(lambda,2))/(power(lambda,2)-
   power(lambda3,2)));
   n1 = sqrt(n1squared);
   n1vector(1,i) = n1;

   n2 = n1*(1+delta);
   n2vector(1,i) = n2;

   V = (2*pi/lambda)*a*n1*sqrt(2*delta);
   Vvector(1,i) = V;

   dy1dx = (-2*G1*power(lambda1,2)*lambda)/
   (power(power(lambda,2)-power(lambda1,2),2));
   dy2dx = (-2*G2*power(lambda2,2)*lambda)/
   (power(power(lambda,2)-power(lambda2,2),2));
   dy3dx = (-2*G3*power(lambda3,2)*lambda)/
   (power(power(lambda,2)-power(lambda3,2),2));

d2y1dx2 = (2*G1*power(lambda1,2)*(3*power(lambda,2)+power(lam
bda1,2)))/(power(power(lambda,2)-power(lambda1,2),3));

d2y2dx2 = (2*G2*power(lambda2,2)*(3*power(lambda,2)+power(lam
bda2,2)))/(power(power(lambda,2)-power(lambda2,2),3));

d2y3dx2 = (2*G3*power(lambda3,2)*(3*power(lambda,2)+power(lam
bda3,2)))/(power(power(lambda,2)-power(lambda3,2),3));
```

```
d2ndx2 = 0.5*(((d2y1dx2+d2y2dx2+d2y3dx2)*power(n1,2)-
0.5*(power(dy1dx+dy2dx+dy3dx,2)))/power(n1,3));

M = (-d2ndx2/c)*lambda;
Mvector(1,i) = M;%row vector

Dw = (-n2*delta)/c*(0.080+0.549*power(2.834-V,2))*(1/
lambda);
Dwvector(1,i) = Dw;

if(i < (((finish-start)/resolution)+1))
   lambdavector(1,i+1) = lambdavector(1,i)+(resolution*1e-9
   );
   lambda = lambdavector(1,i+1);
end
end

plot(lambdavector,Mvector,lambdavector,Dwvector,lambdavector,M
vector+Dwvector);
grid;
```

REFERENCES

1. L.B. Jeunnhomme, *Single-Mode Fiber Optics—Principles and Applications*, Optical Engineering Series, vol. 4, Marcel Dekker, 1983.
2. J.P. Gordon and H. Kogelnik, "PMD Fundamentals: Polarization Mode Dispersion in Optical Fibers," *Proc. Natl. Acad. Sci. USA*, 97(9), 4541–4550, 2000.
3. Corning, Inc., *An Introduction to the Fundamentals of PMD in Fibers*, White Paper, July 2006.
4. A. Galtarossa and L. Palmieri, "Relationship between Pulse Broadening Due to Polarisation Mode Dispersion and Differential Group Delay in Long Single-Mode Fiber," *Elect. Lett.*, 34(5), 1998.
5. J.M. Fini and H.A. Haus, "Accumulation of Polarization-Mode Dispersion in Cascades of Compensated Optical Fibers," *IEEE Photonics Technol. Lett.*, 13(2), 124–126, 2001.
6. A. Carena, V. Curri, R. Gaudino, P. Poggiolini, and S. Benedetto, "A Time-Domain Optical Transmission System Simulation Package Accounting for Nonlinear and Polarization-Related Effects in Fiber," *IEEE J. Select. Areas Commun.*, 15(4), 751–765, 1997.
7. S.A. Jacobs, J.J. Refi, and R.E. Fangmann, "Statistical Estimation of PMD Coefficients for System Design," *Elect. Lett.*, 33(7), 619–621, 1997.
8. I. Morita and S.L. Jansen, "High Speed Transmission Technologies for 100-Gbit/s-Class Ethernet," Presentation at Proceedings of the ECOC 07, Berlin, Germany, September 16–20, 2007.
9. A.F. Elrefaie and R.E. Wagner, "Chromatic Dispersion Limitations for FSK and DPSK Systems with Direct Detection Receivers," *IEEE Photonics Technol. Lett.*, 3(1), 71–73, 1991.
10. A.F. Elrefaie, R.E. Wagner, D.A. Atlas, and A.D. Daut, "Chromatic Dispersion Limitation in Coherent Lightwave Systems," *IEEE J. Lightw. Technol.*, 6(5), 704–710, 1988.
11. L.N. Binh, *Digital Optical Communications*, Taylor & Francis Group, Boca Raton, FL, 2008.

12. L.N. Binh, *Optical Fiber Communications Systems: Theory and Practice with MATLAB® and Simulink® Models (Optics and Photonics)*, Taylor & Francis Group, Boca Raton, FL, 2010.

13. S. Kobayashi, S. Shibata, and T. Izawa, "Refractive-Index Dispersion of Doped Fused Silica," in *Proc. International Conference on Integrated Optics and Optical Fiber Communication*, Tokyo, p. 309, 1977.

14. J. W. Fleming, "Material Dispersion in Lightguide Glasses," *Elect. Lett.*, 14, 11, 326–328, 1978.

8 Guided Wave Optical Transmission Lines

Transfer Functions

Launching optical modulated signals of information channels over a guided wave medium, especially the optical fibers, and their propagation are critical for optical communication systems. The transmittance in terms of the fields or intensity of the guided waves is very critical for determining the quality of signals at the receiving end. Transfer functions are commonly used in electrical systems to interpret a number of quantities as a function of frequency range. Similarly in optical systems, the transfer function is indeed closely related to the transmittance of the guided lightwaves, not only in linear but also in nonlinear dynamics of signals.

This chapter gives an introduction to the transfer functions of optical fibers that can operate in linear and nonlinear regions.

8.1 TRANSFER FUNCTION OF SINGLE-MODE FIBERS

The transfer function, the output over the input in the frequency domain, of any subsystem of an overall system is very important for the designer to evaluate its behavior and effects on the quality of the performance of the system. In optical communication systems, optical fibers play a major role in the guidance of modulated lightwaves over very long distances; therefore, the transfer function of optical fiber as a transmission medium must be known and represented in the frequency domain. That is, the envelope of modulated signals plays a principal role in the characteristics of the double sidebands on both sides of the optical carrier. This section describes the behavior of signals transmitted over the single-mode optical fiber in terms of the transfer function in the frequency domain operating in both linear and nonlinear regions.

8.1.1 LINEAR TRANSFER FUNCTION

The treatment of the propagation of modulated lightwaves through single-mode fiber in the linear and nonlinear regimes has been well documented.[1-6] For completeness of the transfer function of single-mode optical fibers, in this section we restrict our study to the frequency transfer function and impulse responses of the fiber to the linear region of the media. Furthermore, the delay term in the nonlinear Schrodinger equation (NLSE) can be ignored, as it has no bearing on the size and shape of the pulses. From NLSE we can model the fiber simply as a quadratic phase function. This is derived from the fact that the nonlinear term of NLSE can be removed, and the Taylor series approximation around the operating frequency, for example, the central frequency, can be obtained, as well as frequency and impulse responses of the single-mode fiber. The input–output relationship of the pulse can therefore be depicted. Equation (8.1) expresses the time-domain impulse response $h(t)$ and the frequency-domain transfer function $H(\omega)$ as a Fourier transform pair:

$$h(t) = \sqrt{\frac{1}{j4\pi\beta_2}} \exp\left(\frac{jt^2}{4\beta_2}\right) \leftrightarrow H(\omega) = e^{-j\beta_2\omega^2} \tag{8.1}$$

where β_2 is well known as the group velocity dispersion (GVD) parameter. The input function $f(t)$ is typically a rectangular pulse sequence, and β_2 is the GVD parameter of the fiber and is proportional to the length of the fiber. The output function $g(t)$ is the dispersed waveform of the pulse sequence. The propagation transfer function in (8.1) is an exact analogy of diffraction in optical systems (see Papoulis[7,8]). Thus the quadratic phase function also describes the diffraction mechanism in one-dimensional optical systems, where distance x is analogous to time t. The establishment of this analogy affords us to borrow many of the imageries and analytical results that have been developed in the diffraction theory. Thus, we may express the step response $s(t)$ of the system, $H(\omega)$, in terms of Fresnel cosine and sine integrals as follows:

$$s(t) = \int_0^t \sqrt{\frac{1}{j4\pi\beta_2}} \exp\left(\frac{jt^2}{4\beta_2}\right) dt = \sqrt{\frac{1}{j4\pi\beta_2}} [C(\sqrt{1/4\beta_2}\,t) + jS(\sqrt{1/4\beta_2}\,t)] \tag{8.2}$$

with

$$C(t) = \int_0^t \cos\left(\frac{\pi}{2}\tau^2\right) d\tau$$

$$\tag{8.3}$$

$$S(t) = \int_0^t \sin\left(\frac{\pi}{2}\tau^2\right) d\tau$$

where $C(t)$ and $S(t)$ are the Fresnel cosine and sine integrals. The excitation of the step signal into a system is shown in Figure 8.1.

Using this analogy, one may argue that it is always possible to restore the original pattern $f(x)$ by refocusing the blurry image $g(x)$ (e.g., image formation[9]). In the electrical analogy, it implies that it is possible to compensate the quadratic phase media perfectly. This is not surprising. The quadratic phase function $H(\omega)$ in (8.1) is an all-pass transfer function; thus, it is always possible to find an inverse function to recover $f(t)$. One can express this differently in information theory terminology, that the quadratic phase channel has a theoretical bandwidth of infinity; hence, its information capacity is infinite. Shannon's channel capacity theorem states that there

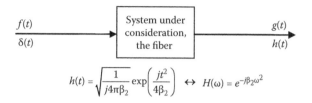

FIGURE 8.1 Representation of a system with input and output signals, especially the impulse response $h(t)$ due to $\delta(t)$, impulse signal as an excitation source.

is no limit on the reliable rate of transmission through the quadratic phase channel. Figure 8.2 shows the pulse and impulse responses of the fiber. Only the envelope of the pulse is shown, and the phase of the lightwave carrier is included as the complex values of the amplitudes. As observed, the chirp of the carrier is significant at the edges of the pulse. At the center of the pulse, the chirp is almost negligible at some limited fiber length; thus, the frequency of the carrier remains nearly the same as at its original starting value. One could obtain the impulse response quite easily, but the pulse response is much more relevant in the investigation of the uncertainty in the pulse sequence detection. Rather, the impulse response is much more important in the process of equalization, in which a convolution in the time domain of the impulse responses of cascaded subsystems would be necessary for designing the equalizer at the receiver. Alternatively, the transfer functions of various cascaded subsystems can be multiplied together, and then with that of the equalizer, to achieve a unity overall transfer function. The impulse and step responses are most critical in the characterization of the propagation medium, the optical fiber.

The uncertainty of the detection depends on the modulation formats and detection process. The modulation can be implemented by manipulation of the amplitude, the phase or the frequency of the carrier, or both the amplitude and phase of multisubcarriers, such as the orthogonal frequency division multiplexing (OFDM). The amplitude detection would be mostly affected by the ripples of the amplitudes of the edges of the pulse. The phase of the carrier is mostly affected near the edge due to the chirp effects. However, if differential phase detection is used, then the phase change at the transition instant is the most important, and the opening of the detected eye diagram. For frequency modulation the uncertainty in the detection is not very critical provided the chirping does not enter into the region of the neighborhood of the center of the pulse in which the frequency of the carrier remains almost constant.

The picture changes completely if the detector/decoder is allowed only a finite time window to decode each symbol. In the convolution coding scheme, for example, it is the decoder's constraint length that manifests due to the finite time window. In the adaptive equalization scheme, it is the number of equalizer coefficients that determines the decoder window length. Since the transmitted symbols have already been broadened by the quadratic phase channel, if they are next gated by a finite time window, the information received could be severely reduced. The longer the fiber, the more the broadening of the pulses is widened, and the more uncertain it becomes in the decoding. It is the interaction of the pulse broadening on one hand

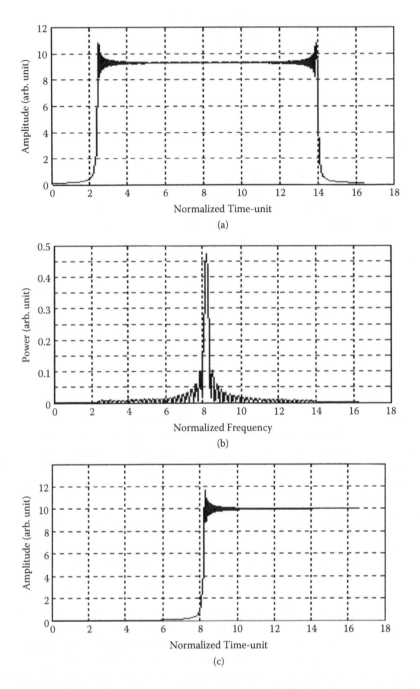

FIGURE 8.2 Rectangular pulse transmission through an SMF: (a) pulse response, (b) frequency spectrum, and (c) step response of the quadratic phase transmittance function. Note that the horizontal scale is in normalized unit of time.

and the restrictive detection time window on the other that give rise to the finite channel capacity.

It is also observed that the chirp occurs mainly near the edge of the pulses when it is in the near-field region, about a few kilometers for standard single-mode fibers. In this near-field distance the accumulation of nonlinear effects is still very weak, and thus these chirp effects dominate the behavior of the single-mode fiber.[2,10] The nonlinear Volterra transfer function presented in the next section would thus have minimum influence. This point is important for understanding the behavior of lightwaves circulating in short-length fiber devices in which both linear and nonlinear effects are to be balanced, such as active mode locked soliton and multibound soliton lasers.[11,12] In the far field the output of the fiber is Gaussian-like for the square pulse launched at the input. In this region the nonlinear effects would dominate over the linear dispersion effects, as they have been accumulated over a long distance.[13,14]

The linear time-variant system such as the single-mode fiber would take a transfer function of

$$H(f) = |H(f)| e^{-j\alpha(f)} \tag{8.4}$$

where $\alpha = \pi^2 \beta_2 L = -\pi D L \lambda^2 / 2c$ is proportional to the length L and the dispersion factor $D(\lambda)$ (s/m^2). The phase of the frequency transfer response is a quadratic function of the frequency; thus, the group delay would follow a linear relationship with respect to the frequency as observed in Figure 8.3. The frequency response in amplitude terms is infinite and is a constant, while the phase response is a quadratic function with respect to the frequency of the baseband signals. The carrier is chirped accordingly as observed in Figures 8.4 and 8.5. The chirping effect is very significant near the edge of the rectangular pulse and almost nil at the center of the pulse, in the near-field region of less than 1 km of standard single-mode fiber. In the far-field region the pulse becomes Gaussian-like. Thus the response of the fiber in the linear region can be seen as shown in Figure 8.6 for a Gaussian pulse input to the fiber. The output pulse is also Gaussian by taking the Fourier transform of the input pulse and multiplying by the fiber transfer function. Thence, an inverse Fourier would indicate the output pulse shape follows a Gaussian profile. Figure 8.3(a–c) illustrates the typical variation of the phase and magnitude responses of SSMF. The phase responses are important for signals under the phase shift keying modulation scheme, as these effects will rotate the constellation of the transmitted signals.[8] The chirping effects due to fiber dispersion are also illustrated in Figure 8.4. It is noted that chirp occurs mostly at the edges of the pulse, and then spreads to the center over a long transmission distance. Finally, the pulse would take a Gaussian profile shape in the far-field region, as shown in Figure 8.5.

This leads to a rule of thumb for consideration of the scaling of the bit rate and transmission distance: given that a modulated lightwave of a bit rate B can be transmitted over a maximum distance L of single-mode optical fiber with a bit error rate (BER) of 10^{-9} error-free level, if the bit rate is halved, the transmission distance can be increased by four times, and vice versa. For example, for 10 Gb/s amplitude shift keying modulation format signals can be transmitted over 80 km of standard

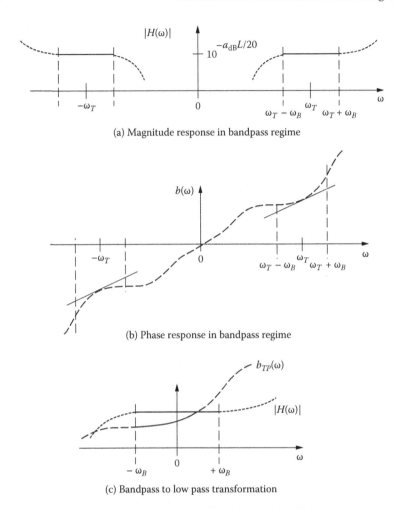

(a) Magnitude response in bandpass regime

(b) Phase response in bandpass regime

(c) Bandpass to low pass transformation

FIGURE 8.3 Frequency response of a single-mode optical fiber: (a) magnitude, (b) phase response in bandpass regime, and (c) baseband equivalence.

single-mode optical fiber; then at 40 Gb/s only 5 km can be transmitted for a BER of 10^{-9} without any forward error coding and under direct detection.

8.1.2 Single-Mode Optical Fiber Transfer Function: Simplified Linear and Nonlinear Operating Regions

In this section, a closed expression of the frequency transfer function of dispersive and nonlinear single-mode optical fibers for broadband operation can be derived, similar to the case under microwave photonics. The expression takes into account both chromatic dispersion and self-phase modulation (SPM) effects and is valid for optical double-sideband modulation, optical single-sideband (SSB) modulation, and chirped optical transmitters.

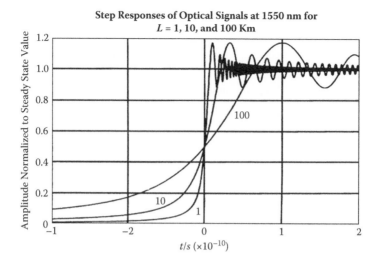

FIGURE 8.4 Carrier chirping effects and step response of a single mode optical fiber of length, $L = 1$, 10, and 100 km.

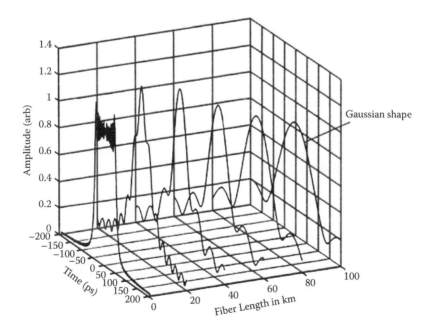

FIGURE 8.5 Pulse response from near field ($\sim$<2 km) to far field (>80 km).

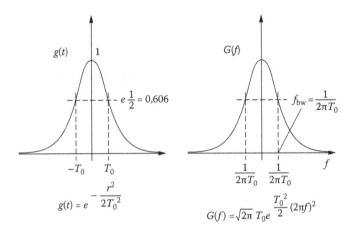

FIGURE 8.6 Fiber response to Gaussian pulse. Gaussian → Gaussian!

The evolution along the propagation path z of the small-signal intensity modulation (IM), or complex power $\bar{p}(\omega,z)$ and phase rotation (PM) $\bar{\phi}(\omega,z)$, during the propagation of the guided mode through the single-mode optical fiber (SMF), taking into account both the chromatic dispersion and the nonlinearity (SPM) effects, is governed by the following set of differential equations[15–18]:

$$\frac{\delta \bar{p}(\omega,z)}{\delta z} = \beta_2 \omega^2 \bar{P}_0 \bar{\phi}(\omega,z); \bar{A}(\omega,z) = \sqrt{\bar{p}(\omega,z)} \tag{8.5}$$

$$\frac{\delta \bar{\phi}(\omega,z)}{\delta z} = -\left[\frac{\beta_2 \omega^2}{4\bar{P}_0} + \gamma e^{-\alpha z} \right] \bar{p}(\omega,z) \tag{8.6}$$

where $\bar{A}(\omega,z)$ and $\bar{\phi}(\omega,z)$ are defined as the normalized complex amplitude and phase, respectively, of the optical field in the Fourier domain; ω is the radial frequency of the radio frequency (RF) or broadband signal; z is the distance along the propagation axis of the fiber; α is the attenuation coefficient in the linear scale of SMF; and β_2 is the first-order dispersion coefficient, i.e., the group delay factor as a function of the optical wavelength given by

$$\beta_2 = -\frac{\lambda^2 D(\lambda)}{2\pi c} \tag{8.7}$$

where c is the velocity of light in vacuum, $D(\lambda)$ is the dispersion factor of the fiber typically taking a value of 17 ps/nm/km for silica SMF at the operating wavelength $\lambda = 1550$ nm, and γ is the nonlinear SPM factor defined by

$$\gamma = \frac{2\pi n_2}{\lambda A_{eff}}; \quad \text{with } A_{eff} = \pi r_0^2 \tag{8.8}$$

where n_2 is the nonlinear coefficient of the fiber, typically $n_2 = 1.3 \times 10^{-23}$ m²/W for the standard SMF, Corning SMF-28, and $A_{eff} = \pi r_0^2$ is the effective area of the fiber, which is the area of the Gaussian mode spot size r_0 of the guided mode in a single-mode optical fiber under the weakly guiding condition.[19] These parameters were described in Chapter 7.

Equations (8.5) and (8.6) are derived from the observer positioned on the moving frame of the phase velocity of the waves, which are normally expressed by the nonlinear Schrodinger equation (NLSE):

$$\frac{\partial A(t,z)}{\partial z} = -\left[\alpha A(t,z) + \frac{j}{2}\beta_2 \frac{\partial A^2(t,z)}{\partial t^2} \right] + j\gamma |A(t,z)|^2 A(t,z) \qquad (8.9)$$

Present coherent optical systems are based on digital signal processing at the transmitter and receivers using digital-to-analog and analog-to-digital converters at the transmitters and receivers, respectively, allowing complex modulation formats imposed on the optical carriers. However, at extremely high speed, the processing algorithms must be simple enough so as not to consume too much processing time for real-time applications. The simplified transfer function of the guided medium must be obtained. Furthermore, phase shift keying is proven to be a very efficient scheme for modulation of the lightwaves. Under the quadrature amplitude modulation (QAM) both the in-phase and quadrature phase components are recovered in the amplitude, which is proportional to the power of the optical signals arriving in the front of the optical waves of the optical hybrid coupler and the balanced photodetector pair. A schematic of the transmission is shown in Figure 8.7(a), in which the transmitter can generate an optical sequence or near-single-frequency sinusoidal waves at frequencies reaching 30 GHz using a Fujitsu DAC sampling rate of 65 GSa/s. The optical modulator is a typical Fujitsu IQ modulator modulated by electrical signals output from the DAC and phase shifted in the RF domain by an electrical phase shifter (PS). The RF phase can be set such that when the signals are $\pi/2$ shifted with respect to each other, the suppression of one of the single sidebands can be achieved at the output spectrum. The main carrier can be suppressed by biasing the "children" Mach–Zehnder intensity modulators (MZIMs) at the minimum transmission point. An insert of the structure of an I/Q modulator is shown in Figure 8.7(b).

By differentiating (8.5) and substituting into (8.6) we obtain

$$\frac{\delta^2 \overline{p}(\omega,z)}{\delta z^2} = -\left[\frac{\beta_2^2 \omega^4}{4} - \beta_2 \omega^2 + \gamma \overline{P}_0 e^{-\alpha z} \right] \overline{p}(\omega,z) \qquad (8.10)$$

subject to the initial conditions given by

$$\overline{p}(\omega,0) = \overline{p}_{in}(\omega) \quad \text{and} \quad \frac{\delta \overline{p}(\omega,0)}{\delta z} = \beta_2 \omega^2 \overline{P}_0 e^{-\alpha z} \phi(\omega,0) = \beta_2 \omega^2 \overline{P}_0 e^{-\alpha z} \phi_{in}(\omega) \quad (8.11)$$

where the subscript *in* indicates the input location, which is at the starting of the propagation of the modulated optical waves through the SMF.

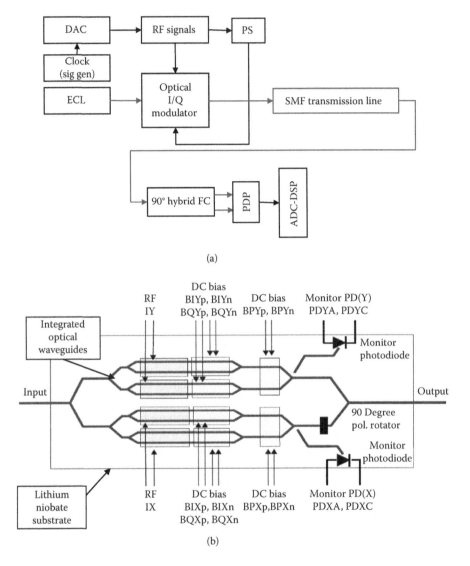

(a)

(b)

FIGURE 8.7 (a) Digital-based optical transmitter and coherent reception with real-time sampling and digital signal processing. DAC = digital-to-analog converter, ADC = analog-to-digital converter, DSP = digital signal processing, PDP = photodetector pair, FC = fiber coupler, I/Q = in-phase/quadrature phase. (b) Structure of an IQ optical modulator in integrated guided waveform (not to scale). RF = radio frequency, PD = photodetector, X = horizontal, Y = vertical, n = negative, p = positive. (See also Chapter 5 for 3D-guided wave structures and modulators.)

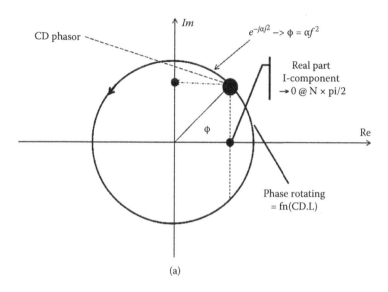

(a)

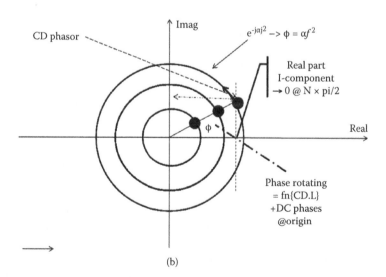

(b)

FIGURE 8.8 Phasor of the amplitude evolution along the propagation path, assuming the amplitude is not affected by attenuation: (a) phasor and (b) phase or phase constellation of M-QAM (M = 16 square QAM), three-amplitude level.

Now by changing some variables with the setting of

$$x = 2\sqrt{B}e^{-\alpha z/2}; \quad B = -\frac{\beta_2 \omega^2 \gamma P_0}{\alpha^2} \tag{8.12}$$

then (8.10) can be rewritten as

$$\left[x\frac{\partial^2}{\partial x^2} + x\frac{\partial}{\partial x} - (x^2 - \upsilon^2)\right]\bar{p}(\omega, z) = 0 \tag{8.13}$$

The solution of this equation is a combination of purely imaginary Bessel functions L and K and is subject to the initial conditions of (8.11). Thus the evolution of the complex amplitude of the modulated optical waves along the propagation path is given by[20,21]

$$\bar{p}(\omega, z) = \frac{2\sinh(\pi\upsilon)}{\pi}\left\{ \begin{array}{l} \sqrt{B}\,\bar{p}_{in}(\omega)\left[\dfrac{\partial L_{i\upsilon}}{\partial x}(2\sqrt{B})K_{i\upsilon}(x) - \dfrac{\partial K_{i\upsilon}}{\partial x}(2\sqrt{B})I_{i\upsilon}(x)\right] \\[2mm] + \dfrac{\alpha B}{\gamma}\phi_{in}(\omega)[K_{i\upsilon}(2\sqrt{B})L_{i\upsilon}(x) - L_{i\upsilon}(2\sqrt{B})K_{i\upsilon}(x)] \end{array} \right\} \tag{8.14}$$

with $\upsilon = -(\beta_2\omega^2/\alpha)$. The first term on the right-hand side (RHS) of (8.14) is the magnitude part, and the second is the phase part, that is, the in-phase and quadrature components of the QAM signal. Thus the in-phase and quadrature parts of the complex magnitude can be expressed as

$$\bar{p}_I(\omega, z) = \frac{2\sinh(\pi\upsilon)}{\pi}\left\{\sqrt{B}\,p_{in}(\omega)\left[\frac{\partial L_{i\upsilon}}{\partial x}(2\sqrt{B})K_{i\upsilon}(x) - \frac{\partial K_{i\upsilon}}{\partial x}(2\sqrt{B})I_{i\upsilon}(x)\right]\right\} \tag{8.15}$$

$$\bar{p}_Q(\omega, z) = \frac{\alpha B}{\gamma}\phi_{in}(\omega)\frac{2\sinh(\pi\upsilon)}{\pi}[K_{i\upsilon}(2\sqrt{B})L_{i\upsilon}(x) - L_{i\upsilon}(2\sqrt{B})K_{i\upsilon}(x)] \tag{8.16}$$

The complex amplitude vector of the lightwaves is rotating as depicted in Figure 8.8. The behavior of the amplitude and phase of the complex power can be explained by referring to Figure 8.9. The in-phase and quadrature phase components move along the horizontal and vertical axes within the ±1 limits, meaning that as the phase rotates around the unit circle, these components oscillate such that when the phase is $(2M + 1)$ $(M = 0, 1, 2, \ldots)$ or an odd number of $\pi/2$, the in-phase component becomes nullified. Likewise for the quadrature phase at $N\pi(N = 0, 1, 2, \ldots)$. In the case of the QAM scheme, e.g., 16 QAM, there would be a three-amplitude level of the phase constellation, and these levels are rotating. The initial phase is set by the initial position of the constellation point of square 16 QAM, but the oscillation and nullified locations would be very much similar to those of Figure 8.9.

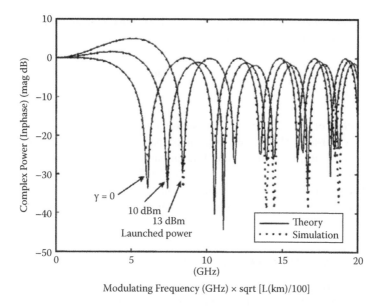

Modulating Frequency (GHz) × sqrt [L(km)/100]

FIGURE 8.9   Variation of the magnitude of the optical field intensity with frequency, the intensity frequency response of SMF-28 of $L = 100$ km (at $z = 100$ km) under coherent detection with normalized amplitude of the intensity of the optical waves under linear ($\gamma = 0$) and nonlinear operating conditions.

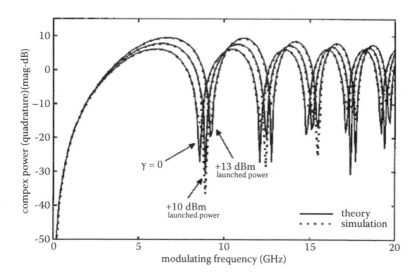

FIGURE 8.10   Variation of the magnitude of phase component with frequency, the QAM in-phase frequency response of the QAM signal SMF of $L = 100$ km (at $z = 100$ km) under coherent detection with normalized amplitude of the intensity of the optical waves under linear ($\gamma = 0$) and nonlinear operating conditions.

Under a linear operating regime, that is, $\gamma = 0$, we can obtain an expression for the complex power amplitude and phase as

$$\bar{p}_l(\omega, z) = \cos\frac{\beta_2\omega^2 z}{2} \tag{8.17}$$

For simulation of the evolution of the modulated lightwave channel over SMF, Appendix 8.1 lists a MATLAB® file for propagating an optical data sequence through an optical fiber transmission line employing the NLSE and the split-step Fourier method (see also Chapter 7, Section 7.5). Furthermore, Appendix 8.2 lists an initialization file for simulating optical signal propagation and parameters of an optical transmission system. This file is to be changed according to desired simulation circumstances.

8.1.3 Nonlinear Fiber Transfer Function

The weakness of most of the recursive methods in solving the NLSE is that they do not provide much useful information to help the characterization of nonlinear effects.[1] The Volterra series transfer function (VSTF) model provides an elegant way for describing a system's nonlinearities, and enables designers to observe clearly where and how the nonlinearity affects the system performance. Although several references[3-6,22-24] have given outlines of the kernels of the transfer function using the Volterra series, it is necessary for clarity and physical representation of these functions; brief derivations are given here on the nonlinear transfer functions of an optical fiber operating under nonlinear conditions.

The VSTF of a particular optical channel can be obtained in the frequency domain as a relationship between the input spectrum $X(\omega)$ and the output spectrum $Y(\omega)$ as

$$Y(\omega) = \sum_{n=1}^{\infty} \int_{-\infty}^{\infty} \ldots \int_{-\infty}^{\infty} H_n(\omega_1, \ldots, \omega_{n-1}, \omega - \omega_1 - \ldots - \omega_{n-1}) \times X(\omega_1) \ldots X(\omega_{n-1}) \tag{8.18}$$

$$X(\omega - \omega_1 - \ldots - \omega_{n-1}) d\omega_1 \ldots d\omega_{n-1}$$

where $H_n(\omega_1, \ldots, \omega_n)$ is the nth-order frequency-domain Volterra kernel, including all signal frequencies of orders 1 to n. The wave propagation inside a single-mode fiber can be governed by a simplified version of the NLS wave equation with only the self-phase modulation effect included as (also given in Chapter 7)

$$\frac{\partial A}{\partial z} = -\frac{\alpha_0}{2} A - \beta_1 \frac{\partial A}{\partial t} - j\frac{\beta_2}{2}\frac{\partial^2 A}{\partial t^2} - \frac{\beta_3}{6}\frac{\partial^3 A}{\partial t^3} + j\gamma|A|^2 A \tag{8.19}$$

where $A = A(t, z)$. The proposed solution of the NLS equation can be written with respect to the VSTF model of up to fifth order as

$$A(\omega,z) = H_1(\omega,z)A(\omega) + \int\limits_{-\infty}^{\infty}\int\limits_{-\infty}^{\infty} H_3(\omega_1,\omega_2,\omega-\omega_1+\omega_2,z)$$

$$\times A(\omega_1)A^*(\omega_2)A(\omega-\omega_1+\omega_2)d\omega_1 d\omega_2$$

$$+\int\limits_{-\infty}^{\infty}\int\limits_{-\infty}^{\infty}\int\limits_{-\infty}^{\infty}\int\limits_{-\infty}^{\infty} H_5(\omega_1,\omega_2,\omega_3,\omega_4,\omega-\omega_1+\omega_2-\omega_3+\omega_4,z)$$

$$\times A(\omega_1)A^*(\omega_2)A(\omega_3)A^*(\omega_4)\times A(\omega-\omega_1+\omega_2-\omega_3+\omega_4)d\omega_1 d\omega_2 d\omega_3 d\omega_4$$

$$(8.20)$$

where $A(\omega) = A(\omega, 0)$, that is, the amplitude envelope of the optical pulses at the input of the fiber. Taking the Fourier transform of (8.3) and assuming $A(t, z)$ is of sinusoidal form, we have

$$\frac{\partial A(\omega,z)}{\partial z} = G_1(\omega)A(\omega,z)\int\limits_{-\infty}^{\infty}\int\limits_{-\infty}^{\infty} G_3(\omega_1,\omega_2,\omega-\omega_1+\omega_2)A(\omega_1,z)A^*(\omega_2,z)$$

$$\times A(\omega-\omega_1+\omega_2,z)d\omega_1 d\omega_2 \qquad (8.21)$$

where

$$G_1(\omega) = -\frac{\alpha_0}{2} + j\beta_1\omega + j\frac{\beta_2}{2}\omega^2 - j\frac{\beta_3}{6}\omega_3 \quad \text{and} \quad G_3(\omega_1,\omega_2,\omega_3) = j\gamma$$

The parameter ω covers the range over the signal bandwidth, and beyond that it can overlap the signal spectrum of other optically modulated carriers. $\omega_1 \dots \omega_3$ are all also taking values over a similar range as that of ω but in different frequency axes. For a general expression the limit of integration is indicted over the entire range to infiniy. Thus the higher the order of the VSTF, the more complex are the numerical solutions due to multiple spectral ranges to be integrated over. Substituting (8.20) into (8.21) and equating both sides, the kernels can be obtained after some algebraic manipulations, and then by equating the first-order terms on both sides, we obtain

$$\frac{\partial}{\partial z}H_1(\omega,z) = G_1(\omega)H_1(\omega,z) \qquad (8.22)$$

The solution for the first-order transfer function (8.22) is then given by

$$H_1(\omega,z) = e^{G_1(\omega)z} = e^{\left(-\frac{\alpha_0}{2} + j\beta_1\omega + j\frac{\beta_2}{2}\omega^2 - j\frac{\beta_3}{6}\omega^3\right)z} \qquad (8.23)$$

This is in fact the linear transfer function of a single-mode optical fiber with the dispersion factors β_2 and β_3, as already shown in the previous section. Similarly, for the third-order terms we have

$$\frac{\partial}{\partial z} \int\limits_{-\infty}^{\infty} \int\limits_{-\infty}^{\infty} H_3(\omega_1,\omega_2,\omega-\omega_1+\omega_2,z) \times A(\omega_1)A^*(\omega_2)A(\omega-\omega_1+\omega_2)\,d\omega_1\,d\omega_2$$

$$= \int\limits_{-\infty}^{\infty} \int\limits_{-\infty}^{\infty} G_3(\omega_1,\omega_2,\omega-\omega_1+\omega_2)H_1(\omega_1,z)A(\omega_1)H_2^*(\omega_2,z)$$

$$\times A(\omega_2)H_1(\omega-\omega_1+\omega_2)A(\omega-\omega_1+\omega_2)\,d\omega_1\,d\omega_2$$

(8.24)

Now, letting $\omega_3 = \omega - \omega_1 + \omega_2$, it follows that

$$\frac{\partial H_3(\omega_1,\omega_2,\omega_3,z)}{\partial z} = G_1(\omega_1-\omega_2+\omega_3)H_3(\omega_1,\omega_2,\omega_3,z)$$

$$+ G_3(\omega_1,\omega_2,\omega_3)H_1(\omega_1,z)H_1^*(\omega_2,z)H_1(\omega_3,z)$$

(8.25)

Thus, the third kernel transfer function can be obtained as

$$H_3(\omega_1,\omega_2,\omega_3,z) = G_3(\omega_1,\omega_2,\omega_3) \times$$

$$\frac{e^{(G_1(\omega_1)+G_1^*(\omega_2)+G_1(\omega_3))z} - e^{G_1(\omega_1-\omega_2+\omega_3)z}}{G_1(\omega_1)+G_1^*(\omega_2)+G_1(\omega_3)-G_1(\omega_1-\omega_2+\omega_3)}$$

(8.26)

The fifth-order kernel can be similarly found but not included here, refer to Binh et al.[25] for the expression. Higher-order terms can be derived with ease if higher accuracy is required. However, in practice such higher order would not exceed the fifth rank. We can understand that for a length of uniform optical fiber, the first- to nth-order frequency spectrum transfer can be evaluated, indicating the linear to nonlinear effects of the optical signals transmitting through it. Indeed, the third- and fifth-order kernel transfer functions based on the Volterra series indicate the optical field amplitude of the frequency components that contribute to the distortion of the propagated pulses. An inverse of these higher-order functions would give the signal distortion in the time domain. Thus, the VSTFs allow us to conduct distortion analysis of optical pulses and an evaluation of the bit error rate of optical fiber communication systems.

The superiority of such a Volterra transfer function expression allows us to evaluate each effect individually, especially the nonlinear effects, so that we can design and manage the optical communication systems under linear or nonlinear operations. Currently this linear–nonlinear boundary of operations is critical for system implementation, especially for optical systems operating at 40 Gbps, where a linear operation and a carrier-suppressed return-to-zero format are employed. As a norm in series expansion, the series need converged to a final solution. It is this convergence that allows us to evaluate the limit of nonlinearity in a system.

8.2 FIBER NONLINEARITY

The linear effects in optical fibers are described in Section 8.3. This section describes the nonlinear effects and their influence on the propagation of optical signals over a long length of fibers. The nonlinear and linear effects in optical fibers can be classified as shown in Figure 8.11.

The fiber refractive index (RI) is dependent on both operating wavelengths and lightwave intensity. This intensity-dependent phenomenon is known as the Kerr effect and is the cause of fiber nonlinear effects.

8.2.1 SPM AND XPM EFFECTS

The power dependence of RI is expressed as

$$n' = n + \bar{n}_2(P / A_{eff}) \tag{8.27}$$

where P is the average optical power of the guided mode, $\bar{n}_2$ is the fiber nonlinear coefficient, and A_{eff} is the effective area of the fiber.

Fiber nonlinear effects include intrachannel SPM, interchannel cross-phase modulation (XPM), four-wave mixing (FWM), stimulated Raman scattering (SRS), and stimulated Brillouin scattering (SBS). SRS and SBS are not the main degrading factors, as their effects get noticeably large only with very high optical power. On the other hand, FWM severely degrades the performance of an optical system with the generation of ghost pulses only if the phases of optical signals are matched with each other. However, with high local dispersions such as in SSMF, effects of FWM

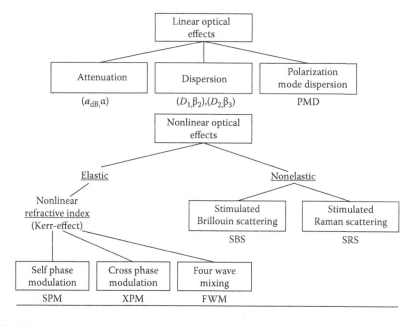

FIGURE 8.11 Linear and nonlinear fiber properties in SMF.

become negligible. In terms of XPM, its effects can be considered to be negligible in a dense wavelength division multiplexing (DWDM) system in the following scenarios: (1) highly locally dispersive system and (2) large channel spacing. However, XPM should be taken into account for optical transmission systems deploying non-zero-dispersion-shifted fiber (NZ-DSF) where local dispersion values are small. Thus SPM is usually the dominant nonlinear effect for systems employing transmission fiber with high local dispersions, e.g., SSMF and DCF. The effect of SPM is normally coupled with the nonlinear phase shift ϕ_{NL}, defined as

$$\phi_{NL} = \int_0^L \gamma P(z)\,dz \; = \; \gamma L_{eff} P$$

$$\gamma = \omega_c \bar{n}_2 / (A_{eff} c) \tag{8.28}$$

$$L_{eff} = (1 - e^{-\alpha L}) / \alpha$$

where ω_c is the lightwave carrier, L_{eff} is the effective transmission length, and α is the fiber attenuation factor, which normally has a value of 0.17 to 0.2 dB/km in the 1550 nm spectral window. The temporal variation of the nonlinear phase ϕ_{NL} results in the generation of new spectral components far apart from the lightwave carrier ω_c, indicating the broadening of the signal spectrum. This spectral broadening $\delta\omega$ can be obtained from the time dependence of the nonlinear phase shift as

$$\delta\omega = -\frac{\partial \phi_{NL}}{\partial T} = -\gamma \frac{\partial P}{\partial T} L_{eff} \tag{8.29}$$

Equation (8.29) indicates that $\delta\omega$ is proportional to the time derivative of the average signal power P. Additionally, the generation of new spectral components occurs mainly at the rising and falling edges of optical pulses; i.e., the amount of generated chirps is larger for an increased steepness of the pulse edges.

The wave propagation equation can be represented as

$$\frac{\partial A(z,t)}{\partial z} + \frac{\alpha}{2} A(z,t) + \beta_1 \frac{\partial A(z,t)}{\partial t} + \frac{j}{2}\beta_2 \frac{\partial^2 A(z,t)}{\partial t^2} - \frac{1}{6}\beta_3 \frac{\partial^3 A(z,t)}{\partial t^3}$$

$$= -j\gamma |A(z,t)|^2 A(z,t) - \frac{1}{\omega_0}\frac{\delta}{\delta t}(|A|^2 A) - T_R A \frac{\delta(|A|^2)}{\delta t} \tag{8.30}$$

in which we have ignored the pure delay factor involving β_1. The last term on the RHS represents the Raman scattering effects.

8.2.2 Modulation Instability

The mutual effect between the nonlinear dispersion effects and the nonlinear effects can lead to the modulation of the lightwave pulses and, thus, unstable states of the

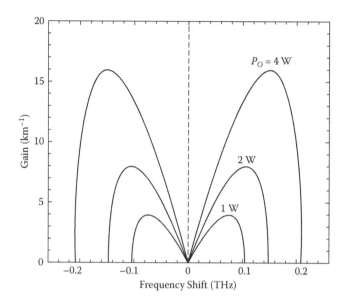

FIGURE 8.12 Spectrum of the optical gain due to modulation instability at three different average power levels in an optical fiber with $\beta_2 = 20$ ps^2/km and $\gamma = 2$ W/km.

optical pulses. This phenomenon is usually called the modulation instability and is normally observed in soliton lasers. The gain spectrum of the modulation instability is shown in Figure 8.12.[26]

8.2.3 EFFECTS OF MODE HOPPING

So far we have assumed that the source center emission wavelength is unaffected by the modulation. In fact, when a short current pulse is applied to a semiconductor laser, its center emission wavelength may hop from one mode to its neighbor, a longer wavelength. In the case where a multilongitudinal mode laser is used, this hopping effect is negligible; however, it is very significant for a single longitudinal mode laser. Currently external cavity lasers can offer very narrow linewidths of about 100 kHz and high stability without any mode-hopping effects. These lasers are employed in digital coherent transmission systems operating at 100 G and beyond bit rates.

8.3 NONLINEAR FIBER TRANSFER FUNCTIONS AND APPLICATION IN COMPENSATIONS

Nonlinear effects have been considered in the previous section, in which the SPM effects play the major role in the distortion of modulated signals, besides the linear chromatic dispersion effects. We have also seen that the nonlinear Schrodinger equation has been used extensively in modeling the lightwave signals propagating through optical fiber links in which both linear and nonlinear effects are included.

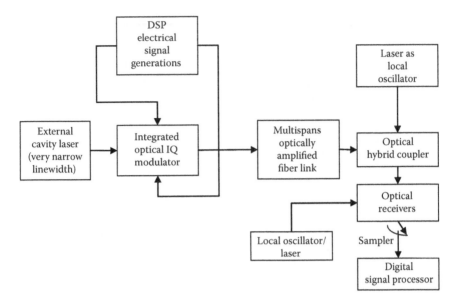

FIGURE 8.13 Schematic form of an optical receiver based on a digital signal processor using coherent detection in a modern optically amplified fiber link transmission system.

In practice we have seen many optical components, such as the fiber Bragg gratings, dispersion compensating fibers (DCFs), or optical fiber filter structures,[27] compensate for chromatic dispersion effects in the optical domain, as described in Sections 8.2. Nonlinear dispersion compensation can also be compensated in the optical domain by phase conjugators,[28,29] but these require being placed exactly at the midway of optical fiber links, which would be hard to be determined. However, under current high-speed optical communication technology, electronic digital signal processing of received signals occurs in the electronic domain after the coherent receivers. Thus it is possible to compensate for both the linear and nonlinear dispersions if algorithms can be found to do the reverse dispersion processes in the electronic domain to minimize the signal distortion. These algorithms would be developed if such transfer functions of the fibers operating in linear and nonlinear regions could be simplified so as to cost the least number of processing steps for processors working at ultra-high speed.[30,31] The schematic of the optical coherent receiver in the long-haul optical fiber communication system is shown in Figure 8.13. Both transmitters and receivers can integrate digital signal processors before and after the optically amplified multispan optical fiber transmission link. The fiber link can be represented by a canonical form of transfer functions. It is noted here that the sampler must operate at very high rate, normally at about 56 or 64 GSa/s;[32] thus, the DSP would have minimum memory banks and processing speed must be high enough so that real-time processing of signals can be achieved. Hence, algorithms must be very efficient and cost minimum time.

This section is thus dedicated to some of the recent developments in representing the transfer functions of optical fibers for signal propagation and compensations, with applications especially in the electronic domain.

8.3.1 CASCADES OF LINEAR AND NONLINEAR TRANSFER FUNCTIONS IN TIME AND FREQUENCY DOMAINS

In order to reduce computational requirements at the receiver and assuming that the nonlinear phase rotation on the optical carrier can be separable from the linear phase effects, one can represent the transfer functions of the propagation of the optical pulse sequence over a length L by a cascade of linear and nonlinear phase rotation as[33]

$$E_n(t,z+h) = E(t,z)e^{j\gamma h |E(t,z)|^2} \tag{8.31}$$

and

$$\tilde{E}(\omega,z+h) = \tilde{E}_n(\omega,z+h)e^{-j\left(\frac{\alpha}{2}+\frac{\beta_2}{2}\omega^2\right)h} \tag{8.32}$$

where the nonlinear phase is multiplied to the signals envelope at the input of a fiber length. This nonlinear phase is estimated under a number of considerations, so that it is valid under certain constraints. h is the step size, as we have assumed in previous section, but it can also take a much larger distance—thus allowing reduction of computational resources. $\tilde{E}(\omega,z+h)$ and $\tilde{E}_n(\omega,z+h)$ are the approximated optical fields at the input and output of the fiber over a step of order n. Clearly from (8.31) we can observe that the phase accumulated over the distance step h is contributed to the rotation of the phase of the carrier, while (8.32) represents the rotation of the phase of the carrier after propagating through h by the linear GVD effect evaluated in the spectral domain. Thus the transfer function of the linear dispersion effect given in (8.4) can be employed together with the nonlinear phase contribution as shown in Figure 8.14 over the whole transmission link of N spans or cascades of span by span over the whole link.

The assumptions and observations through experiments of the nonlinear phase effects on transmission of signals are as follows:

- Amplitude-dependent phase rotation to improve system performance has been demonstrated in Kikuchi et al.[34] and Charlet et al.[35] over short fiber spans with nearly perfect chromatic dispersion (CD) compensation per span.
- The received signal has a spiral-shaped constellation as reported in Lau and Kahn.[36] It is possible to exploit the correlation between the received amplitude and nonlinear phase shift to reduce nonlinear phase noise variance as shown in Pina et al.[37] under simulation using the split-step Fourier method (SSFM). This spiral rotation leads to conclusions that under possibly weak nonlinear effects the phase can be superimposed on the modulated signals as an additional phase components—thus, the cascade of the nonlinear phase superposition and linear transfer function.

The effective length of each step must be evaluated so as not to overcompensate. This effective length can be estimated as given in (8.28), which is typically about

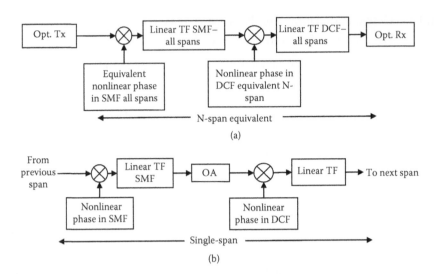

FIGURE 8.14 Representation of optical signals propagating through optical fibers with separable nonlinear and linear transfer functions: (a) equivalent for all spans and (b) equivalent for each span.

22 km for standard single-mode optical fiber (SSMF) with a nonlinear coefficient of 2.1e-20 m/W. The rotation of the constellation of a quadrature phase shift keying (QPSK) signal sequence is shown in Figure 8.15, indicating the effects of nonlinear rotation when the linear chromatic dispersion is completely compensated.

Once the nonlinear phase noises can be represented as a phase superposition on the signals, under coherent detection the optical field would be detected and presented as an electronic current or voltages at the output of an optical receiver whose signals are

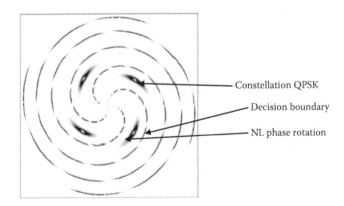

FIGURE 8.15 Received signal constellation of QPSK under coherent detection over 5000 km of SSMF under nonlinear (NL) effects and linear dispersion with decision boundary (spiral lines) for detection. (From A.P.T. Lau and J.M. Kahn, *IEEE J. Lightw. Technol.*, 25(10), 3008, 2007. With permission.)

then sampled by a high-speed sampler to covert to the digital domain and processed by a digital signal processor (DSP). The compensation of nonlinear phase noises is then conducted in the digital domain, and thus a back propagation algorithm is required. This algorithm can be implemented by forward propagation with a nonlinear coefficient of sign opposite to that of the transmission fiber. The numerical implementation of such a transfer function and phase superposition given in (8.31) and (8.32) is quite straightforward and numerically effective, as the phase over the propagation step can be over one span or sections of spans, or even the whole transmission link.[33] However, the compensation may be too much, and thus distortion does also happen. In this case there must be an adaptive technique to monitor the compensation process so that when the nonlinear phase distortion is just completely compensated, the process must be finished.[33]

8.3.2 VOLTERRA NONLINEAR TRANSFER FUNCTION AND ELECTRONIC COMPENSATION

As described in Appendix 8.1, the wave propagation inside a single-mode fiber (SMF) can be governed by a simplified version of the NLSE, in which only SPM is affected. $A = A(t, z)$ is the electric field envelope of the optical signal, β_2 is the second-order dispersion parameter, α is the fiber attenuation coefficient, and γ_L is the nonlinear coefficient of the fiber. The solution of the NLSE can be written with VSTFs of kernels of the fundamental order and $(2N + 1)$th-order as described in Peddanarappagari and Brandt-Pearce.[38] Shown in Figure 8.16a is the transfer function VSTF of a span of transmission fiber by a parallel combination of a linear and two nonlinear kernels. The whole transmission link can be thus represented by a cascade of the VSTFs and optical amplification stage as in Figure 8.16b. However, up to the third order is sufficient to represent the weakly nonlinear effects in the slowly varying amplitude of the guide wave propagating in a single-mode weakly guiding fiber; the frequency domain of the amplitude along the transmission line is given as[25]

$$A(\omega,z) = \begin{cases} H_1(\omega,z)A(\omega) + \int\limits_{-\infty}^{\infty}\int\limits_{-\infty}^{\infty} H_3(\omega_1,\omega_2,\omega,z) \\ \times A(\omega_1)A^*(\omega_2)A(\omega-\omega_1+\omega_2)d\omega_1 d\omega_2 \end{cases} \tag{8.33}$$

$$H_1(\omega,z) = e^{-\alpha z/2}e^{-j\omega^2\omega_2 z/2} \tag{8.34}$$

$$H_3(\omega_1,\omega_2,\omega,z) = \left\{ -\frac{j\gamma}{4\pi^2}H_1(\omega,z)\times\frac{1-e^{-(\alpha+j\beta_2(\omega_1-\omega)(\omega_1-\omega_2))z}}{\alpha+j\beta_2(\omega_1-\omega)(\omega_1-\omega_2)} \right. \tag{8.35}$$

where $A(\omega) = A(\omega,z = 0)$ represents the optical pulse at the input of the fiber in the frequency domain. ω_1, ω_2, and ω are the dummy variables acting as parameters and indicating the cross-interactions of the lightwaves at different frequencies, i.e.,

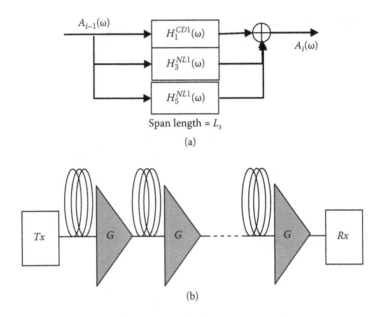

FIGURE 8.16 (a) Representation of a fiber span by VSTFs of first and higher order (up to fifth order); (b) cascade of optically amplified N_s-span fiber link without DCF.

intra- or interchannels, especially the interchannel interaction effects. The range of these spectral variables changes from $(-\infty, +\infty)$. Thus we can observe that (ω_1, ω_2) form a plane of the angular frequency components, and the angular frequency ω can be scanned across all regions to see the interactions of the nonlinear effects. We can distinguish the regions on this plane, different nonlinear effects, after the propagation of the lightwaves in the nonlinear regime.

There are regions where there are cross-terms indicating the interaction of different and nonidentical frequency components of the signal spectra. These cross-terms are the intermodulation terms, i.e., due to XPM, as commonly known. The term $j\beta_2(\omega_1 - \omega)(\omega_1 - \omega_2)$ accounts for the waveform distortion within a single span. Higher-order kernels, for example, the fifth-order kernel $H_5(\omega_1, \omega_2, \omega_3, \omega_4, \omega)$, can be used if higher accuracy is required. These nonlinear transfer functions indicate the nonlinear distortion effects on the linear transfer part; thus, they are the power penalty or distortion noise that degrades the channel capacity.

8.3.3 SPM AND INTRACHANNEL NONLINEAR EFFECTS

Under consideration of only the SPM of all the nonlinear effects on the optical signals transmitting through a dispersive transmission link, we can drop all the cross-coupling terms but $\Delta\Omega$ in (8.25); the nonlinear effect is thus contributed by additional intrachannel effects with ω_1, ω_2 taking the values with the spectra of the optical signal and not crossing over the spectra of other adjacent channels. With the substituting of the fundamental order transfer function (8.22) we arrive at

$$H_3(\omega_1,\omega_2,\omega)_{SPM,inter} = -\frac{j\gamma_L}{4\pi^2}(e^{-j\omega^2 N_s \beta_2 L_s/2})$$

$$(1-e^{-(\alpha+j\beta_2\Delta\Omega)L_s})(L_s\sqrt{\alpha^2+\beta_2^2\Delta\Omega^2}\,e^{j\tan^{-1}\frac{\alpha}{\beta_2\Delta\Omega}})\sum_{k=0}^{N_s-1}e^{-jk\beta_2 L_s\Delta\Omega}$$

(8.36)

The nonlinear distortion noises contributed to the signals when operating under the two regimes of large and negligible dispersion are given by Tang,[3] depending on the dispersion factor of the fiber spans.

The nonlinear transfer function H_3 indicates the power penalty due to nonlinear distortion and can be approximated as[25]

$$H_3^i(\omega_1,\omega_2,\omega) \approx \begin{cases} j\gamma_L e^{-[\alpha/2-j\beta_2(\omega_1-\omega_2)(\omega-\omega_2)]} \\ \times[L_s^{eff}-j\beta_2(\omega_1-\omega_2)(\omega-\omega_2)]\left(\dfrac{L_s-L_s^{eff}}{\alpha}-L_s L_s^{eff}\right) \end{cases}$$

(8.37)

Thus if the ASE noise of the in-line optical amplifier is weak compared with signal power, then we can obtain the nonlinear distortion noises for highly dispersive fiber spans (e.g., G.652 SSMF) as

$$K_{N,\beta}(\omega_0) = N_s\left[Q(\omega_0)+2\left(\frac{\gamma_L}{2\pi}\right)^2\frac{\Omega^2}{\alpha^2}\frac{P^3}{\Delta\omega_c^3}\partial\left(\frac{\omega_0}{\Delta\omega_c},\frac{\beta\Omega^2}{\alpha}\right)\right]$$

(8.38)

and for mildly dispersive fiber spans (e.g., G.655 LEAF fiber spans):

$$K_{N,\beta\ll20}(\omega_0) \approx N_s\left[Q(\omega_0)+2\left(\frac{\gamma_L}{2\pi}\right)^2\frac{\Omega^2}{\alpha^2}\frac{P^3}{\Delta\omega_c^3}\partial\left(\frac{\omega_0}{\Delta\omega_c},0\right)\right]$$

(8.39)

with

$$\partial_\chi(x,\xi) = \int_{-\infty}^{\infty}dx_1\int_{-\infty}^{\infty}dx_2 *\left(\frac{(1+e^{-\alpha L})-2e^{-\alpha L}\cos[\alpha L\xi(x-x_1)(x_1-x_2)]}{1+\xi^2(x-x_1)^2(x_1-x_2)^2}\right) \\ *\eta(x_1)\eta(x_2)\eta(x-x_1+x_2)$$

and

(8.40)

$$\eta(x) = \begin{cases} 1 & \text{for } x = [-1/2,1/2] \\ 0 & \text{elsewhere} \end{cases}$$

The nonlinear power penalty thus consists of the linear optical amplifier noises; the second is the SPM noises from the input signal and nonlinear interference between the input and optical amplifier noises, which may be ignored when the ASE is weak. Equations (8.38) and (8.39) show the variation of the penalty and, hence, channel capacity of dispersive fibers, operating under nonlinear effects and in multispans with optically amplified fiber spans of dispersion parameters from 0 to -20 ps^2/km. The nondispersive fiber has restricted the channel capacity to about 3–4 bps/Hz but 6–9 bps/Hz with 4 and 32 spans, respectively, for a dispersion parameter of -20 ps^2/km with 100 DWDM channels of 50 GHz spacing between the channels and an optical spectral noise density of 1 μW/GHz.[3] The length of each fiber span is 80 km.

By the definition of the nonlinear threshold determined at 1 dB degradation from the linear optical signal-to-noise ratio (OSNR), the contribution of the nonlinear noise term, from (8.38), we can obtain the maximal launched power at which there is an onset of the degradation of the channel capacity as

$$\max P = \sqrt{0.1 \frac{\omega_c^3}{2N_s \left(\dfrac{\gamma_L}{2\pi}\right)^2 \dfrac{\Omega^2}{\alpha^2}}} \qquad (8.41)$$

An example of the estimation of the maximum level of power per channel to be launched to the fiber before reaching the nonlinear threshold 1 dB penalty level follows: for an overall 100 channels of 150 GHz spacing $\Omega_T \approx 200$ nm, then $P_{th} \approx 58$ μW/GHz, or for 25 GHz bandwidth, we have the threshold power level at $P_{th\,\beta\,low} = 0.15$ mW per channel.

For highly dispersive and eight-wavelength channels we have $P_{th\,\beta\,low} \rightarrow 7$–10 $P_{th\,\beta\,low}$, which may reach 1.5 mW/channel threshold level. The estimations given here, as an example, are consistent with the analytical expression obtained in Equation (8.41). Thus, this shows clearly: (1) Dispersive multispan long-distance transmission under a coherent ideal receiver would lead to better channel capacity than a low-dispersive transmission line. (2) If a combination of low- and high-dispersive fiber spans is used, then we expect that from our analytical Volterra approach, the penalty would reach the same level of threshold power so that a 1 dB penalty on the OSNR is reached. Note that this approach relies on the average level of optical power of the lightwave-modulated sequence. This may not be easy to estimate if a simulation model is employed. (3) However, under simulation, the estimation of average power cannot be done without costing extremely high time; thus, the instantaneous power is commonly estimated at the sampled time interval of a symbol. This sampled amplitude and the instantaneous power can be deduced. Thence the nonlinear phase is estimated and superimposed on the sampled complex envelope for further propagation along the fiber length. This may create some differences between the analytical estimates and numerical simulated results, as we can observe from the published results given in Tang.[3] The sequence high–low-dispersive spans would offer slightly better performance than a low–high combination. This can be due to the fact that for low-dispersive fiber the output optical pulse would be higher in amplitude, which is to be launched into the high-dispersive fibers; thus this would suffer higher nonlinear effects because the instantaneous power launched into the fiber would be different— even the average power would be the same for both cases.

The argument in (3) can be further strengthened by representing a fiber span by the VSTFs, as shown in Figure 8.17. Any swapping of the sequence of low- and high-dispersion fiber spans would offer the same power penalty due to nonlinear phase distortion, except the accumulated noises contributed from the ASE noises of the in-line optical amplifiers of all spans. Thus we could see that the noise figure (NF) of both configurations can be approximated the same. This is in contrast to the simulated results reported in Pina et al.[37] We believe that the difference in the power penalty in different order of arrangement of low- and high-dispersion fiber spans reported in Papoulis[14] is due to a numerical error, as possibly the SSFM was employed and the instantaneous amplitude of the complex envelope was commonly used. This does not indicate the total average signal power of all channels. Therefore, we can conclude that the simulated nonlinear threshold power level would suffer an additional artificial OSNR penalty due to the instantaneous power of the sampled complex amplitude of the propagating amplitude.

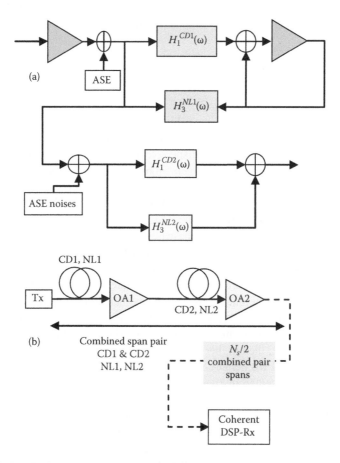

FIGURE 8.17 (a) System of concatenation of fiber spans consisting of a pair of different CDs and NL; (b) optically amplified N_s-span fiber link without DCF.

Tang[3] reported the variation of the channel capacity against the input power/channel with dispersion as a parameter −2 to −20 ps²/km with a noise power spectral density of 10 μW/GHz over 4 spans, and that for 4 and 32 spans of dispersive fibers of 0 and −20 ps2/km with a channel spacing of 50 GHz, 100 channels, noise spectral density of 10 μW/GHz. The deviation of the capacity is observed at the onset of the power per channel of 0.1, 2, and 5 mW.

Further observations can be made here. The noise responses indicate that the nonlinear frequency transfer function of a highly dispersive fiber link is related directly to the fundamental linear transfer function of the fiber link. When the transmission is highly dispersive, the linear transfer function acts as a low-pass filter, and thus, all the energy concentrates in the passband of this filter, which may be lower than that of the signal at the transmitting end. This may lower the nonlinear effects, as given in Equation (8.36). For lower-dispersive fiber this transfer function would represent a low-pass filter with 3 dB roll-off frequency, much higher than that of a dispersive fiber. For example, the G.655 would have a dispersion factor about three times lower than that of the G.652 fiber. This wideband low-pass filter will allow the nonlinear effects of intrachannels and interchannel interactions. The dispersive accumulation term

$$\sum_{k=0}^{N_s-1} e^{-jk\beta_2 L_s \Delta\Omega}$$

dominates when the number of spans is high.

Simulation results given in Binh,[39] using the Volterra series transfer function, consider this kind of arrangement of dispersive fiber spans. We expect from our analytical expression (8.36) that the arrangement of alternating positions between G.655 and G.652 would not create any penalty. The simulation reported in Binh et al.[25] indicates a 1.5 dB difference at 10×2 spans (SSMF + non-DCF) and no difference at 20×2 spans. The contribution by the ASE noises of the optical amplifiers at the end of each span would influence the phase noises, and hence, the effects on the error vector magnitude (EVM) of the sampled signal detected constellation.

From the transfer functions, including both linear and nonlinear kernels of the dispersive fibers, we could see that if the noises are the same, then the nonlinear effects would not be different regardless of whether high- or low-dispersive fiber spans were placed at the front or back. However, if the nonlinear noises are accounted for, and especially the intrachannel effects, we could see that if less dispersive fibers are placed in the front, then higher noises are expected, and thus a lower nonlinear threshold (at which a 1 dB penalty is reached on the OSNR). This is opposite to the simulation results presented in Tang.[6] However, these accumulated noises are much smaller than the average signal power. Under simulation, depending on the numerical approach used to solve the NLSE, the estimation of signal power at the sampled instant is normally obtained from the sampled amplitude at this instant, and thus, it is different from the average launched power into the fiber span. This creates discrepancies in the order of the high- or low-dispersion fiber spans as argued in Binh, Liu, and Li (2012).[25] Thus there are possibilities that the peak is of above-average amplitude of the very dispersive pulse sequence at some instants along the propagation path at which there is a superposition of several pulses. This amplitude may reach a level much higher than the nonlinear threshold and, thus, create a different distortion penalty due to nonlinear effects.

We have simulated the optical transmission systems over alternating cascading of high- and low-dispersion fiber spans with the number of spans in order of at least five consecutive of one type and then the other. The modulation format is non–return-to-zero quadrature phase shift keying (NRZ-QPSK), and the fibers are SSMF as high-dispersion type and TWC as low-dispersion type. The performance BER versus OSNR is obtained as depicted in Binh, Liu, and Li,[25] in which we can also observe the difference between the arrangements of the order of low- and high-dispersive spans in the transmission link:

a. For high launched power, regardless of whether high- or low-dispersion sequence, the BER versus OSNR performances are almost identical—see 1, 2, and 3 dBm launched power curves for 5 and 10 SSMF + 5 and 10 TWC, respectively.
b. For weaker nonlinearity with launched power of –2 and –3 dBm, we observe that there are some differences in the power penalty between the arrangements of the consecutive sequences of fiber spans of high and low dispersion.

We believe that this is due to the numerical modeling of the NLSE by SSFM as also observed. The power penalty due to nonlinear effects, whether by self-phase modulation or cross-phase modulation or intra- and interchannels or four-wave mixing (FWM), will degrade the channel capacity of an optical transmission system. PDM and QPSK modulation formats with pulse-shaping NRZ or RZ will likely be deployed transmission systems in the near future. Estimation of channel capacity under this environment system operation is critical.

This report has arrived at analytical expressions of the power penalty or departure from the ideal Shannon channel capacity. VSTFs are also derived for the relationship between the linear and nonlinear contributions to the power penalty. Thus conclusions on the structures of high- and low-dispersive fiber arrangements are made in order to obtain the least dispersive nonlinear effects on the OSNR.

It is expected from analytical analyses using the perturbation approach and Volterra series transfer functions up to the third order that the order of arrangement of the low- and high-dispersive fibers would not make any difference in the power OSNR penalty. However, under a numerical simulation model, this would be different due to the instantaneous power at the sampled instant. There are possibilities that superimposed amplitudes at different instants along the highly dispersive pulse sequence produce highly complex amplitudes, and high nonlinear phase distortion and, therefore, additional penalty on OSNR.

The Volterra transfer function offers better accuracy and covers a number of SPM and parametric scattering, but suffers costs of computing resources due to two-dimensional fast Fourier transform (FFT) for the SPM and XPM. This model should be employed when such extra nonlinear phase noises are required, such as in the case of superchannel transmission.

8.4 CONCLUDING REMARKS

As an introduction of the concept of transfer function of optical signals propagating over the guided medium, this chapter describes the behavior of the envelope of transmitted lightwaves when the magnitude is in the linear or nonlinear regions. Also, the chirping of the carriers in the near and far fields is also described. Analytical forms of the transfer functions of optical fibers can be derived and intuitively approximated so that they can be employed in the design of the optical transmission systems or used for the development of algorithms in the processing and compensation of nonlinear phase noises during the propagation of signal channels in nonlinear regions. The transfer functions are play important roles in modern-day ultra-high-capacity ultra-long-reach optical communication systems.

APPENDIX 8.1: PROGRAM LISTINGS OF SPLIT-STEP FOURIER METHOD (SSFM) WITH NONLINEAR SPM EFFECT AND RAMAN GAIN DISTRIBUTION

```
function output = ssprop_matlabfunction_raman(input)

nt = input(1);
u0 = input(2:nt+1);
dt = input(nt+2);
dz = input(nt+3);
nz = input(nt+4);
alpha_indB = input(nt+5);
betap = input(nt+6:nt+9);
gamma = input(nt+10);
P_non_thres = input(nt+11);
maxiter = input(nt+12);
tol = input(nt+13);
%Ld = input(nt+14);
%Aeff = input(nt+15);
%Leff = input(nt+16);

tic;
%tmp = cputime;

%- - - - - - - - - - - - - - - - - - - - - - - - - - - - - - - -
% This function ssmf solves the nonlinear Schrodinger equation
for
% pulse propagation in an optical fiber using the split-step
% Fourier method.
%
% The following effects are included in the model: group
velocity
% dispersion (GVD), higher order dispersion, loss, and self-
phase
% modulation (gamma). Raman gain is treated as a distributed
amplification.
```

```
%
% USAGE
%
% u1 = ssprop(u0,dt,dz,nz,alpha,betap,gamma);
% u1 = ssprop(u0,dt,dz,nz,alpha,betap,gamma,maxiter);
% u1 = ssprop(u0,dt,dz,nz,alpha,betap,gamma,maxiter,tol);
%
% INPUT
%
% u0 - starting field amplitude (vector)
% dt - time step - [in ps]
% dz - propagation stepsize - [in km]
% nz - number of steps to take, ie, ztotal = dz*nz
% alpha - power loss coefficient [in dB/km], need to convert
to linear to have P = P0*exp(-alpha*z)
% betap - dispersion polynomial coefs, [beta_0... beta_m] [in
ps^(m-1)/km]
% gamma - nonlinearity coefficient [in (km^-1.W^-1)]
% maxiter - max number of iterations (default = 4)
% tol - convergence tolerance (default = 1e-5)
%
% OUTPUT
%
% u1 - field at the output
%- - - - - - - -
% Convert alpha_indB to alpha in linear domain
%- - - - - - - -
alpha = 1e-3*log(10)*alpha_indB/10;       % alpha (1/km) - see
Agrawal p57
%- - - - - - - -
%P_non_thres = 0.0000005;

ntt = length(u0);
w = 2*pi*[(0:ntt/2-1),(-ntt/2:-1)]'/(dt*nt);
%t = ((1:nt)'-(nt+1)/2)*dt;
gain = numerical_gain_hybrid(dz,nz);

for array_counter = 2:nz+1
  grad_gain(1) = gain(1)/dz;
  grad_gain(array_counter) = (gain(array_counter)-gain(array_
  counter-1))/dz;
end
gain_lin = log(10)*grad_gain/(10*2);

clear halfstep
  halfstep = -alpha/2;
    for ii = 0:length(betap)-1;
      halfstep = halfstep - j*betap(ii+1)*(w.^ii)/
      factorial(ii);
    end
```

```
    square_mat = repmat(halfstep, 1, nz+1);
    square_mat2 = repmat(gain_lin, ntt, 1);
    size(square_mat);
    size(square_mat2);
    total = square_mat + square_mat2;
clear LinearOperator
    % Linear Operator in Split Step method
    LinearOperator = halfstep;
    halfstep = exp(total*dz/2);

u1 = u0;
ufft = fft(u0);

% Nonlinear operator will be added if the peak power is
greater than the
% Nonlinear threshold
iz = 0;
while (iz < nz) && (max((gamma*abs(u1).^2 + gamma*abs(u0).^2))
> P_non_thres)
  iz = iz+1;

  uhalf = ifft(halfstep(:,iz).*ufft);

  for ii = 1:maxiter,
    uv = uhalf.* exp((-j*(gamma)*abs(u1).^2 +
    (gamma)*abs(u0).^2)*dz/2);
    ufft = halfstep(:,iz).*fft(uv);
    uv = ifft(ufft);

    if (max(uv-u1)/max(u1) < tol)
      u1 = uv;
      break;
    else
      u1 = uv;
    end

  end
% fprintf('You are using SSFM\n');
  if (ii = = maxiter)

  fprintf('Failed to converge to%f in%d
  iterations',tol,maxiter);
end

  u0 = u1;
end

if (iz < nz) && (max((gamma*abs(u1).^2 + gamma*abs(u0).^2)) <
P_non_thres)
  % u1 = u1.*rectwin(ntt);
  ufft = fft(u1);
```

```
ufft = ufft.*exp(LinearOperator*(nz-iz)*dz);
u1 = ifft(ufft);

%fprintf('Implementing Linear Transfer Function of the Fiber
Propagation');
end

%toc;

output = u1;
```

APPENDIX 8.2: PROGRAM LISTINGS
OF AN INITIALIZATION FILE

```
% This file initialization file - declaring all parameters and
data required for
% Simulink model and Split Step Fourier - this file should be
incorporated in Simulink% model via the use of model
properties.
% This "initialization" program is to be modified to match
parameters employed for
% any specific optical transmission systems.

clear all
close all

% Constants
c = 299792458;               % speed of light (m/s)in vacuum
% NUMERICAL PARAMETERS

numbitspersymbol = 1
P0 = 0.003;                  % peak power (W)
FWHM = 25                    % pulse width FWHM (ps)
%halfwidth = FWHM/1.6651     % for Gaussian pulse
halfwidth = FWHM            % for square pulse

bitrate = 1/halfwidth;       % THz
baudrate = bitrate/numbitspersymbol;
signalbandwidth = baudrate;
%%%%%%%%%%%%%%%%%%%%%%%%%%%%%%%%%%%%%%%%
% biasing condition on optical modulator for Differential
Phase Sshift Keying
Vpi = 5;
halfVpi = Vpi/2;
twoVpi = Vpi*2;

% nt = 2^8;                  % number of points in FFT
PRBSlength = 2^5;

% Make sure : FFT time window (= nt*dt) = PRBSlength * FWHM...
```

```
% FFTlength nt = PRBSlength/block * numbersamples/bit =
PRBSlength * (FWHM/dt)
% num_samplesperbit = FWHM/dt should be about 8 - 16 samples/
bit
num_samplesperbit = 32;% should be 2^n
dt = FWHM/num_samplesperbit; % sampling time(ps);% time step
(ps)
nt = PRBSlength*num_samplesperbit; % FFT length

% nt = 2^9;
% nt = num_samplesperbit;

dz = 0.2;                       % distance stepsize (km)
nz = 500;

%melbourne to gippsland: transmission distance of the link as
an example
%170km two spans
nz_MelbToGipps = 500;

%undersea link - as a part of the overall link with Raman
amplification or
% scattering effects
% total undersea distance = 290km over which Raman pump is
employed; nz is the number% of distance in steps of
propagation
nz_Raman = 250;
nz_undersea = 950;
nz_DCF = 145;

%George Town to Hobart - ANOTHER transmission link span
EMPLOYING Raman amplification
nz_GtownToHobart = 500;

% number of z-steps
maxiter = 10;            % max number of iterations
tol = 1e-5;              % error tolerance for convergence
                         determination

% OPTICAL PARAMETERS

nonlinearthreshold = 0.010;% 10mW— % Nonlinear Threshold Peak
                                   Power in mW
lambda = 1550;                     % operating wavelength (nm)
                                   of channel under
                                   considerations
optical_carrier = c/(lambda*1e-9); % convert wavelength to
                                   frequency
alpha_indB = 0.17;                 % fiber loss (dB/km)
D = 18.5;% GVD (ps/nm.km); if anomalous dispersion(for
compensation),D is negative
```

```
beta3 = 0.06;                        % GVD slope (ps^3/km)
ng = 1.46;                           % group index
n2 = 2.6e-20;                        % nonlinear index (m^2/W)
Aeff = 76;                           % effective area (um^2)

% CALCULATED QUANTITIES

T = nt*dt;                           % FFT window size (ps)
-Agrawal: should be about 10-20 times of the pulse width
alpha_loss = log(10)*alpha_indB/10;% alpha (1/km)
beta2 = -1000*D*lambda^2/(2*pi*c); % beta2 (ps^2/km);

%- - - - - - - - - - - - - - - - - - - - - - - - - -
% beta 3 can be calculated from the Slope Dispersion (S) as
follows:]
% Slope Dispersion
% S = 0.092;                 % ps/(nm^2.km)
% beta31 = (S - (4*pi*c./lambda.^3))./(2*pi*c./lambda.^2)
%- - - - - - - - - - - - - - - - - - - - - - - - - -
gamma = 2e24*pi*n2/(lambda*Aeff); % nonlinearity coef (km^-1.
W^-1)
t = ((1:nt)'-(nt+1)/2)*dt; % vector of t values (ps)
t1 = [(-nt/2+1:0)]'*dt;    % vector of t values (ps)
t2 = [(1:nt/2)]'*dt;       % vector of t values (ps)

w = 2*pi*[(0:nt/2-1),(-nt/2:-1)]'/T; % vector of w values
(rad/ps)
v = 1000*[(0:nt/2-1),(-nt/2:-1)]'/T; % vector of v values
(GHz)
vs = fftshift(v);          % swap halves for plotting
v_tmp = 1000*[(-nt/2:nt/2-1)]'/T;

% STARTING FIELD

% P0 = 0.001              % peak power (W)
% FWHM = 20               % pulse width FWHM (ps)
%halfwidth = FWHM/1.6651     % for Gaussian pulse

%For square wave input, the FWHM = Half Width
%halfwidth = FWHM;

L = nz*dz

Lnl = 1/(P0*gamma)         % nonlinear length (km)
Ld = halfwidth^2/abs(beta2)    % dispersion length (km)
N = sqrt(abs(Ld./Lnl))     % governing the which one is
dominating: dispersion or Non-linearities
ratio_LandLd = L/Ld           % if L << Ld- > NO
Dispersion Effect
ratio_LandLnl = L/Lnl         % if L << Lnl- > NO
Nonlinear Effect
```

```
% Monitor the broadening of the pulse with relative the
Dispersion Length
% Calculate the expected pulsewidth of the output pulse
% Eq 3.2.10 in Agrawal "Nonlinear Fiber Optics" 2001 pp67
FWHM_new = FWHM*sqrt(1 + (L/Ld)^2)

% N<<1- > GVD ; N >>1- -> SPM
Leff = (1 - exp(-alpha_loss*L))/alpha_loss
expected_normPout = exp(-alpha_loss*2*L)
NlnPhaseshiftmax = gamma*P0*Leff

betap = [0 0 beta2 beta3]';

% Constants for ASE of EDFA
% PSD of ASE: N(at carrier freq) = 2*h*fc*nsp*(G-1) with nsp =
Noise
% Figure 2 (assume saturated gain)
%*************** Standdard Constant ************************
*******
h = 6.626068e-34; %Plank's Constant
%*******************************************
```

REFERENCES

1. G.P. Agrawal, *Fiber Optic Communication Systems*, Academic Press, New York, 2002.
2. A.F. Elrefaie, R.E. Wagner, D.A. Atlas, and D.G. Daut, "Chromatic Dispersion Limitations in Coherent Lightwave Transmission Systems," *IEEE J. Lightw. Technol.*, 6(6), 704–709, 1998.
3. J. Tang, "The Channel Capacity of a Multispan DWDM System Employing Dispersive Nonlinear Optical Fibers and an Ideal Coherent Optical Receiver," *IEEE J. Lightw. Technol.*, 20(7), 1095–1101, 2002.
4. B. Xu and M. Brandt-Pearce, "Comparison of FWM- and XPM-Induced Crosstalk Using the Volterra Series Transfer Function Method," *IEEE J. Lightw. Technol.*, 21(1), 40–54, 2003.
5. J. Tang, "The Shannon Channel Capacity of Dispersion-Free Nonlinear Optical Fiber Transmission," *IEEE J. Lightw. Technol.*, 19(8), 1104–1109, 2001.
6. J. Tang, "A Comparison Study of the Shannon Channel Capacity of Various Nonlinear Optical Fibers," *IEEE J. Lightw. Technol.*, 24(5), 2070–2075, 2006.
7. A. Papoulis, *Systems and Transforms with Applications in Optics*, Krieger Publ. Co., New York, 1981, item 1, Table 2-1, p. 14.
8. J.G. Proakis, *Digital Communications*, 4th ed., McGraw-Hill, New York, 2001, pp. 185–213.
9. Papoulis, 1981, item 5, Table 2-1, p. 14.
10. R.C. Srinivasan and J.C. Cartledge, "On Using Fiber Transfer Functions to Characterize Laser Chirp and Fiber Dispersion," *IEEE Photonics Technol. Lett.*, 7(11), 1327–1329, 1995.
11. L.N. Binh, *Digital Optical Communications*, Taylor & Francis, Boca Raton, FL, 2009.
12. L.N. Binh and N. Nguyen, "Generation of High-Order Multi-Bound-Solitons and Propagation in Optical Fibers," *Opt. Commun.*, 282, 2394–2406, 2009.
13. A.V.T. Cartaxo, B. Wedding, and W. Idler, "Influence of Fiber Nonlinearity on the Fiber Transfer Function: Theoretical and Experimental Analysis," *IEEE J. Lightw. Technol.*, 17(10), 1806–1813, 1999.

14. A. Papoulis, *Systems and Transforms with Applications in Optics*, McGraw Hill, New York, 1968, p. 14.

15. A.V.T. Cartaxo, B. Wedding, and W. Idler, "New Measurement Technique of Nonlinearity Coefficient of Optical Fibre Using Fibre Transfer Function," in *Proceedings of ECOC98*, Madrid, Spain, 1998, pp. 169–170.

16. A.V.T. Cartaxo, B. Wedding, and W. Idler, "Influence of Fiber Nonlinearity on the Phase Noise to Intensity Noise Conversion in Fiber Transmission: Theoretical and Experimental Analysis," *IEEE J. Lightw. Technol.*, 16(7), 1187, 1998.

17. F. Ramos and J. Martí, "Frequency Transfer Function of Dispersive and Nonlinear Single-Mode Optical Fibers in Microwave Optical Systems," *IEEE Photonics Technol. Lett.*, 12(5), 549, 2000.

18. G. Agrawal, *Nonlinear Fiber Optics*, 2nd ed., Academic Press, San Diego, 1995.

19. L.N. Binh, *Guided Wave Photonics*, Taylor & Francis, Boca Raton, FL, 2012.

20. F. Ramos and J. Martí, "Frequency Transfer Function of Dispersive and Nonlinear Single-Mode Optical Fibers in Microwave Optical Systems," *IEEE Photonics Technol. Lett.*, 12(5), 549, 2000.

21. T.M. Dunster, "Bessel Functions of Purely Imaginary Order, with an Application to Second-Order Linear Differential Equations Having a Large Parameter," *SIAM J. Math. Anal.*, 21(4), 995–1018, 1990.

22. K.V. Peddanarappagari and M. Brandt-Pearce, "Volterra Series Approach for Optimizing Fiber-Optic Communications System Designs," *IEEE J. Lightw. Technol.*, 16(11), 2046–2055, 1998.

23. M.B. Brilliant, *Theory of the Analysis of Nonlinear Systems*, Technical Report 345, MIT Research Laboratory of Electronics, March 3, 1958.

24. A. Samelis, D.R. Pehlke, and D. Pavlidis, "Volterra Series Based Nonlinear Simulation of HBTs Using Analytically Extracted Models," *Elect. Lett.*, 30(13), 1098–1100, 1994.

25. L.N. Binh, L. Liu, and L.C. Li, "Volterra Transfer Functions in Optical Transmission and Nonlinear Compensation," in *Nonlinear Optical Systems: Principles, Phenomena, and Advanced Signal Processing*, Ed. L.N. Binh and D.V. Liet, Taylor & Francis Group, Boca Raton, FL, 2012, Chapter 10.

26. G.P. Agrawal, *Nonlinear Fiber Optics*, 3rd ed., Academic Press, San Diego, 2001.

27. L.N. Binh, *Photonic Signal Processing*, Taylor & Francis, Boca Raton, FL, 2007.

28. D.M. Pepper, "Applications of Optical Phase Conjugation," *Scientific American*, January 1986.

29. P. Minzioni, V. Pusino, I. Cristiani et al., "Optical Phase Conjugation in Phase-Modulated Transmission Systems: Experimental Comparison of Different Nonlinearity-Compensation Methods, *Opt. Exp.*, 18(17), 18119, 2010.

30. X. Liu and D.A. Fishman, "A Fast and Reliable Algorithm for Electronic Pre-Equalization of SPM and Chromatic Dispersion," Presentation at Optical Fiber Communications Conference, OFC2007, 2007.

31. E. Ip and J.M. Kahn, "Compensation of Dispersion and Nonlinear Impairments Using Digital Back Propagation," *IEEE J. Lightw. Technol.*, 26(20), 3416, 2008.

32. Fujitsu, Inc., "Reality Check: Challenges of Mixed-Signal VLSI Design for High-Speed Optical Communications," Presentation at ECOC2011, Geneva, 2011.

33. A. Dochhan, R. Rath, C.C. Hebebrand, J.J. Leibrich, and W.W. Rosenkranz, "Evaluation of Digital Back-Propagation Performance Dependent on Stepsize and ADC Sampling Rate for Coherent NRZ- and RZ-DQPSK Experimental Data," Presentation at ECOC 2011, Geneva, 2011.

34. K. Kikuchi, M. Fukase, and S. Kim, "Electronic Post-Compensation for Nonlinear Phase Noise in a 1000-km 20-Gb/s Optical QPSK Transmission System Using the Homodyne Receiver with Digital Signal Processing," Presentation at Proceedings of Optical Fiber Communications Conference (OFC'07), Los Angeles, 2007, Paper OTuA2.

35. G. Charlet, N. Maaref, J. Renaudier, H. Mardoyan, P. Tran, and S. Bigo, "Transmission of 40 Gb/s QPSK with Coherent Detection over Ultralong Distance Improved by Nonlinearity Mitigation," Presentation at Proceedings of the European Conference on Optical Communication (ECOC 2006), Cannes, France, 2006, Paper Th4.3.4.
36. A.P.T. Lau and J.M. Kahn, "Signal Design and Detection in Presence of Nonlinear Phase Noise," *IEEE J. Lightw. Technol.*, 25(10), 3008–3016, 2007.
37. J. Pina, C. Xia, A.G. Strieger, and D.V.D. Borne, "Nonlinear Tolerance of Polarization-Multiplexed QPSK Transmission over Fiber Links," Presentation at ECOC2011, Geneva, 2011.
38. V. Peddanarappagari and M. Brandt-Pearce, "Volterra Series Transfer Function of Single-Mode Fibers," *J. Lightw. Technol.*, 15(12), 2232–2241, 1997.
39. L.N. Binh, "Linear and Nonlinear Transfer Functions of Single Mode Fiber for Optical Transmission Systems," *JOSA A*, 26(7), 1564–1575, 2009.

9 Fourier Guided Wave Optics

ABBREVIATIONS

Term	Full Name
AWG	Array waveguide grating
DWDM	Dense wavelength division multiplexing
PLC	Planar lightwave circuit
GW	Guided wave
DFT or FFT	Discrete/fast Fourier transform
DWT	Discrete wavelet transform
IFFT	Inverse FFT
oFFT	Optical FFT
oDWT	Optical DWT

9.1 INTRODUCTION

Photonic signal processing has emerged as one of the most essential techniques in modern optical communication systems beyond 100 Gb/s,[1–6] especially in the intense effort to increase the transmission capacity per wavelength lightwave channel to terabits/s. In order to increase the spectral efficiency defined as the number of bits/s that are accommodated in 1 Hz, advanced modulation formats are employed, such as differential quadrature phase shift keying (D-QPSK), quadrature amplitude modulation (QAM) in the amplitude, and phase plane or orthogonal frequency division multiplexing as multisubcarrier modulation.[7] If it is possible to generate orthogonal channels in the optical domain, then this would offer significant advantages to the preservation of the subchannels via the orthogonality of the channels after transmission over dispersive a fiber channel, and hence the processing of the received signals at the end of the transmission link.

Fourier transform offers the orthogonality needed for such an operation due to the summation of the harmonic terms of sine and cosine terms. Fourier optics offers an excellent technique for processing optical signals. The idea of transformation in optics was investigated some decades ago. The Fourier transform of a continuous and coherent spatial distribution or image can be evaluated physically to a high degree of accuracy by use of one or more simple lenses plus free-space light propagation, leading to the well-established technology of Fourier optics as described in texts by Goodman,[8] Papoulis,[9] and Gaskill.[10] This can be applied to a sequence of modulated lightwave signals carried by several optical subcarriers so that these channels can be positioned to be orthogonal to each other. This is one of the main features of the discrete Fourier transform (DFT).

In modern optical communications in the 21st century, digital and coherent detection techniques are employed for detection and processing of transmitted symbol sequences over several thousand single-mode optical fibers without using any dispersion-compensating fibers (DCFs). The optical signals are normally sampled, and so in the optical domain the signals are considered sampled complex values. That is, the phase of the embedded lightwaves is contained in the complex term of the amplitude. In this chapter we treat the optical signals in the discrete domain.

This chapter introduces the fundamental principles of Fourier optics and its implementation in guided wave structures using either fiber or integrated planar or channel waveguide forms. Both discrete Fourier and wavelet transforms are given as the transformations that generate orthogonal components in the spectral domain. The optical structures of these transforms can take either serial or parallel forms. From a fundamental 2×2 optical coupler as a 2-point discrete Fourier transformer, one can build up a DFT of Nth order. Thus compact optical DFT and optical inverse DFT (IDFT) devices can be designed. Alternatively, parallel waveguide paths can be employed with appropriate delay length corresponding to specific spectral resolution, the array waveguide gratings. Another method that can also be used is multimode interference (MMI) waveguides to obtain resolution in the spatial domain and, hence, distribution of the guided waves into different spatial and frequency resolutions. However, this structure is strongly dependent on the interference between the guided high-order mode and the fundamental mode to achieve the time resolution, hence the frequency spacing. This interference makes this MMI unstable. On the other hand, an optical wavelet packet transformer requires only precise, within reasonable tolerance, coupling coefficients in a number of splitting/coupling stages and no phase tuning; it thus offers significant flexibility in the selectivity of the filter passbands and combining channels in the inverse. Discrete wavelet transformation (DWT) offers some significant advantages in the reduction of accumulated and interference noises; some 10 dB improvement in the optical signal-to-noise ratio (OSNR) is expected for long-haul nondispersion compensation transmission systems.

The applications of optical DFT and IDFT as well as DWT are presented in advanced optical transmission systems operating at speeds on the order of terabits/s. It is noted that the difference between these two technologies is the path length, which can be very short in guided wave structures and only in moderate terms for fiber structures. If the operating speed is tens of Gb/s, then integrated optical structures must be used, the guided wave Fourier optics.

This chapter treats DFT, specifically fast Fourier transform (FFT), of discrete signals in an optical transmission system or network, thus leading to the prospective concept of fiber Fourier optics. In this approach a discrete set of coherently related optical input amplitudes a_n are fed into a lossless fiber or integrated optic device through a corresponding set of single-mode optical waveguides or single-polarization input fibers, and the DFT b_n of this sequence is taken out through N similar output optical waveguides. This device can be termed the Fourier optical circuit (FOC).[11] Modern technology on planar lightwave circuits (PLCs) using silica on silicon[12] offers such an implementation for an integrated Fourier optical transformer.

Similarly, another orthogonal transformation, DWT, is also analyzed and implemented using guided wave components.

The chapter is organized as follows. Section 9.2 gives an introduction to discrete Fourier transform. Section 9.3 then introduces the guided wave technology that offers the practical implementation of DFT devices, including channel and planar optical waveguides, array waveguide gratings (AWGs), and cascaded 2-point DFT devices to form Nth-order DFT components. These components can also be implemented in fiber structures. Alternatively, the MMI device structures can also be designed to provide the functionalities of a DFT device. Some fundamental designs and analyses of the planar waveguides and channel waveguides are given in Section 9.3.3—in particular, the dispersion characteristics of the waveguides, so that one can select the number of modes to be supported, and then realization with geometrical and index values. We then describe the relationship between mathematical representations of DFT, DWT, and optical realization in both serial and parallel forms using guided wave optical components in a discrete integrated form. The design features of these transformers and their implementations employing guided wave devices are given in Sections 9.2.4.1 and 9.3.3, respectively. Applications of these DFTs in terabits/s optical transmission are given with focus on the optical signal processing, although some basic optical transmission system concepts employing the DFTs are also given for the sake of completeness. We avoid using any jargon terms related to transmission performance, but put emphasis on terms of guide wave phenomena.

Optical processing is always conducted in the analog domain. However, the discrete Fourier and wavelet transformation terms are used due to the equivalent delay path length and the sampling time. Thus, the delay transform z can be used to represent and simplify the mathematical expressions.

The term *discrete* is used in this chapter to imply duality of the sampling in the optical domain and the sampling interval as commonly defined in the digital domain, that is, a delay of the optical carrier by an amount that is equivalent to the delay time of a symbol of optical modulated signals[13] or the time interval between samples. For example, for a 25 GSym/s modulated signal the symbol period is 40 ps. If such a digital symbol is sampled at 50 GSamples/s, then the sample period is 20 ps. So a unit delay time is 20 ps and a delay length of an optical fiber is $L \approx 3e^8/1.5 \times 20e^{-12} = 20$ μm, assuming that the effective refractive index of the guided mode is close to 1.5. Indeed, this index depends on the order of the guided mode and the composition of the core and cladding regions, as well as the materials used in the fabrication of the guided wave structure.[14] With the very short length for high-frequency operation, it is only possible to use integrated optic technology, and hence the guided wave optics term used for this chapter.

9.2 BACKGROUND: FOURIER TRANSFORMATION

9.2.1 Basic Transform

In the standard approach to the discrete Fourier transform, consider an N-term complex-valued input discrete sequence a_n as a complex sequence of the continuous signal $a(t)$:

$$a_n(t) = \sum_{n=-\infty}^{+\infty} a(t)\delta(t-nT) \tag{9.1}$$

Then the DFT A_m of this discrete sequence is given by

$$A_k = \sqrt{\frac{1}{N}} \sum_{n=0}^{N-1} a_n e^{-jkn\frac{2\pi}{N}} \quad k = 0,1,\ldots N-1 \tag{9.2}$$

And the IDFT is given as

$$a_n = \sqrt{\frac{1}{N}} \sum_{k=0}^{N-1} A_k e^{+jkn\frac{2\pi}{N}} \quad n = 0,1\ldots N-1 \tag{9.3}$$

The summation in (9.2) can be split into even $2k$ and odd $2k+1$ terms, leading to the sequence coefficients of the even and odd DFT terms for the upper and lower halves of the output components, given by

for $0 \le q \le \dfrac{N}{2}-1$ upper-half number of coefficients

$$A_{m=q} = \sqrt{\frac{1}{N}}\left(\sum_{k=0}^{\frac{N}{2}-1} a_{2k} e^{-jqk\frac{\pi}{N}} + e^{-jq\frac{2\pi}{N}}\sum_{k=0}^{\frac{N}{2}-1} a_{2k+1} e^{-jqk\frac{\pi}{N}}\right)$$
$$\tag{9.4}$$

and for $0 \le q \le \dfrac{N}{2}-1$ lower-half number of coefficients

$$A_{m=q+\frac{N}{2}} = \sqrt{\frac{1}{N}}\left(\sum_{k=0}^{\frac{N}{2}-1} a_{2k} e^{-jqk\frac{\pi}{N}} - e^{-jq\frac{2\pi}{N}}\sum_{k=0}^{\frac{N}{2}-1} a_{2k+1} e^{-jqk\frac{2\pi}{N}}\right)$$

The index q is scanning across all values. The summations on the right-hand sides of (9.4) can be recognized as simply $(N/2)$th-order DFTs on the even and odd components of the Nth-order array, with some additional phase shifts of value $\exp(-jq(2\pi/N))$ added to half of the transformed elements after the transformations. Thus depending on the order N of the transform when q and N equate to a ratio matching 2π would lead to a no phase shifting at all.

9.2.2   OPTICAL CIRCUITRY IMPLEMENTATION

9.2.2.1   2 × 2 Asymmetric Coupler

The optical blocks of the coupler and the phase shifter can be realized without much difficulty in integrated lightwave circuits such as silica on silicon, InP, and even in

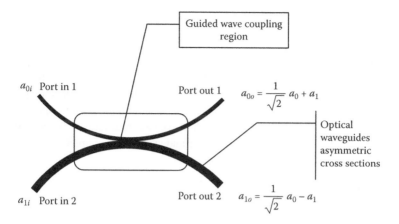

FIGURE 9.1 A 3 dB 2 × 2 asymmetric guided wave coupler (AGWC) represented as a second-order Fourier transformer.

LiNbO$_3$-integrated photonic circuitry as given in Figure 9.1. Phase modulation or shifting can be implemented by using the electro-optic effect via an applied voltage on the lumped (for low frequency) or traveling wave type (for high frequency). The interpretation of the mathematical representation of the DFT is described as follows, and corresponding to the assignments of the order of discrete input samples as shown in Figure 9.3.

The summations on the right-hand sides of (9.4) can be recognized as simply (N/2)th-order DFTs on the even and odd components of the Nth-order array, with some additional phase shifts of value $e^{-jqk(\pi/N)}$ added to half of the transformed elements after the transformations. We note that a phase shift in the frequency domain is equivalent to a delay in the time domain. We will see later that this delay is indeed an optical path difference in the implementation of an integrated optical structure. Furthermore, (9.4) illustrates the general principle that one can evaluate an Nth-order DFT by structuring the two (N/2)th-order transforms and combining the results with appropriate phase shifts. This provides the foundation for the FFT algorithm,[15,16] which is universally employed for the numerical evaluation of DFTs. If this same procedure is applied again to the (N/2)th-order DFTs, they can each be separated into two (N/4)th-order transforms. If the original order N is a power of 2 so that N = 2^M, applying this procedure (M − 1) times reduces the initial Nth-order DFT to M/2 second-order DFTs. A second-order DFT is thus simply given as

$$b_0 = \frac{1}{\sqrt{2}}(a_0 + a_1)$$

$$b_1 = \frac{1}{\sqrt{2}}(a_0 - a_1)$$

(9.5)

This transformation requires addition and subtraction, and no multiplication is involved. In addition, a scaling factor of $1/\sqrt{2}$ is required. This factor is indeed the coupling factor in the optical field term of a 3 dB optical coupler or splitter. The other splitting port would then involve a complex term of $-j$. So a 2×2 Fourier transform is simply a 3 dB coupler (or 50:50),[11] as shown in Figure 9.1. It is noted that such a coupler can be implemented in bulk, integrated, or fiber optics. The principal motivation for implementation-integrated optics is the minimization of the insertion loss and alignment difficulty. This is essential when the order of DCT is increased much higher. The inputs a_{0i} and a_{1i} represent the field amplitude of the input signals injected into ports 0 and 1 of the coupler. The transfer matrices of the optical fields of a 50:50 2×2 lossless asymmetric coupler and symmetric coupler are given by

$$T_s = \begin{pmatrix} s_{11} & s_{12} \\ s_{21} & s_{22} \end{pmatrix} = \frac{1}{\sqrt{2}} \begin{pmatrix} 1 & j \\ j & 1 \end{pmatrix}; \text{ symmetric coupler}$$

$$(9.6)$$

$$T_{as} = \begin{pmatrix} s_{11} & s_{12} \\ s_{21} & s_{22} \end{pmatrix} = \frac{1}{\sqrt{2}} \begin{pmatrix} 1 & 1 \\ 1 & -1 \end{pmatrix}; \text{ asymmetric coupler}$$

In general the transfer matrix or the transmittance matrix involving the optical fields between the input and output ports of the *asymmetric coupler* is given by

$$[T]_{as} = \frac{1}{\sqrt{k}} \begin{pmatrix} 1 & 1 \\ 1 & -1 \end{pmatrix} \tag{9.7}$$

where k is the intensity coupling coefficient. The optical fields at the output ports of a 3 dB asymmetric coupler are given by

$$\begin{pmatrix} a_{0i} \\ a_{1i} \end{pmatrix} = \begin{pmatrix} 1 & 1 \\ 1 & -1 \end{pmatrix} \begin{pmatrix} a_{0o} \\ a_{1o} \end{pmatrix} \rightarrow a_{0o} = \frac{1}{\sqrt{2}}(a_{0i} + a_{1i}); \ a_{1o} = \frac{1}{\sqrt{2}}(a_0 - a_1) \quad (9.8)$$

An asymmetric guided wave coupler can be considered to have formed by two nonidentical core optical waveguides being placed side by side. The tunneling of the guided waves from one to the other is not symmetrical, and hence the phase shifts are imposed on one optical path with respect to the other.

9.2.2.2 Exemplar Models

We describe here two exemplar models for 4×4 and 8×8 Fourier transformers and their implementation in the optical domain employing asymmetric optical couplers and phase shifters.

9.2.2.2.1 4 × 4 Transformation Optical Circuitry

For a 4×4 transform we can obtain the coefficients of the stages of the couplers and phase shifters as follows. Referring to Equation (9.4), we have

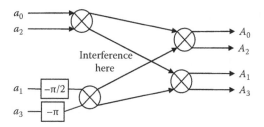

FIGURE 9.2 Optical circuit for a 4×4 optical Fourier transformer.

for $0 \leq q \leq 1 \ldots A_0 A_1$

$$A_{m=q} = \sqrt{\frac{1}{4}} \left(\sum_{k=0}^{1} a_{2k} e^{-jqk\frac{\pi}{4}} + e^{-jq\frac{2\pi}{4}} \sum_{k=0}^{1} a_{2k+1} e^{-jqk\frac{\pi}{4}} \right)$$

$$\rightarrow A_0 = \frac{1}{2}[(a_0 + a_2) + (a_1 + a_3)]; \quad A_1 = \frac{1}{2}[(a_0 + a_2) + e^{-j\frac{\pi}{2}}(a_1 + a_3 e^{-j\frac{\pi}{4}})]$$

(9.9)

and for $0 \leq q \leq 1 \ldots A_2 A_3$

$$A_{m=q+2} = \sqrt{\frac{1}{4}} \left(\sum_{k=0}^{1} a_{2k} e^{-jqk\frac{\pi}{N}} - e^{-jq\frac{2\pi}{4}} \sum_{k=0}^{1} a_{2k+1} e^{-jqk\frac{2\pi}{4}} \right)$$

$$\rightarrow A_2 = \frac{1}{2}[(a_0 + a_2) - (a_1 + a_3)]; \quad A_3 = \frac{1}{2}[(a_0 + a_2) - e^{-j\frac{\pi}{2}}(a_1 + a_3 e^{-j\frac{2\pi}{4}})]$$

Hence we have the transfer matrix and the inputs and outputs as

$$\begin{bmatrix} A_0 \\ A_1 \\ A_2 \\ A_3 \end{bmatrix} = \frac{1}{2} \begin{bmatrix} 1 & 1 & 1 & 1 \\ 1 & e^{-j\frac{\pi}{2}} & 1 & e^{-j\frac{\pi}{2}} e^{-j\frac{\pi}{4}} \\ 1 & 1 & -1 & -1 \\ 1 & -e^{-j\frac{\pi}{2}} & 1 & -e^{-j\frac{\pi}{2}} e^{-j\frac{2\pi}{4}} \end{bmatrix} \begin{bmatrix} a_0 \\ a_1 \\ a_2 \\ a_3 \end{bmatrix}$$

(9.10)

The optical circuit representing the 4×4 transform is given in Figure 9.2 directly interpreted from (9.9) and (9.10). Note that an alternative structure can be formed with the position of the phase shifter.

9.2.2.2.2 8×8 Transformation Optical Circuitry

Equation (9.4) has been written in the composition of the odd and even parts. Hence, the DFT is formed by the order of even and odd of the input ports and output ports. Equation (9.4) can be realized by using an optical splitter and phase shifting circuit of an 8×8 transfer transmittance matrix, as shown in Figure 9.3; the matrix coefficients are given by

for $q = 0,1,2,3,\ldots \rightarrow A_0, A_1, A_2, A_3$

$$A_{m=q} = \sqrt{\frac{1}{8}}\left(\sum_{k=0}^{3} a_{2k}e^{-jqk\frac{\pi}{8}} + e^{-jq\frac{2\pi}{8}}\sum_{k=0}^{3}a_{2k+1}e^{-jqk\frac{\pi}{8}}\right)$$

(9.11)

and for $q = 0,1,2,3. \rightarrow A_4, A_5, A_6, A_7$

$$A_{m=q+4} = \sqrt{\frac{1}{8}}\left(\sum_{k=0}^{3} a_{2k}e^{-jqk\frac{\pi}{8}} - e^{-jq\frac{2\pi}{8}}\sum_{k=0}^{3}a_{2k+1}e^{-jqk\frac{2\pi}{8}}\right)$$

Similar to the design of the 4×4 optical Fourier transformer, the transfer matrix can be obtained as given in (9.12).

$$\begin{bmatrix} A_0 \\ A_1 \\ A_2 \\ A_3 \\ A_4 \\ A_5 \\ A_6 \\ A_7 \end{bmatrix} = \frac{1}{2^{3/2}} \begin{bmatrix} 1 & 1 & 1 & 1 & 1 & 1 & 1 & 1 \\ 1 & 1e^{-j\frac{2\pi}{8}} & 1e^{-j\frac{\pi}{8}} & 1e^{-j\frac{2\pi}{8}}e^{-j\frac{\pi}{8}} & 1e^{-j\frac{2\pi}{8}} & 1e^{-j\frac{2\pi}{8}}e^{-j\frac{2\pi}{8}} & 1e^{-j\frac{3\pi}{8}} & 1e^{-j\frac{2\pi}{8}}e^{-j\frac{3\pi}{8}} \\ 1 & 1e^{-j\frac{2\times2\pi}{8}} & 1e^{-j\frac{2\pi}{8}} & 1e^{-j\frac{2\times2\pi}{8}}e^{-j\frac{\pi}{4}} & 1e^{-j\frac{4\pi}{8}} & 1e^{-j\frac{2\times2\pi}{8}}e^{-j\frac{2\pi}{4}} & 1e^{-j\frac{6\pi}{8}} & 1e^{-j\frac{2\times2\pi}{8}}e^{-j\frac{3\pi}{8}} \\ 1 & 1e^{-j\frac{3\times2\pi}{8}} & 1e^{-j\frac{3\pi}{8}} & 1e^{-j\frac{3\times2\pi}{8}}e^{-j\frac{3\pi}{8}} & 1e^{-j\frac{3\times2\pi}{8}} & 1e^{-j\frac{3\times2\pi}{8}}e^{-j\frac{6\pi}{8}} & 1e^{-j\frac{3\times3\pi}{8}} & 1e^{-j\frac{3\times2\pi}{8}}e^{-j\frac{3\times3\pi}{8}} \\ 1 & -1 & 1 & -1 & 1 & -1 & 1 & -1 \\ 1 & -1e^{-j\frac{2\pi}{8}} & 1e^{-j\frac{1\pi}{8}} & -1e^{-j\frac{2\pi}{8}}e^{-j\frac{2\pi}{8}} & 1e^{-j\frac{2\pi}{8}} & -1e^{-j\frac{2\pi}{8}}e^{-j\frac{2.2\pi}{8}} & 1e^{-j\frac{3\pi}{8}} & -1e^{-j\frac{2\pi}{8}}e^{-j\frac{3.2\pi}{8}} \\ 1 & -1e^{-j\frac{2\times2\pi}{8}} & 1e^{-j\frac{2\pi}{8}} & -1e^{-j\frac{2\times2\pi}{8}}e^{-j\frac{2\times2\pi}{8}} & 1e^{-j\frac{2\pi}{8}} & -1e^{-j\frac{2\times2\pi}{8}}e^{-j\frac{2\times2\pi}{8}} & 1e^{-j\frac{3\times2\pi}{8}} & -1e^{-j\frac{2\times2\pi}{8}}e^{-j\frac{3\times2\times2\pi}{8}} \\ 1 & -1e^{-j\frac{3\times2\pi}{8}} & 1e^{-j\frac{3\pi}{8}} & -1e^{-j\frac{3\times2\pi}{8}}e^{-j\frac{3\times2\pi}{8}} & 1e^{-j\frac{3\times2\pi}{8}} & -1e^{-j\frac{3\times2\pi}{8}}e^{-j\frac{3\times2\times2\pi}{8}} & 1e^{-j\frac{3\times3\pi}{8}} & -1e^{-j\frac{3\times2\pi}{8}}e^{-j\frac{3\times3\times2\pi}{8}} \end{bmatrix} \begin{bmatrix} a_0 \\ a_1 \\ a_2 \\ a_3 \\ a_4 \\ a_5 \\ a_6 \\ a_7 \end{bmatrix}$$

(9.12)

From this matrix we can obtain the transform optical circuit as shown in Figure 9.3. Note that a minus sign can be replaced by a phase shift of π-rads and j is a phase shift of $\pi/2$. One can trace the summation via the propagation of the optical fields of the input ports through the asymmetric couplers.

Equation (9.12) shows the symmetry of the coupling with additional phase shifting at some specific positions due to the phase components of the second half of the transform equation, especially the minus sign, which is equivalent to a π phase shift. The matrix (9.12) can be rearranged so that the order of the elements would be a_0, $a_4\ldots a_2$, $a_6\ldots a_1$, $a_5\ldots a_3$, a_7. This arrangement will lead to the coupling into the summation and subtraction resultant outputs, and then phase shifting to the final time-domain outputs shown in Figure 9.3.

It is noted that such a 2×2 coupler can also be used as a 1×2 coupler with one of the input ports left unused.

Note the similar phase terms in the pairs of the rows of the matrix. By examining the row of the matrix given in (9.12), we notice the following:

- Row $(a_0 + a_4)$
- Row $(a_1 - a_5)e^{-j2\pi/8}$
- Row $(a_2 - a_6)e^{-j4\pi/8}$
- Row $(a_3 - a_7)e^{-j6\pi/8}$

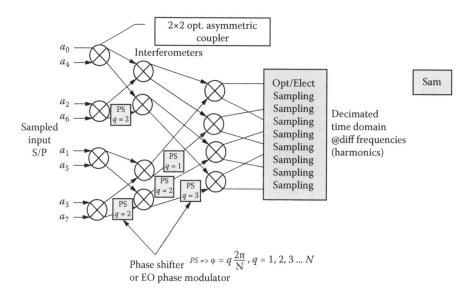

FIGURE 9.3 Signal flow of optical guided waves and operation for implementation of Nth-order optical DFT formed by couplers and phase shifters (PS; $N = 8$ for this diagram). (Note: For numbering of the order of the input signal and the output ports, the first half of input ports are even, and the second half odd. Scaling coefficients for amplitude are not shown. Samplers are employed for sampling the output waveform in a time domain.)

These rows are for the propagation of the transformation via the couplers and phase shifting; the minus signs are extracted in output port 2 of the asymmetric coupler. These are displayed in the optical structure depicted in Figure 9.3.

The appendix at the end of this chapter tabulates the specifications and illustrates the optical transmittance characteristics of a commercial array waveguide grating (AWG); the spectral characteristics indicate the principles of the FFT of such AWG in which the spectra of individual channels are overlapped and orthogonal to adjacent channels, that is, going to zero at the maximum of the others.

9.2.3 OPTICAL DFT BY MACH–ZEHNDER DELAY INTERFEROMETERS (MZDIs)

The DFT or FFT equation (9.4) can be rewritten in the form that would be implemented by a set of delay interferometers. For a continuous input signal $x(t)$, the output $X_m(t)$ can be expressed as

$$X_m(t) = \frac{1}{N} \sum_{n=0}^{N-1} e^{-jn\frac{m}{N}2\pi} \delta\left(t - n\frac{T}{N}\right) * x(t) \tag{9.13}$$

where δ is the impulse function. Now taking the Fourier transform of (9.13), we have

$$\tilde{X}_m(\omega) = \frac{1}{N} \sum_{n=0}^{N-1} e^{-jn\frac{T}{N}\omega} e^{-jn\frac{m}{N}2\pi} \tilde{x}(\omega) \tag{9.14}$$

$\tilde{X}_m(\omega)$, $\tilde{x}(\omega)$ are the Fourier transforms on the output and input signals in the spectral domain. Referring to Figure 9.3, the outputs of an 8×8 optical DFT can be formed with the input sampled signals $a_0 \ldots a_7$ coupled to a three-stage (2^3) coupling system consisting of asymmetric couplers and phase shifters. As seen from (9.4) the couplers are arranged in such a way that the even and odd sampled inputs are coupled and phase shifted according to the phase shift coefficients given in this DFT relationship.

Alternatively, (9.14) can be manipulated to give

$$H_m(\omega) = \frac{2}{N} \sum_{n=0}^{\frac{N}{2}-1} e^{-j\frac{2n}{N}(m2\pi + \omega T)} \frac{1}{2} \left(1 + e^{-j\left(\omega\frac{T}{N} + \frac{2\pi m}{M}\right)} \right) \tag{9.15}$$

By inspecting Equation (9.15) we can see that the first part represents the DFT of order $N/2$ and the term in the bracket represents the transfer function in the frequency domain of a delay interferometer of a delay τ and a phase shift φ_m and 3 dB 2 $\times 2$ couplers at the input and output ports. The delay time and phase shift are given as

$$\tau = \frac{T}{N} = \frac{T}{2^p}; p = \text{order of FFT points} \tag{9.16}$$

$$\varphi_m = \frac{2\pi m}{N} - \pi$$

So we can see that using an FFT of order $N/2$ can be obtained by cascading a number of MZDIs. This will be described in the next section in the implementation of the DFT or the wavelet packet transformer in the photonic domain.

9.2.4 FOURIER TRANSFORM SIGNAL FLOW AND OPTICAL IMPLEMENTATION

Based on (9.2) and (9.3), we can arrive at the signal flow graph shown in Figure 9.3. Note that the 2×2 asymmetric coupler is assigned as a cross-coupling symbol instead of the non-cross flow described in NTT Electronic Labs.[12] Only couplers and phase shifters are required where the phase shift amount depends on the order of the FFT. An active switching device can be used instead of the 3 dB coupler so that one can change the order of FFT as required.

9.2.4.1 Practical Integrated Guided Wave Structure

As we can observe, Equation (9.4) is composed of the followings parts, and hence operations, as outlined in Figure 9.4. The mathematical operations in guided wave systems include the following:

$$A_{k=q} = \sqrt{\frac{1}{N}} \left[\sum_{k=0}^{N/2-1} a_{2k} e^{-jqk\frac{\pi}{N}} + e^{-jq\frac{2\pi}{N}} \sum_{k=0}^{N/2-1} a_{2k+1} e^{-jqk\frac{\pi}{N}} \right] for\_0 \le q \le \frac{1}{2} - 1 .. upper\_half$$

and

$$A_{k=q+\frac{N}{2}} = \sqrt{\frac{1}{N}} \left[\sum_{k=0}^{N/2-1} a_{2k} e^{-jqk\frac{\pi}{N}} - e^{-jq\frac{2\pi}{N}} \sum_{k=0}^{N/2-1} a_{2k+1} e^{-jqk\frac{\pi}{N}} \right] for\_0 \le q \le \frac{1}{2} - 1 .. lower\_half$$

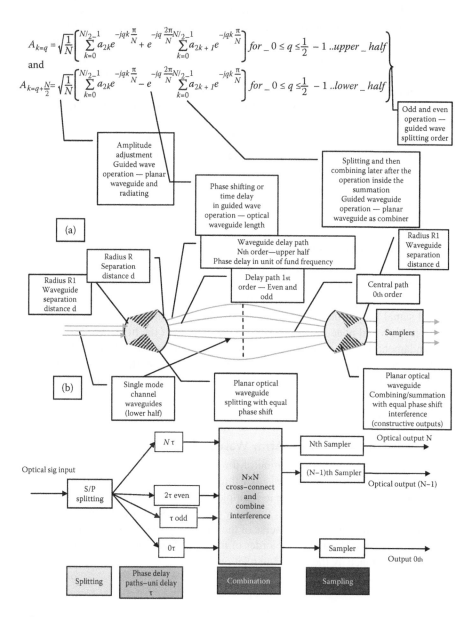

FIGURE 9.4 (a) DFT mathematical operation representation and (b) equivalent operations by guided wave components in parallel form.

- Incoming optical waves (possibly modulated optical waves) are guided into a wave splitting region, normally a planar waveguide with receiving waveguides positioned where they can receive the maximum power distribution.
- Forming asymmetric coupling systems and appropriate phase shifters for the odd and even sampled discrete components in the guided wave structure.
- The number of stages depends on the total number of Fourier transform order, N, and appropriate number of couplers and phase shifters. The coupling coefficients and asymmetry of the coupling regions must be precise in order to obtain the low loss and phase shifting according to the coefficients of the DFT given by (9.4).
- Cascade of asymmetric couplers or MZDI stages can be implemented and offer the same DFT structures. However, the MZDI can be used to offer a more simplified structure.

Design details of the guided wave optical transformers are briefly given in Section 9.6. More details are also described in Chapter 5.

9.2.4.2   Cascade—Series Formation

In optical processing the frequency downsampling components of the Fourier transform are realized by using a delay path whose propagation time is equivalent to the inverse of the frequency component. We can see from Figure 9.5 that an original optical signal is split into two equal-intensity field outputs that are then delayed by Δt in one path and none in the other. Thus the outputs of the first stage MZDI are in the fundamental order of the DFT, and then at the output ports of the 3 dB couplers the frequency is in the range of the fourth order. The higher the order, the shorter is the delay path. Figure 9.6 depicts a possible structure that can be implemented by integrated optical waveguides using both planar and channel waveguides described in Chapters 3 and 4.

9.2.4.3   Parallel Formation—Array Waveguide Grating

An example of the parallel formation of the FFT representing Equation (9.4) is the array waveguide grating that consists of three stages. The first stage is the split by radiation from a single waveguide to multiple output waveguides via a planar waveguide section. The middle stage is the delay paths in optical waveguides with a difference in the time equal to the inverse of the frequency spacing between the lines of the highest order as desired or the order of the FFT. The time delay is equivalent to the phase difference by the following relationship:

$$z^{-1} = e^{j\beta\Delta L} \equiv e^{j\omega\Delta t} \tag{9.17}$$

where β is the effective propagation constant of the channel waveguide, ΔL is the path difference between the two consecutive paths, ω is the angular frequency of the lightwaves, and z is the symbol of the commonly known z-transform employed in digital signal processing.[17] Once we know the order of the DFT, we can estimate the frequency of the highest-order spectral components, then the delay time with a

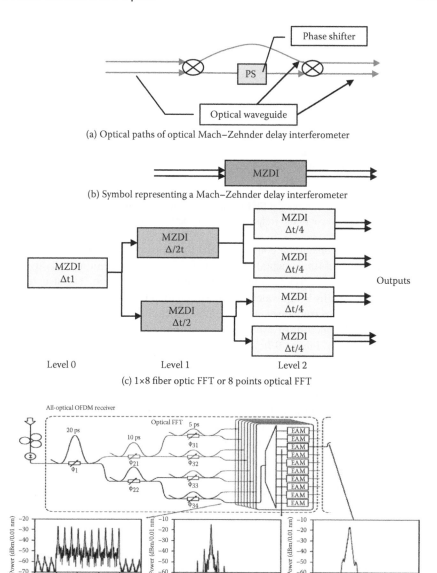

FIGURE 9.5 Operations by guided wave components using fiber optics. (a–b) Guided wave optical path of a Mach–Zehnder delay interferometer (MZDI) or asymmetric interferometer with phase delay tunable by thermal or electro-optic effects. (c) Block diagram representation. (d) Implementation of optical FFT using cascade stages of fiber optical MZDI structure. EAM = electroabsorption modulator used for demultiplexing in time domain. Note also phase shifters employed in MZDIs between stages. Inserts are spectra of optical signals at different stages, as indicated by the optical FFT (serial type). (Extracted from Hillerkuss, R. et al., *Nature Photonics*, 5, 364, 2011.)

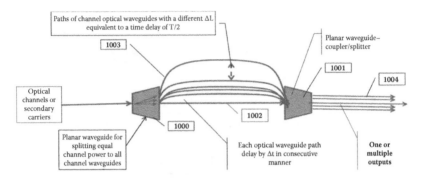

FIGURE 9.6 Schematic of optical FFT using array waveguide grating; the delay difference is equivalent to the inverse of the frequency spacing between individual spectral lines.

known fabricated propagation constant of the lightwave, that is, the effective index of the guided wave of fundamental order. The final stage is combining and directing lightwave channels to individual outputs of each frequency region. The structure of the parallel delay paths is equivalent to the summation of a number of MZDI pairs; each is equivalent to a frequency component—thence the DFT operation.

A geometrical structure of a multiple input multiple output AWG is shown in Figure 9.7(a) and (b). Figure 9.7(c) illustrates the radiation of the lightwaves from the input to the optical parallel waveguide paths, and then the inteference of the outputs to other outputs of the device via another planar waveguide section. It is the interference that would ensure the guidance of a particular frequency component of the original signals. On the other hand, if a number of lightwaves of different frequencies are launhched from the outputs back to the input side of the AWG, we have a multiplexing of the time-domain signals and hence the inverse Fourier transform operation. The spectra of a commercial AWG with 50 and 100 GHz spacing are shown in Figure 9.8. Note that the spectrum of the lines overlaps at about −11 dB, meaning that the passband of each filter rolls off and crosses over at this level. The operating principles of the AWG are given as follows. An AWG wavelength multi/demultiplexer combines and splits optical signals of different wavelengths for use in WDM systems. The heart of the device, the AWG, consists of a number of arrayed channel waveguides that act together like a diffraction grating in a spectrometer. The grating offers high-wavelength resolution, thus attaining narrow-wavelength channel spacing such as 0.8 nm in International Telecommunication Union (ITU) channel allocation. Moreover, the multiplexer's extreme stability eliminates the negative effects caused by mechanical vibration; in addition, it delivers long-term reliability because it is composed of silica-based planar lightwave circuits.

A number of AWGs based on PLC have been fabricated and made available by NTT NEL Co. Ltd. With 25, 50, and 100 GHz channel spacing the total number of input and output ports can reach 256. The spectra of both transverse electric (TE) and transverse magnetic (TM) modes coincide with each other, which makes the PDM modulated signals demuxed or go through an optical DFT without loss of misalignment of the spectra. A typical spectrum of the AWG is shown in Figure 9.9.

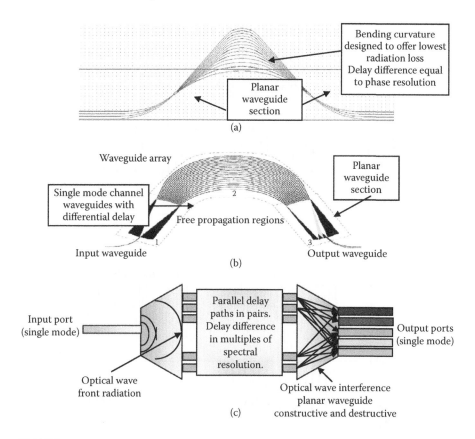

FIGURE 9.7 A geometrical design of (a) 4 × 4 and (b) 1 × 4 array waveguide grating.[19] (c) Radiation and interference of lightwave rays in planar sections of the AWG (not to scale).

9.2.5 AWG Structure and Characteristics

The configuration of a 1 × N AWG multiplexer is shown in Figure 9.4(b). The multiplexer consists of 1/N input/output waveguides or possibly N × N, two focusing slab waveguides, and arrayed waveguides with a constant path length difference ΔL between neighboring waveguides. The length of the difference is determined by the inverse of the fundamental frequency of the transform. The input light is launched into the first slab waveguide and then excites the arrayed waveguides with equal phase distribution. This is effectively the diffraction of the lightwave from the end of the single-mode waveguide when entering into the slab waveguide. After traveling through the arrayed waveguides, the light beam interferes *constructively* at one focal point in the second slab. The location of the focal point depends on the signal wavelength λ because the relative phase delay in each arrayed waveguide is given by $\Delta L/\lambda$. The slab and array waveguides act as a lens and grating, respectively, as shown in Figure 9.4(a–c).

We can consider the principle of the AWG in more detail as follows. In the first slab region, input waveguide separation is D_1, the arrayed waveguide separation is d_1,

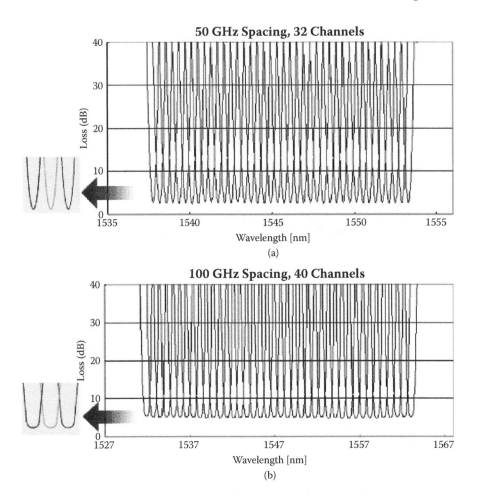

FIGURE 9.8   Spectrum of a commercial AWG[21] of (a) 50 GHz and (b) 100 GHz spacing between frequency components. (Extracted from Binh, L. N., *Guided Wave Photonics*, Taylor & Francis, Boca Raton, FL, 2011, Chapter 3.)

and the curvature radius is R_1. The waveguide parameters in the first and the second slab regions may differ. In the second slab region the output waveguide separation is D, the arrayed waveguide separation is d, and the curvature radius is R. The input light at position x_1 (x_1 is arbitrary and measured in a counterclockwise direction from the center of the input waveguides) is radiated to the first slab and then excites the arrayed waveguides. The amplitude profile in each arrayed waveguide usually has a Gaussian distribution. After traveling through the arrayed waveguides, the light beams constructively interfere at one focal point x (x is measured in a counterclockwise direction from the center of the output waveguides) in the second slab. Let us consider the phase retardation for the two light beams passing through the $(i-1)$th and ith arrayed waveguides. The difference between the total phase retardations for the two light beams passing through the $(i-1)$th and ith arrayed waveguides must be

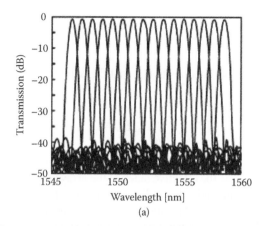

(a)

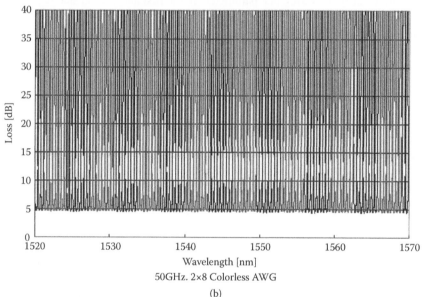

50GHz. 2×8 Colorless AWG

(b)

FIGURE 9.9 NTT NEL Co. Ltd. SiO$_2$:Si PLC AWG device with TE and TM modes, 25–50 GHz channel spacing: (a) Single band 225 GHz AWG; (b) cyclic colorless AWG characteristics.

an integer multiple of 2π in order that the two beams constructively interfere at focal point x. Therefore we have the interference condition as

$$\beta_s(\lambda_0)\frac{d_1 x_1}{R_1} - \beta_s(\lambda_0)\frac{dx}{R} + \beta_c(\lambda_0)\Delta L = 2m\pi \qquad (9.18)$$

where β_s, β_c are propagation constants in a slab region and an arrayed waveguide, m is an integer, and λ_0 is the center wavelength of the multiwavelength array system. When the condition $\beta_c(\lambda_0)\Delta L = 2m\pi$ or

$$\lambda_0 = \frac{n_{eff} \Delta L}{m} \tag{9.19}$$

is satisfied for λ_0, the light input position x_1 and the output position x should satisfy the condition

$$\frac{d_1 x_1}{R_1} = \frac{dx}{R} \tag{9.20}$$

In Equation (9.19), n_{eff} is the effective index of the arrayed waveguide ($n_{eff} = \beta_c/k$; k = wave number in vacuum) and m is the diffraction order. The above equation means that light is coupled into the input position x_1 and the output position x is determined by (9.20). Usually the waveguide parameters in the first and second slab regions are the same; they are $d_1 = d$ and $R_1 = R$. Therefore, the input and output distances are the same as $x_1 = x$. The dispersion of the focal position x with respect to the wavelength λ for the fixed light input position x_1 is given by differentiating (9.18) to λ as

$$\frac{\Delta x}{\Delta \lambda} = \frac{N_c}{n_s} \frac{R \Delta L}{d \lambda_0} \tag{9.21}$$

where n_s is the effective index in the slab region, and N_c is the group index of the effective index n_c of the arrayed waveguide ($N_c = n_c - \lambda \, d \, n_c/d\lambda$). The input and output waveguide separations are $\Delta x = D$ when $\Delta \lambda$ is the channel spacing of the WDM signal. Putting these relations into Equation (9.4), the wavelength spacing $\Delta \lambda$ for the fixed light input position is given by

$$\Delta \lambda = -\frac{n_c}{N_c} \frac{dD\lambda_0}{R \Delta L} \tag{9.22}$$

The path length difference ΔL is also obtained from (9.19). The spatial separation of the mth and $(m + 1)$th-focused beams for the same wavelength is obtained from (9.18) as

$$X_{FSR} = x_m - x_{m+1} = \frac{\lambda_0 R}{d n_s} \tag{9.23}$$

X_{FSR} represents the free spectral range of AWG. The number of available wavelength channels N_{ch} is given by X_{FSR}/D, where D is the output waveguide separation.

In silica-based AWGs, the equivalent refractive index n_c is different for the orthogonal polarization modes due to stress-induced birefringence. The focusing position is therefore different for the TE and TM polarizations. The TE/TM mode conversion method, in which a thin half-waveplate is inserted at the center of the arrayed waveguides, is the simplest and most practical way to correct for the different focal positions.

9.3 GUIDED WAVE WAVELET TRANSFORMER

Among the transformation techniques that produce orthogonality of the time domain, waveforms would minimize the deleterious effects of dispersion. Wavelet transformation offers the use of time–frequency plane and finite impulse response (FIR), so implementation in the optical domain may offer some advantages. This section considers the wavelet transformation and its implementation using guided wave structures. Similar to the FFT and IFFT described above, the mathematical representation can be expressed in serial and parallel forms; the discrete wavelet transform can also be presented in these structures.

9.3.1 WAVELET TRANSFORMATION AND WAVELET PACKETS

Cincotti et al. have recently[21,22] proposed photonic architectures that perform the fast Fourier transform, the discrete wavelet transform (DWT), and the wavelet packet (WP) decomposition of an optical signal. The same architecture is also proposed to implement a full optical encoder–decoder that generates a set of the orthogonal codes simultaneously. WP decomposition is an appealing technique for processing signals with time-varying spectra due to its remarkable property to describe frequency content along with time localization. Wavelets have a large number of applications, such as image compression, signal denoising, human vision, and radar, and in many fields, such as mathematics, quantum physics, electrical engineering, and seismic geology.

In optical communications, wavelets have been proposed for time–frequency multiplexing[23] in order to minimize the linear dispersion effects. This section gives a short description of the features of the DWT and WP decomposition.

9.3.1.1 Cascade Structure

The DWT of a discrete sequence can be numerically evaluated via recursive discrete convolutions with a low-pass and a high-pass filter, followed by a subsampling of factor 2, according to the Mallat's pyramidal decomposition algorithm,[24,25] as

$$c_1[n] = \sum_k h[2n - k]s[k]$$

$$d_1[n] = \sum_k g[2n - k]s[k]$$

(9.24)

where $c_1[n]$, $d_1[n]$ are the scaling and details coefficients, respectively, at the level decomposition or resolution of 2. The coefficients of the resolution at 2^l are iteratively generated starting from the scaling coefficients at the previous resolution 2^{l-1} as

$$c_l[n] = \sum_k h[2n - k]c_{l-1}[k]$$

$$d_l[n] = \sum_k g[2n - k]c_{l-1}[k]$$

(9.25)

On the other hand, the reconstruction of the input sequence can be generated by

$$s[n] = \sum_k h[2n-k]c_l[k] + g[2n-k]d_l[k]$$

(9.26)

The scaling and detail coefficients are the orthogonal projections of the input sequence $s[n]$ onto two complementary spaces $\mathbb{Q}_l, \mathbb{N}_l$, which are, respectively, spanned and scaled versions of the scaling function

$$\varphi_{l,k}(t) = 2^{-l/2}\varphi(2^{-l}t - k)$$

(9.27)

and the wavelet function

$$\psi_{l,k}(t) = 2^{-l/2}\psi(2^{-l}t - k)$$

(9.28)

These two functions must satisfy the dilation relationships:

$$\varphi(t) = 2^{1/2}\sum_k h[k]\varphi(2t - k\tau)$$

$$\psi(t) = 2^{1/2}\sum_k g[k]\varphi(2t - k\tau)$$

(9.29)

with τ determined by the inverse of the free spectral range defined as the frequency region in which all frequency components are included. The two infinite impulse response filters (FIRs) of length M have the frequency responses given by

$$H(\omega) = \frac{1}{\sqrt{2}}\sum_{k=0}^{M-1} h[k]e^{-jk\omega\tau}$$

$$G(\omega) = \frac{1}{\sqrt{2}}\sum_{k=0}^{M-1} g[k]e^{-jk\omega\tau}$$

(9.30)

These two frequency responses are related by

$$G(\omega) = e^{-jk\omega\tau}H^*\left(\omega + \frac{\pi}{\tau}\right)$$

$$|H(\omega)|^2 + \left|H\left(\omega + \frac{\pi}{\tau}\right)\right|^2 = 1$$

(9.31)

Hence,

$$|H(\omega)|^2 + |G(\omega)|^2 = 1$$

(9.32)

where the superscript * denotes the complex conjugation. The filters $G(\omega)$, $H(\omega)$ are the quadrature mirror filter (QMF) and half-band filter (HB), respectively.[26] They can be implemented by a lossless two-port coupler[27] with the conservation of energy given by (9.32). In the complex plane z, the QMF and HB responses of (9.31) can be rewritten as

$$G(z) = -z^{-(M-1)}H_*(-z)$$

$$H(z)H_*(z) + H(-z)H_*(-z) = 1$$

(9.33)

where $z = e^{j\omega\tau} = e^{j\beta L}$ with the delay variable length L, β is the propagation constant of the guided mode, and L is the delay length equivalent to the time τ.[17,28] The subscript * denotes the Hermitian conjugation, i.e., $H_*(z) = H(1/z^*)$. We can see that the wavelet transfer can be implemented by fiber couplers and delay lines whose lengths can be tailored to match the sampling rate in the time domain, and hence the frequency spectrum of the base band signals that are carrying the optical waves in guided wave structures such as integrated optics or planar lightwave circuits. The z-transform expression can be implemented using optical circuits without much difficulty.[29] A unit delay is represented by z^{-1} equivalent to a phase shift of propagation length L and the guided wave constant β.

If a wavelet filter is a half-band filter, we can follow the design procedure described in Cincotti.[22] An optical FIR filter of length M can be formed by cascading M two-port lattice structures with M optical delay lines, M phase shifters, and ($M + 1$) directional couplers. However, if the filter can be simplified to HB type, then the number of Mach–Zehnder delay interferometers (MZDIs) can be halved. An HB filter of order $2N$ can be realized by cascading an MZDI with a delay τ and ($N - 1$) MZDI with a delay time of 2τ, as depicted in Figure 9.10(a). Figure 9.10(b) depicts the equivalent circuit by a 3 dB coupler MZDI and a network transfer function $S(z)$ of $N - 1$ cascaded MZDI with coupling coefficients k_2, ..., k_N and phase shifters ϕ_1, ..., ϕ_N.

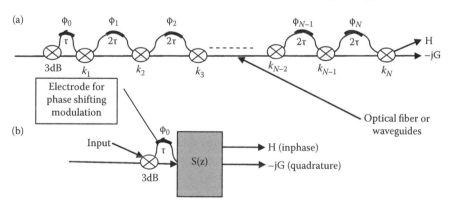

FIGURE 9.10 Structure of an optical wavelet filter (i.e., half-band filter [HB]) made up by cascading of MZDIs of length t and 2τ and phase shifters and a 3 dB 2×2 coupler cascaded with $N - 1$ couplers of amplitude coupling coefficients k. (a) N coupler structure; (b) equivalent by one MZDI and a network $S(z)$ of identical $(N - 1)$ 2×2 couplers with coefficients 2 to k_N. $H(z)$ is the HB and $G(z)$ is the QMF. Note: 2×2 coupler is symmetric in this structure.

9.3.1.2 Parallel Structure

The DWT decomposes a signal $s(t)$ from a subspace V_o into nonoverlapping frequency subbands by means of orthogonal projections $W_j^m f$ given as

$$f(\zeta) = \sum_{j=0}^{2^m-1} W_j^m f(\zeta) \quad m = 1, 2 \ldots \tag{9.34}$$

with

$$W_j^m f(\zeta) = \sum_{k \in Z} \langle f(\zeta), w_{m,k,j}(\zeta) \rangle_{L^2} w_{m,k,j}(\zeta) \equiv \sum_{k \in Z} w_k^{m,j} w_{m,k,j}(\zeta) \tag{9.35}$$

where $\{w_k^{m,j}\}$ are the wavelet coefficients. The wavelet molecules $w_{m,k,j}$ are recursively determined starting from the low-pass filter H and high-pass filter G, i.e., into mapping the input signal sequence into its low- and high-frequency parts by means of two orthogonal projections:

$$w_{m,k,2j}(\zeta) = \sum_{n \in Z} h_{n-2k} w_{m-1,n,j}(\zeta)$$

$$\tag{9.36}$$

$$w_{m,k,2j+1}(\zeta) = \sum_{n \in Z} g_{n-2k} w_{m-1,n,j}(\zeta)$$

The wavelet molecules are characterized by three parameters: the frequency order j, the scale m, and the position k. The time-varying harmonics in the input signals are detected from the position and scale of high-amplitude wavelet coefficients. Note that in DWT only the scaling coefficients $\{c_k^m\}$ are recursively filtered. In full decomposition, both the scaling and detail coefficient vectors are recursively decomposed into two parts with the scheme illustrated in Figure 9.11.

9.3.2 Fiber Optic Synthesis

Figure 9.12 depicts an optical circuit following the decomposed structure of Figure 9.11 using guided wave devices, including an asymmetric MDI incorporating a phase shifter to resolve the frequency components of a serial optical pulse sequence input. We note that unlike for the DFT optical circuit described in Section 9.2.4.1, the decomposition for DWT no-phase shifters is required; thus there is no need for tuning the phase shifting by either electro-optic or pyroelectric effects. The Haar wavelet transform of first order is given in Cincotti.[22]

The transfer matrix of a lattice of an asymmetric MDI can be written as

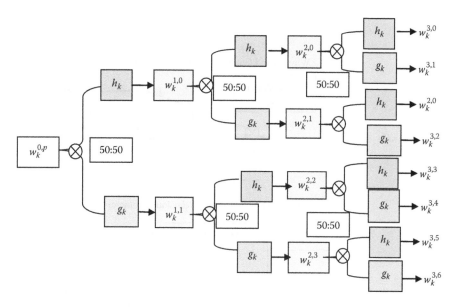

FIGURE 9.11 Pyramidal WP decomposition scheme.

$$S_l = \begin{pmatrix} e^{-j\omega\Delta\tau/2} & 0 \\ 0 & z^{+j\omega\Delta\tau/2} \end{pmatrix} = \begin{pmatrix} z^{-1/2} & 0 \\ 0 & z^{1/2} \end{pmatrix} \tag{9.37}$$

The overall transfer matrix of a network of lattice of MZDI can be formed by multiplication of appropriate matrices to give

$$S_l = \begin{pmatrix} H(z) & -F_*(z) \\ F_*(z) & H_*(z) \end{pmatrix} \tag{9.38}$$

$$F_*(z)F(z) + H(z)H_*(z) = I$$

The condition for a unitary matrix is required to satisfy the conservation of energy in the coupling of the optical fields from one lattice to the other.

In the case that there are N input channels in parallel, the DWT would look like the structure depicted in Figure 9.13. The outputs (c_0^2, d_0^2) have the spectrum consisting of all optical passbands and their quadrature counterparts.

The coefficients c_k^{lev} and d_k^{lev}, with the subscript indicating the order and the superscript the level of propagation, can be evaluated as[22]

$$c_k^l = \sum_{n \in Z} h_{n-2k} c_n^{l-1}; \ d_k^l = \sum_{n \in Z} g_{n-2k} c_n^{l-1} \tag{9.39}$$

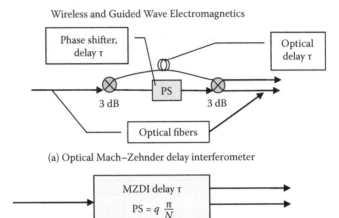

(a) Optical Mach–Zehnder delay interferometer

(b) Symbol of an asymmetric Mach–Zehnder delay interferometer with a delay line τ as a phase shifter. This makes the H or G filters

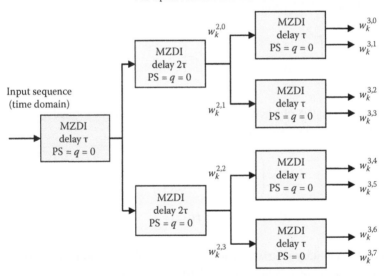

(c) Optical guided wave network for the decomposition of a serial optical signal with Haar (Daubechies of order $M = 2$) using asymmetric Mach–Zehnder delay interferometer with delay lines and no phase shifter.

FIGURE 9.12 Optical guided wave network for the decomposition of a serial input optical signal with Haar (Daubechies of order $M = 1$) using asymmetric Mach–Zehnder delay interferometers with delay lines and no phase shifter. (a) Asymmetric MZDI with phase shifter (PS); (b) symbol; (c) guided wave optical circuit of a Haar wavelet (Daubechies of order $M = 1$).

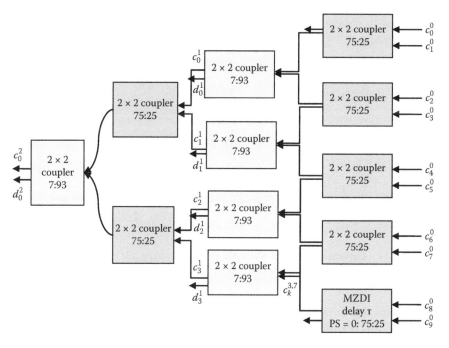

FIGURE 9.13 Optical guided wave network for the decomposition of a serial input optical signal with Haar (Daubechies of order $M = 2$) in four stages using a guided wave coupler 2×2 with different coupling coefficients as indicated in the coupler notation.

These coefficients are given by the transfer matrices of the filterd H and G as

For $M = 1$, first-order Haar wavelet ($l = 1$):

$$\begin{pmatrix} H_1 \\ G_1 \end{pmatrix} = \frac{1}{\sqrt{2}} \begin{pmatrix} 1 & 1 \\ 1 & -1 \end{pmatrix} \tag{9.40}$$

For second-order Haar wavelet ($l = 2$):

$$\begin{pmatrix} H_2 \\ G_2 \end{pmatrix} = \frac{1}{4\sqrt{2}} \begin{pmatrix} 1-\sqrt{3} & 3-\sqrt{3} & 3+\sqrt{3} & 1+\sqrt{3} \\ 1+\sqrt{3} & -3-\sqrt{3} & 3-\sqrt{3} & -1+\sqrt{3} \end{pmatrix} \tag{9.41}$$

9.3.3 Synthesis Using Multimode Interference Structure

Planar optical waveguides are normally considered guided wave optical devices that are infinitely long in the lateral direction, y, and have restricted confinement in the vertical direction, x. It is normally assumed that the field distribution in the lateral direction is uniform, and there is an oscillating profile in the vertical direction.

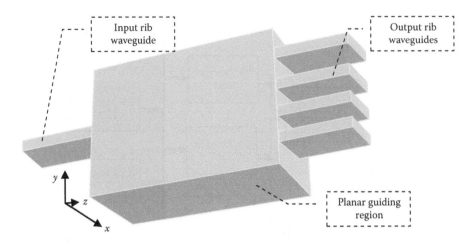

FIGURE 9.14 General structure of MMI coupler, input waveguide radiating field in planar structure and distributed with mode oscillating form into output channel waveguides; substrate not shown.

Assume now that the planar waveguide is supporting only one guided mode in the x-direction and a multimode in the y-direction and that the input waveguide and output waveguides are single channel waveguides. Thus one can consider that light emitting from this kind of structure is similar to the diffraction from a slit to multiple slots or an antenna radiating an oscillating field in the planar waveguide region and distributed to different locations. This is illustrated in Figure 9.14. The field of guided modes of the planar waveguide with some restriction on the lateral dimension, that is, multimode guiding in the lateral direction, would be oscillating and thus distributed according to the high- and low-field regions.

Figure 9.3 shows typical field intensity distribution in a multimode planar waveguide in the lateral direction. The interference of these lateral modes forms the patent of maximum and minimum intensities as shown in Figure 9.15, depending on the order of the modes of the interference. This interference effect can be considered as overlapping imaging effects.[30,31] Figure 9.15 shows the planar waveguide interference section. The length of this section is critical for the interferences of the lateral

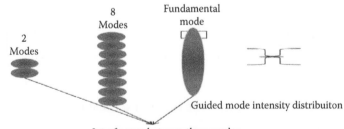

FIGURE 9.15 Lateral modes of the multimode planar waveguide.

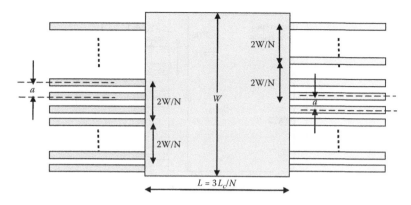

FIGURE 9.16 General $N \times N$ MMI coupler with an arbitrary access number of waveguides of N. (After Zhou, J., *IEEE Photonic Technol. Lett.*, 22(15), 1093–1097, 2010.[34])

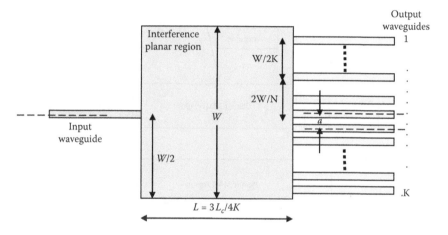

FIGURE 9.17 Symmetric interference $1 \times K$ coupler. (After Siegmen, A.E., *Opt. Lett.*, 26(16), 1215–1217, 2001.)

distribution. Thus its accuracy in the fabrication or thermal dependence would play a major part in the splitting and combining of the optical spectral distribution. Note that the number of the guided modes in the vertical direction is one. Figure 9.14 shows sketches of the confined modes of the planar waveguides. Thus a number of general $N \times N$ MMI couplers can be formed as shown in Figures 9.16, 9.17, 9.18, and 9.19. For the detailed design of multimode planar waveguides we can refer to the appendix. We would design the waveguide such that it is multimode in the lateral direction and single mode in the vertical direction. So an effective index technique would be computationally efficient.

9.3.4 REMARKS

It is noted that the principal differences between the Fourier and wavelet transformations are that the scaled and detailed versions under wavelets all have the same number

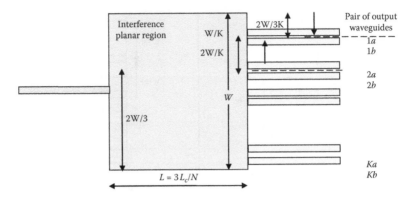

FIGURE 9.18 Pair interference $2 \times K$ MMI coupler. (After Siegmen, A.E., *Opt. Lett.*, 26(16), 1215–1217, 2001.)

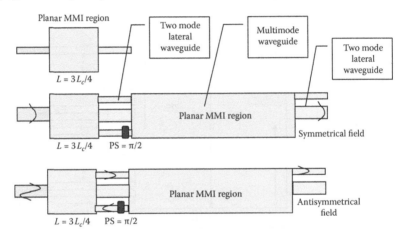

FIGURE 9.19 Overlapping image MMI coupler to achieve a mode splitter or combiner. The phase shifter PS (π rad.) is used for reversing the phase. (After Siegmen, A.E., *Opt. Lett.*, 26(16), 1215–1217, 2001.)

of oscillations. If this were traditional Fourier analysis, scaling coefficients would correspond to the Fourier coefficients, and detail coefficients (wavelet functions) would correspond to $\sin(n\omega t)$ and $\cos(n\omega t)$. The more important differences between the wavelet expansion and the Fourier expansion are that the wavelet coefficients are localized in both time and frequency, contrary to sine and cosine waves, which are completely delocalized in time, and that to describe finer details in time, wavelet expansion uses scaled basis functions, contrary to Fourier analysis, which uses higher frequencies.

In summary the photonic implementation of an optical wavelet transform can be decomposed into either cascade or parallel forms using guided wave structures of input and output waveguides, waveguide inerferometers, and phase shifters. Tunable phase shifting can be integrated. However, if MZDIs with tunable optical delay lines are used, these phase shifters can be eliminated. Switchable phase delay lines can be used to reconfigure the optical transformer, or a different order can be implemented in active guided wave structures such as the electro-optic efficient crystal with Ti-diffused waveguides and phase modulation and switching.[32] The guided wave structures can be in fiber optic form, with planar lightwave technology including guided wave channel and planar waveguides and multimode interferometers. With the base band signal frequency or speed reaching several GSymbols/s, we expect devices in planar waveguide technology will prevail in the near future.

On implementation, the principal difference between photonic DWT and DFT transformers is that the frequency bands of the DWT can be selectively assigned with the preservation of the orthogonality of the signals in the time and frequency domains. The assignment of the spectral bands allows simplification of computing resources and is much more suitable for optical implementation.

9.4 OPTICAL ORTHOGONAL FREQUENCY DIVISION MULTIPLEXING

The current trend in the speed of optical transmission is developed toward terabits/s from a single laser source.[1–5] The principles of such Tbps optical transmission systems are as follows:

- The original lightwave source, the primary laser, is operating in the continuous wave (CW). Ultra-short pulse with a repeating frequency can then be generated in a mode-locked fiber ring resonator.[33]
- These short pulse sequences are then launched into a highly nonlinear fiber (HNLF), and the nonlinear interactions of the pulses would then generate a comb-like feature of several subcarriers, the secondary carriers.
- These secondary carriers are demultiplexed into individual secondary carriers. Each is then modulated by optical modulators.
- The modulated secondary carriers are then combined by an optical DFT and launched into an optical fiber transmission line.
- At the output of the transmission line the subcarrier channels are then demultiplexed by an optical inverse DFT and individually coherently detected by optical receivers and processed in the electronic domain and by algorithms in digital signal processors.
- DFT channels can be interleaved and decomposed into odd and even channels at the transmitter and receiver sides so that the frequency spacing between channels can have larger spatial tolerances and, hence, less channel cross talk.

A generic block diagram of the optical transmission system shown in Figure 9.20 summarizes the steps given above. A detailed structure of the optical receivers is shown in Figure 9.21, in which an optical FFT is employed to recover the transmission channels in the electronic processor after the optical processing stages, including an optical FFT to demultiplex the individual optical transmitted channels and mixing with local oscillators of individual channels to recover the signals in the electronic domain and digitalized for digital signal processing. The spacing between the channels is about 50 GHz or 0.4 nm. The principal function of an optical FFT is to separate the channels

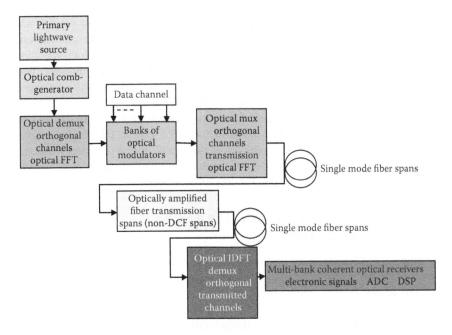

FIGURE 9.20 Transmission block diagram employing optical DFT and IDFT as an application of optical DFT for generating optical OFDM multicarrier modulation in the optical domain.

with assurance of the orthogonality of the channels after transmitting through a very long and dispersive fiber transmission line without dispersion compensation. Advanced optical fiber communication systems extensively exploit coherent receiving techniques in which the optical fields of the arriving signals are mixed with a local laser oscillator via a hybrid coupler with polarization diversity. Both polarized mixed fields are then separately detected by balanced optical receivers. The output electronic signals are then sampled by analog-to-digital converters (ADCs) to digital domain signals that are then processed by digital signal processors to compensate for linear dispersion and nonlinear distortion effects, to recover the carrier phase, etc. The orthogonality of the optical channels before transmitting through long distance is very critical, and so the roles of the optical DFT and IDFT. The insert in Figure 9.21(c) shows the constellation of the 16-quadrature amplitude modulation (QAM) as decoded by a real-time sampling oscilloscope positioned after the coherent balanced receiver. These sampled digital signals are then processed by a digital signal processor (DSP) to recover the original data sequence. In this type of optical transmission system, the orthogonality property of the multiplexed multiwavelength channels is very important, so that even overlapping spectral regions can be decoded by optical FFT or a discrete wavelet transformer (DWT).

9.5 NYQUIST ORTHOGONAL CHANNELS FOR TBPS OPTICAL TRANSMISSION SYSTEMS

Optical DFT and DWT can be employed in the organization of data channels modulated at the Nyquist rate so that the total capacity for transmission over a single-mode

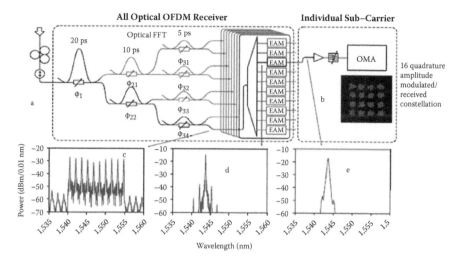

FIGURE 9.21 An all-optical OFDM transmitter and receiver. (a) Input pulse sequence from output of a comb generator generates a frequency comb of 75 equidistant subcarriers that are individually demultiplexed and individually modulated with 16 QAM, then recombined and polarization multiplexed to generate an OFDM signal. (b) All-optical OFDM receiver with an optical DFT consisting of a three-stage DI cascade, one branch of which was actually implemented, two cascaded single-cavity 1 nm filters, and an EAM gate, after which the signal is received by an optical modulation analyzer (OMA). (a) Optical circuit. (b) Receiving subsystem. (c) Optical spectrum of all channels. (d) Spectrum of a single channel passing through the optical FFT. (e) Close view of the spectrum of a single channel. (From Hillerkuss, D. et al., *Nature Photonics*, 5, 364, 2011; Zhou, J., *IEEE Photonic Technol. Lett.*, 22(15), 1093, 2010.)

optical fiber can reach several Tbps over the C-band. This section considers the design of such optical transmission systems for Tbps with orthogonal channels operating at 56 GSym/s and an I/Q modulation scheme, the QAM.

A proposed optical transmission is shown in Figure 9.22 consisting of

- A comb generator generating multiple equally spaced subcarriers from a single-frequency lightwave source.
- The comb subcarriers are then demultiplexed by an array waveguide grating of equal spacing to that of the comb components.
- These individual subcarriers are modulated by the data channels encoded with certain symbols of a specific modulation format such as Mary-QAM.
- The modulated subcarrier channels are then arranged and multiplexed through an optical wavelet packet filter that gives the in-phase and quadrature mirror channels. Both outputs can be combined and transmitted through the optical transmission line.
- The single-frequency characteristics of the subcarriers of the comb lightwave source are critical for the spectral properties of the channels orthogonal after transmitting over a long-distance transmission line.
- When the channels arrive at the receiver, an optical processing front end is employed to extract and select a group of particular desired channels.

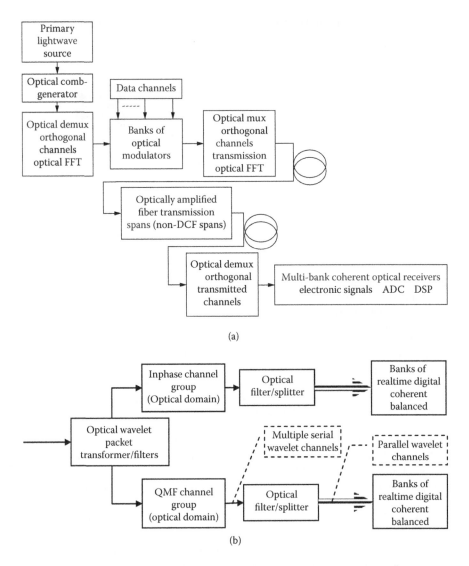

FIGURE 9.22   Optical transmission employing orthogonal channels in the optical/photonic domain: (a) system schematic and (b) multibank wavelet packet realtime digital receiver.

The photonic/optical processing front end takes the form/structure given in Figure 9.13 or Figure 9.17 to give the in-phase group of wavelets and quadrature mirror filter group of channels, which are then demuxed into individual channels by using an AWG. Note that a number of filter channels can be selectively switched into either the in-phase or quadrature outputs.

- Each individual channel is then coherently mixed with a local oscillator laser and detected by a balanced receiver via a hybrid optical coupler. The electrical signals at the output of the optical receivers are then sampled by a dual pair of ADCs and processed in the digital domain by DSPs.

9.6 DESIGN OF OPTICAL WAVEGUIDES FOR OPTICAL FFT AND IFFT

As derived in the previous section, the entire sequence of operations involved in successively subdividing an input array of N pixels and applying the added phase shifts to calculate its Nth-order FFT can be implemented in a lossless optical fiber (or other) network with nothing more than $N/2 \log_2(N)$ couplers or 3 dB beam splitters plus a number of in-line optical phase shifts. The implementation of such a structure in an integrated optical circuit is critical and would allow compact integration of several 2×2 couplers and phase shifters into a high-order Fourier optical transformer. Indeed, one can form a signal flow graph,[17] and the implementation is quite straightforward.

In the implementation of this coupler, guided wave optical devices are commonly used, with the planar lightwave circuit on a silicon substrate with silica, with pure silica as the guiding material. The waveguide can be formed with geometrical structures such as rib waveguide or doped materials, as described in Chapters 5 and 6.

As an example, we consider a dielectric bar of index immersed in a medium with index n_2, as shown in Figure 9.23, with uniform refractive indices in the regions surrounding the channel waveguiding region. To facilitate comparison, we define the

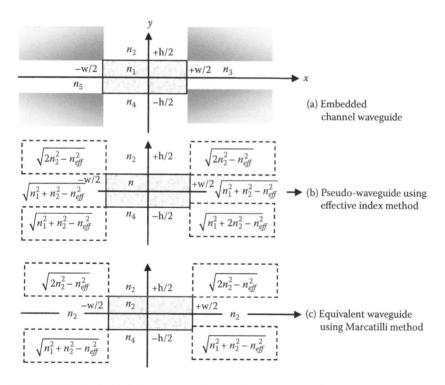

FIGURE 9.23 Embedded channel optical waveguide: (a) waveguide structure and its representation using (b) the effective index method and (c) Marcatilli's method with index profiles in different regions.

normalized frequency parameter V and the normalized guide index b, or normalized propagation constant, in terms of n_1, n_2, h.

$$V = kh\sqrt{n_1^2 - n_2^2} \simeq \frac{2\pi}{\lambda} hn\sqrt{2\Delta}$$

$$b = \frac{\beta^2 - k^2 n_2^2}{k^2(n_1^2 - n_2^2)}$$

(9.42)

The V-parameter can be approximated as given in (9.42) provided that the difference in the refractive index between the guiding region and cladding is small enough, usually less than a few percent.

Thus the normalized effective refractive index can be evaluated as a function of the normalized frequency parameter V to give the dispersion curves as shown in Figure 9.24 in which the curves obtained from the finite element method (FEM) and Marcatilli's methods[35] and the effective index technique confirm their agreement to within a tolerable accuracy.

A numerical evaluation for silica doped with a GeO_2 waveguide and cladding region is pure silica. The relative refractive index of the core and the pure silica cladding is 0.3 or 0.5%; then using the single-mode operation given in Figure 9.24, we can select $V = 1$, and using (9.42), then the cross section of the rectangular waveguide

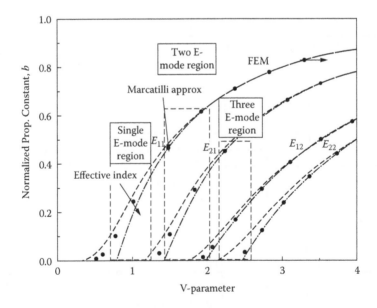

FIGURE 9.24 Dispersion characteristics, dependence of the normalized propagation constant of the guided modes as a function of the parameter V, the normalized frequency: comparison of three numerical, analytical methods for rectangular optical waveguides consisting of uniform core and cladding. Shown in boxes are regions of supported guided modes. (From Zhou, J., *IEEE Photonic Tech. Lett.*, 22, 15, 2010, p. 1093.)

is 3×3 μm^2 for 0.5% relative refractive index, for 0.3% the dimension is 6×6 μm^2, and the refractive index of pure silica is 1.448 for an operating wavelength of 1550 nm. Extending this dispersion curve for the lateral region we can design multimode planar waveguides with a single mode in the vertical direction and support a few modes in the lateral region by determining the value of the V-parameter for the lateral direction and the lateral length of the waveguide. This type of laterally few-mode waveguide can be employed in the multimode interference (MMI) of the modes for optical field splitting and combining as described in Section 9.4. These MMI structures can be formed in cascade or parallel to create a discrete Fourier or wavelet optical transformer.

In practice most of the optical channel waveguides are fabricated using buried rectangular channel waveguides; for example, a ridge silica on silicon structure is first formed on a silicon substrate and then covered with another layer of pure silicon by chemical vapor deposition. However, for Fourier optics and advanced optical communications, accurate position and orthogonality of the channels are very critical. This requires high precision in waveguide fabrication and, hence, high accuracy in the design of waveguides. Modern fabrication technology can offer such required precision.

This type of rectangular dielectric waveguide has been extensively investigated by Kumar's method.[36] This method offers higher accuracy on the estimation of the mode propagation constant, especially near the cutoff limit. The dispersion curve for the square of the normalized propagation constant and the parameter B is given as

$$B = \frac{2}{\pi} V_2 \text{ with } V_2 = \frac{k_0 h}{(n_c^2 - n_{cl}^2)^{1/2}}$$

$$n_c = n_1 = \text{core refractive index}$$

$$n_{cl} = n_2 = n_3 = n_4$$

$$(9.43)$$

It is noted that although the perturbation method reported by Kumar et al.[36] would give more accuracy near the cutoff region of the modes, in practice, due to fabrication tolerance, we can expect that both the Kumar method and effective approaches can be used to design and fabricate without any problems. Further details of this perturbation technique can be found in Moreolo et al.[22] In summary, the procedures for the design of a waveguide, planar, or channel structure, for supporting single or multimodal regions of the E- or H-fields, are

- Based on the dispersion curves of both the E- and H-field modes, that is, the curves representing the V-parameter and the normalized propagation index, determine the desired number of modes in either polarization, then go to the curve to get the corresponding normalized propagation constant index. The effective refractive index of the guided modes can be determined from this index and also the expected propagation time of the guided mode over the length of the waveguide. This is important for the design of the multimode interference waveguide.

- Continue for other polarization directions, and then combine the two guided solutions to obtain the approximated analytical values for the combined mode.
- One can of course draw a circle with the value V selected, and the intersection of this circle and the dispersion curves of the modes would give the vertical value of b, the normalized propagation constant index from Figure 9.24. Thus the propagation constants of the guided modes could be estimated. These effective indices of the guided modes give the propagation velocity of the lightwaves confined in the optical waveguides.

9.7   CONCLUDING REMARKS

This chapter has outlined the principles of optical forward and inverse FFT using guided wave techniques. Some simple and basic mathematical representations of Fourier transform in the discrete mode are given. In summary, the techniques include the following steps:

- Determine the final frequency range and spacing between the spectral components using either DFT or DWT and associated decomposition structures.
- From the fundamental frequency deduce the number of stages and the sampling rate required.
- Decide on the serial or parallel form, then the design of individual MZDIs for the serial form and the delay path between waveguide arrays for the parallel form.
- Now design single-mode channel waveguides for the spectral regions of the frequency range. Use an effective index or perturbation techniques with the guide of the dispersion curves leading to a specific dimension of the waveguide.
- Alternatively, use MMI.
- Design the planar waveguide for radiating and combing of the spectral channels.
- Employ any commercial design packages if available. Otherwise, use the graphical techniques and the dispersion curves given in Section 9.6. Determine the number of modes to be supported by the waveguide, and then obtain the V-parameter for the core guiding and cladding regions. It is recommended that the maximal value of the propagation is selected, and then the V-parameter.
- Estimate the delay time and length of the waveguide as well as the coupling coefficient and separation distance between the waveguides of the guided wave directional coupler.
- Once these geometrical structures are determined, simulate detailed behavior of the lightwaves propagating through the optical DFT or DWT using a simulation technique such as the beam propagation method.
- Finally, perform fabrication of the transformers and characterization using integrated photonic techniques such as the planar lightwave circuit (PLC) technology.

Readers can explore many commercial packages to design the geometry and index profile for specific fabrication platforms, such as silica on silicon of polymeric material systems. Figure 9.25 shows the spectrum of an array waveguide grating (AWG) in which the spectrum of each channel is orthogonal to the two adjacent channels. This type of AWG can be employed as an optical FFT at the transmitter and receiver of the terabits per second (Tbps) transmission system depicted in Figure 9.26. A typical specification of the AWG can be found in the appendix of this chapter. (Note that no optical DWT transmission systems have been reported to date, to the best of our knowledge.) The implementation of the optical DWT is much simpler than that for the DFT, as 2×2 couplers with predetermined coupling coefficients are sufficient. Furthermore, we understand that there are other transformation techniques that can also offer orthogonality of the channels, and even higher/faster than the Nyquist (FTN) speed,[37] which is another orthogonal transform, but with a sampling speed higher than that required by Nyquist. The system of FTN requires, naturally, a channel with memory or associated with digital signal processing in the electronic domain, with memory equalization such as the maximum likelihood sequence estimation (MLSE) algorithm. Thus we can see that an interaction between the photonic and electronic domains can be possible to implement FTN transmission at extremely high speed, thanks to the generation of orthogonal spectra of optical channels by guided wave optics.

APPENDIX 9.1

Performance Specification

AWGM 1 Parameters	Min.	Typical	Max.	Unit
Available Channel Frequency Range	191.85		196.20	THz
Channel Spacing		100		GHz
Number of Channels		32, 40 or 44		
Channel Passband		±12.5		GHz
Insertion Loss [1]		5.0	6.0	dB
Uniformity		< 1.5		dB
Ripple		0.5	0.75	dB
Polarization Dependent Loss			0.5	dB
Adjacent Channel Crosstalk	25			dB
Non-adjacent Channel Crosstalk	35			dB
Total Crosstalk	22			dB
Return Loss	40			dB
Directivity	50			dB
Power Consumption		4		W
Maximum Power Handling			300	mW
Fiber Type		G.652D		

(a)

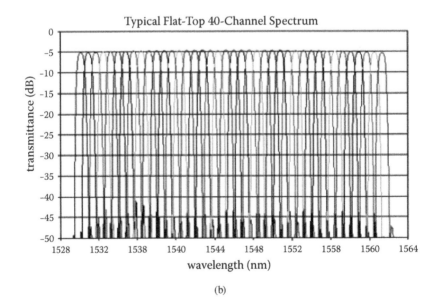

(b)

FIGURE 9.25 Optical spectra of AWG Enablence PM-DWDM 100 GHz spacing and interleaved to 50 GHz orthogonal channel spacing → 2 × 40 channels. (a) Typical parameters. (b) AWG spectra.

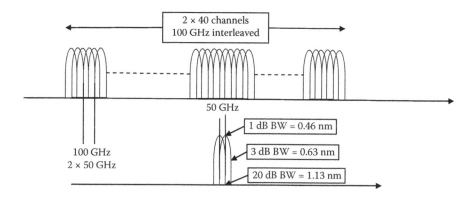

FIGURE 9.26 Illustration of the spectra given in the specifications.

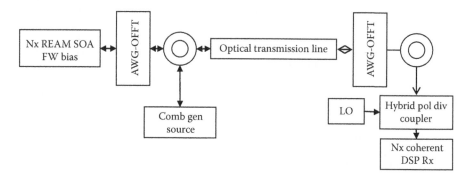

FIGURE 9.27 Tbps transmitter using SOA-REAM array.

REFERENCES

1. M. Nakazawa, "Giant Leaps in Optical Communication Technologies towards 2030 and Beyond," Presentation at Proceedings of European Conference on Optical Communications, ECOC 2011, Geneva, September 2011.
2. T. Kobayashi, A. Sano, E. Yamada, E. Yoshida, and Y. Miyamoto, "Over 100 Gb/s Electro-Optically Multiplexed OFDM for High-Capacity Optical Transport Network," *IEEE J. Lightw. Technol.*, 27(16), 2009.
3. I. Morita, "Optical OFDM for High-Speed Transmission," Presentation at Proceedings of European Conference on Optical Communications, ECOC 2011, Geneva, September 2011.
4. Y. Ma, Q. Yang, Y. Tang, S. Chen, and W. Shieh, "1-Tb/s Single-Channel Coherent Optical OFDM Transmission with Orthogonal-Band Multiplexing and Subwavelength Bandwidth Access," *IEEE J. Lightw. Technol.*, 28(4), 308, 2010.
5. X. Liu and S. Chandrasekhar, "High Spectral-Efficiency Transmission Techniques for Systems Beyond 100 Gb/s," Presentation at Proceedings of Conference on Signal Processing in Photonic Communications, SPPCom-2011, Toronto, Canada, 2011, paper SPMA1.
6. G. Bosco, V. Curri, A. Carena, P. Poggioliniand, and F. Forghieri, "On the Performance of Nyquist-WDM Terabit Superchannels Based on PM-BPSK, PM-QPSK, PM-8QAM or PM-16QAM Subcarriers," *IEEE J. Lightw. Technol.*, 29(1), 53, 2011.

7. J. Proakis, *Digital Communications*, 4th ed., McGraw-Hill, New York, 2006.
8. J.W. Goodman, *Introduction to Fourier Optics*, McGraw-Hill, New York, 1968.
9. A. Papoulis, *Systems and Transforms with Applications in Optics*, McGraw-Hill, New York, 1968.
10. J.D. Gaskill, *Linear Systems, Fourier Transforms, and Optics*, Wiley, New York, 1978.
11. A.E. Siegmen, "Fiber Fourier Optics," *Opt. Lett.*, 26(16), 1215–1217, 2001.
12. NTT Electronic Labs, *Optical OFDM Demultiplexer Using PLC Based Optical FFT Circuit*, report on planar lightwave circuits, 2009.
13. L.N. Binh, *Photonic Signal Processing: Techniques and Applications*, Taylor & Francis, Boca Raton, FL, 2008, Chapter 1.
14. L.N. Binh, *Guided Wave Photonics*, Taylor & Francis, Boca Raton, FL, 2011, Chapter 3.
15. J.W. Cooley and J.W. Tukey, "An Algorithm for the Machine Computation of Complex Fourier Series," *Comput. Math.*, 19, 297–301, 1965.
16. E.O. Brigham, *The Fast Fourier Transform*, Prentice-Hall, Englewood Cliffs, NJ, 1974.
17. L.N. Binh, *Photonic Signal Processing,* Taylor & Francis, Boca Raton, FL, 2008.
18. D. Hillerkuss, R. Schmogrow, T. Schellinger, et al., "26 Tbit/s Line-Rate Super-Channel Transmission Utilizing All-Optical Fast Fourier Transform Processing," *Nature Photonics*, 5, 364, 2011.
19. I. Adam, M. Haniff Ibrahim, N. Mohd Kassim, A. B. Mohammad, and A. S.Mohd Supa'at, "Design of Arrayed Waveguide Grating (AWG) for DWDM/CWDM Applications Based on BCB Polymer," *Elektrika*, 10(2), 17–21, 2008.
20. NTT Electronics. http://www.ntt-electronics.com/en/products/photonics/awg_mul_d.html
21. G. Cincotti, "Fiber Wavelet Filters," *IEEE J. Quant. Elect.*, 38(10), 1420–1427, 2002.
22. M.S. Moreolo, G. Cincotti, and A. Neri, "Synthesis of Optical Wavelet Filters," *IEEE Photonic Technol. Lett.*, 16(7), 1679–1681, 2004.
23. T. Olson, D. Healy, and U. Österberg, "Wavelets in Optical Communications," *IEEE Computing Sci. Eng.*, 1, 51–57, 1999.
24. S. Millat, "A Theory for Multi-Resolution Signal Decomposition: The Wavelet Representation," *IEEE Trans. Pattern Anal. Machine Intellig.*, 11, 674–693, 1989.
25. I. Daubechies, *Ten Lectures on Wavelets*, SIAM, Philadelphia, PA, 1992.
26. K. Jinguji and M. Oguma, "Optical Half-Band Filters," *IEEE J. Lightw. Technol.*, 18, 252–259, 2000.
27. L. Binh, *Guided Wave Photonics,* Taylor & Francis, Boca Raton, FL, 2011.
28. K. Jinguji and M. Kawachi, "Synthesis of Coherent Two Port Lattice Form," *IEEE J. Lightw. Technol.*, 13, 73–82, 1995.
29. Binh, 2008, Chapter 2.
30. M. Bachmann, P.A. Besse, and H. Melchior, "Overlapping-Image Multimode Interference Couplers with a Reduced Number of Self Images for Uniform and Non-Uniform Power Splitting," *Appl. Opt.*, 34(30), 6898–6910, 1995.
31. O. Bryngdahl, "Image Formation Using Self Imaging Technique," *J. Opt. Soc. Am.*, 63, 416–419, 1973.
32. Binh, 2011, Chapter 9.
33. L.N. Binh, *Ultra-Fast Fiber Lasers: Principles and Applications with MATLAB® Models*, Taylor & Francis, Boca Raton, FL, 2010.
34. J. Zhou, "All-Optical Discrete Fourier Transform Based on Multimode Interference Couplers," *IEEE Photonic Tech. Lett.,* 22(15), 1093–1095, 2010.
35. E.A.J. Marcatilli, "Dielectric Rectangular Waveguide and Directional Coupler for Integrated Optics," *Bell Syst. Technol. J.*, 48, 2071–2102, 1969.
36. A. Kumar, K. Thyagarajan, and A.K. Ghatak, "Analysis of Rectangular-Core Dielectric Waveguides: An Accurate Perturbation Approach," *Opt. Lett.*, 8(1), 63–65, 1983.
37. J.E. Mazo, "Faster-Than-Nyquist Signaling," *Bell Syst. Technol. J.*, 54(8), 1451–1462, 1975.

Appendix: Vector Analysis

A.1 DEFINITION

The use of vector analysis in the study of electromagnetic (EM) field theory results in a real economy of times and thoughts. Even more important, the vector forms help to give a clear understanding of the physical laws that mathematics describe. *Scalar* is a quantity that is characterized only by the magnitude and algebraic sign. *Vector* is the quantity defined by the scalar quantity as well as its direction. They are represented by the sign ~ underneath, for example, vectors. Thus a vector can be represented geometrically by an arrow whose direction is appropriately chosen and whose length is proportional to the magnitude of the vector.

A.2 OPERATIONS

A.2.1 RIGHT-HANDED COORDINATE SYSTEM

A vector should be represented in a right-handed coordinate system, as shown in Figure A.1, in which the labeling of the axes follows a right-hand rule, or if we rotate the x-axis to the y-axis in the clockwise of the right hand, then the screwing direction must be the z-direction.

A.2.2 SUM AND DIFFERENCE OF TWO VECTORS

Given that the vector is specified by

$$\underset{\sim}{A} = (\ A_x \quad A_y \quad A_z\ ) = A_x \underset{\sim}{a}_x + A_y \underset{\sim}{a}_y + A_z \underset{\sim}{a}_z = \underset{\sim}{i} A_x + \underset{\sim}{j} A_y + \underset{\sim}{k} A_z$$

$$\underset{\sim}{a}_x, \underset{\sim}{a}_y, \underset{\sim}{a}_z \quad \text{or} \quad \underset{\sim}{i}, \underset{\sim}{j}, \underset{\sim}{k} \triangleq \text{unit vector in } x, y, z \text{ direction}$$

(A.1)

Then for vector $\underset{\sim}{B}$ the summation of the two vectors $\underset{\sim}{A}, \underset{\sim}{B}$ is given as

$$\underset{\sim}{A} + \underset{\sim}{B} = (\ A_x \quad A_y \quad A_z\ ) + (\ B_x \quad B_y \quad B_z\ )$$

$$= (\ A_x + B_x \quad A_y + B_y \quad A_z + B_z\ )$$

(A.2)

A.2.3 MULTIPLICATION OF A SCALAR AND VECTOR AND DOT PRODUCTS

A vector can be scaled by the multiplication of a constant.

A scalar product of two vectors $\underset{\sim}{A}, \underset{\sim}{B}$ is expressed by the dot product of the two vectors to give a scalar quantity:

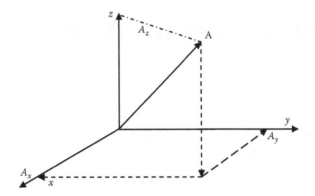

FIGURE A.1 Right-handed coordinate system, that is, if rotating x to y in the clockwise direction, then the screwing direction must be z using the right hand.

$$A \cdot B = \begin{pmatrix} A_x & A_y & A_z \end{pmatrix} \begin{pmatrix} B_x \\ B_y \\ B_z \end{pmatrix} = A_x B_x + A_y B_y + A_z B_z \qquad (A.3)$$

$$= |A||B|\cos\theta = B \times A$$

where θ is the angle between the two vectors.

The cross-product of the two vectors is a vector presented by

$$A \times B = \begin{vmatrix} i & j & k \\ A_x & A_y & A_z \\ B_x & B_y & B_z \end{vmatrix} = \begin{pmatrix} A_y B_z - A_z B_y \\ A_z B_x - A_x B_z \\ A_x B_y - A_y B_x \end{pmatrix} \quad A \times B = -B \times A \qquad (A.4)$$

A.3   CURL, DIVERGENCE, AND GRADIENT OPERATIONS

If V is a scalar function, then the gradient ∇V is given by

$$\nabla V = \begin{pmatrix} \dfrac{\partial V}{\partial x} \\[2ex] \dfrac{\partial V}{\partial y} \\[2ex] \dfrac{\partial V}{\partial z} \end{pmatrix} = i\frac{\partial V}{\partial x} + j\frac{\partial V}{\partial y} + k\frac{\partial V}{\partial z} \qquad (A.5)$$

If $V(x,y,z)$ is a vector function, then the divergence $\nabla \cdot A$ is given as

$$\nabla \cdot \underset{\sim}{V} = \frac{\partial V_x}{\partial x} + \frac{\partial V_y}{\partial y} + \frac{\partial V_z}{\partial z} \triangleq div\underset{\sim}{V} \qquad (A.6)$$

If $\underset{\sim}{V}(x,y,z)$ is a vector function, then the curl of $\underset{\sim}{V}$, $\nabla \times \underset{\sim}{V}$, is given as

$$\nabla \times \underset{\sim}{V} = \begin{vmatrix} \dfrac{\partial}{\partial x} & \dfrac{\partial}{\partial y} & \dfrac{\partial}{\partial z} \\ V_x & V_y & V_z \\ \underset{\sim}{i} & \underset{\sim}{j} & \underset{\sim}{k} \end{vmatrix} = \begin{pmatrix} \dfrac{\partial V_z}{\partial y} - \dfrac{\partial V_y}{\partial z} \\ -\dfrac{\partial V_z}{\partial x} + \dfrac{\partial V_x}{\partial z} \\ \dfrac{\partial V_y}{\partial x} - \dfrac{\partial V_x}{\partial y} \end{pmatrix} \qquad (A.7)$$

This is the curl operation.

A.3.1 IDENTITY

Given that $\underset{\sim}{A}, V$ are the vector and scalar functions, respectively, we have the following identities:

$$div(curl\underset{\sim}{A}) = \nabla(\nabla \times \underset{\sim}{A}) = 0$$

$$curl(grad\underset{\sim}{A}) = (\nabla \times \nabla V) = 0 \qquad (A.8)$$

$$div(grad\underset{\sim}{A}) = (\nabla \cdot \nabla V) = \nabla^2 V = \frac{\partial^2 V_x}{\partial x^2} + \frac{\partial^2 V_y}{\partial y^2} + \frac{\partial^2 V_z}{\partial z^2} \triangleq \text{Laplacian}$$

A.3.2 PHYSICAL INTERPRETATION OF GRADIENT, DIVERGENCE, AND CURL

Gradient: The gradient of any scalar function is the maximum spatial rate of change of that function. If the scalar function represents the temperature, then $grad(V)$ is the temperature gradient or the rate of change of the temperature with distance. It is evident that although the temperature is a scalar function, $grad(V)$ is a vector quantity, with its direction being that in which the temperature changes most quickly.

Divergence: As a mathematical tool, vector analysis finds great usefulness in simplifying the expressions of the relations that exist in a three-dimensional field. A consideration of fluid flow motion gives a direct interpretation of divergence and curl. First consider an incompressible fluid, e.g., water, then the rectangular parallel pipe shown in Figure A.2.

If the fluid is flowing through this small volume, then due to the non-compression of the fluid, it is expected that there will be the same amount of volume of the fluid at the output of the small volume. Thus, there is no divergence of the fluid after flowing through the volume, and the divergence

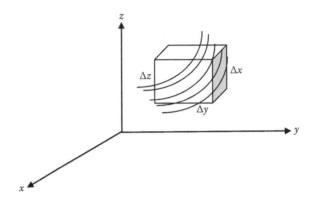

FIGURE A.2   An infinitesimal volume rectangular parallel pipe within a fluid medium.

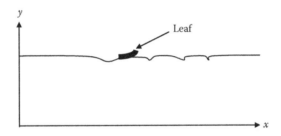

FIGURE A.3   Flow of a leaf on the surface of water.

of the fluid is zero. However, if the fluid can be compressed, then the rate of flow of the fluid through this infinitesimal volume would change, and thus the rate of flow would be different at the output of the volume compared with that at the input. Thence, the divergence of the flow is finite.

Curl: The concept of curl or rotation of a vector quantity can be illustrated in a stream of flow problems, for example, a leaf flowing on the surface of water, as shown in Figure A.3. If the leaf rotates about an axis parallel to the z-axis, then the velocity of flow of the leaf would be different, and the curl of the velocity $\underset{\sim}{v}$ denoted as $\nabla \times \underset{\sim}{v}$ is given as

$$\nabla \times \underset{\sim}{v} = \begin{vmatrix} \dfrac{\partial}{\partial x} & \dfrac{\partial}{\partial y} & \dfrac{\partial}{\partial z} \\ v_x & v_y & v_z \\ \underset{\sim}{i} & \underset{\sim}{j} & \underset{\sim}{k} \end{vmatrix} = \begin{pmatrix} \dfrac{\partial v_z}{\partial y} - \dfrac{\partial v_y}{\partial z} \\ -\dfrac{\partial v_z}{\partial x} + \dfrac{\partial v_x}{\partial z} \\ \dfrac{\partial v_y}{\partial x} - \dfrac{\partial v_x}{\partial y} \end{pmatrix} \qquad (A.9)$$

The rotation velocity of the leaf about the z-axis is given as the change of the velocity of the leaf flow in the x-direction minus that of the y-direction.

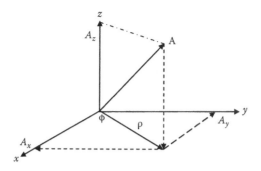

FIGURE A.4 A cylindrical coordinate system.

A.4 VECTOR RELATION IN OTHER COORDINATE SYSTEMS

A.4.1 CYLINDRICAL COORDINATES

Refer to the labels of the cylindrical coordinates assigned in Figure A.4; we have the following expressions for the gradient, divergence, and curl operators:

$$\nabla \underset{\sim}{V} = \begin{pmatrix} \dfrac{\partial V}{\partial \rho} \\[2mm] \dfrac{1}{\rho}\dfrac{\partial V}{\partial \phi} \\[2mm] \dfrac{\partial V}{\partial z} \end{pmatrix} \tag{A.10}$$

If $\underset{\sim}{A}(x,y,z)$ is a vector function, then the divergence $\nabla \cdot \underset{\sim}{A}$ is given as

$$\nabla \cdot \underset{\sim}{A} = \frac{1}{\rho}\frac{\partial(\rho A_\rho)}{\partial \rho} + \frac{1}{\rho}\frac{\partial A_\phi}{\partial y} + \frac{\partial A_z}{\partial z} \triangleq div\underset{\sim}{A} \tag{A.11}$$

If $\underset{\sim}{V}(x,y,z)$ is a vector function, then the curl of $\underset{\sim}{V}$, $\nabla \times \underset{\sim}{V}$, is given as

$$\nabla \times \underset{\sim}{V} = \begin{pmatrix} \left(\dfrac{1}{\rho}\dfrac{\partial(V_z)}{\partial \phi} - \dfrac{\partial V_\phi}{\partial z}\right)\underset{\sim}{a}_\rho \\[3mm] \left(-\dfrac{\partial V_z}{\partial \rho} + \dfrac{\partial V_\rho}{\partial z}\right)\underset{\sim}{a}_\phi \\[3mm] \dfrac{1}{\rho}\left(\dfrac{\partial(\rho V_\phi)}{\partial \rho} - \dfrac{\partial V_\rho}{\partial \phi}\right)\underset{\sim}{a}_z \end{pmatrix} \tag{A.12}$$

The Laplacian in the cylindrical coordinates is then given by

$$\nabla^2 V = \frac{1}{\rho}\frac{\partial}{\partial \rho}\left(\rho\frac{\partial V}{\partial \rho}\right) + \frac{1}{\rho^2}\frac{\partial^2 V}{\partial \phi^2} + \frac{\partial^2 V}{\partial z^2} \triangleq \text{Laplacian} \tag{A.13}$$

A.4.2 Spherical Coordinates

Referring to the spherical coordinate systems depicted in Figure A.5 we can write

$$\nabla V = a_r \frac{\partial V}{\partial z} + a_\theta \frac{1}{r}\frac{\partial V}{\partial \theta} + a_\phi \frac{1}{r\sin\theta}\frac{\partial V}{\partial \phi} \tag{A.14}$$

If $A(x,y,z)$ is a vector function, then the divergence $\nabla \cdot A$ is given as

$$\nabla \cdot A = \frac{1}{r^2}\frac{\partial(r^2 A_r)}{\partial r} + \frac{1}{r\sin\theta}\frac{\partial(A_\theta \sin\theta)}{\partial \theta} + \frac{1}{r\sin\theta}\frac{\partial A_\phi}{\partial \phi} \tag{A.15}$$

If $A(x,y,z)$ is a vector function, then the curl of A, $\nabla \times A$, is given as

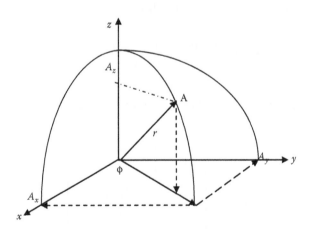

FIGURE A.5   A spherical coordinate system.

$$\nabla \times \underset{\sim}{A} = \begin{pmatrix} curl_r \underset{\sim}{A} \\ curl_\theta \underset{\sim}{A} \\ curl_\phi \underset{\sim}{A} \end{pmatrix} = \begin{pmatrix} \dfrac{1}{r\sin\theta}\left[\dfrac{\partial(A_\phi \sin\theta)}{\partial\theta} - \dfrac{\partial A_\theta}{\partial\phi}\right] \\ \left[\dfrac{1}{r\sin\theta}\dfrac{\partial(A_r)}{\partial\phi} - \dfrac{1}{r}\dfrac{\partial(rA_\phi)}{\partial r}\right] \\ \left[\dfrac{1}{r}\dfrac{\partial(A_\theta)}{\partial r} - \dfrac{\partial A_r}{\partial\theta}\right] \end{pmatrix} \quad (A.16)$$

The Laplacian of a scalar function V in the spherical coordinate system is given as

$$\nabla^2 V = \frac{1}{r^2}\frac{\partial}{\partial r}\left(r^2\frac{\partial V}{\partial r}\right) + \frac{1}{r^2\sin\theta}\frac{\partial}{\partial\theta}\left(\sin\theta\frac{\partial V}{\partial\theta}\right) + \frac{1}{r^2\sin\theta}\frac{\partial^2 V}{\partial\phi^2} \triangleq \text{Laplacian} \quad (A.17)$$

Index